T0358570

Asymptotic Time Decay
in Quantum Physics

Asymptotic Time Decay in Quantum Physics

Domingos H. U. Marchetti
Walter F. Wreszinski

Universidade de São Paulo, Brazil

World Scientific

NEW JERSEY · LONDON · SINGAPORE · BEIJING · SHANGHAI · HONG KONG · TAIPEI · CHENNAI

Published by

World Scientific Publishing Co. Pte. Ltd.

5 Toh Tuck Link, Singapore 596224

USA office: 27 Warren Street, Suite 401-402, Hackensack, NJ 07601

UK office: 57 Shelton Street, Covent Garden, London WC2H 9HE

British Library Cataloguing-in-Publication Data
A catalogue record for this book is available from the British Library.

ASYMPTOTIC TIME DECAY IN QUANTUM PHYSICS

ISBN 978-981-4383-80-6

Typeset by Stallion Press
Email: enquiries@stallionpress.com

Printed in Singapore.

Dedicated to

Carla & Verena

Preface: A Description of Contents

Decay of various quantities (return or survival probability, correlation functions) in time are the basis of a multitude of important and interesting phenomena in quantum physics, ranging from spectral properties, resonances, return and approach to equilibrium, to dynamical stability properties and irreversibility and the "arrow of time". This monograph is devoted to a clear and precise, yet (hopefully) pedagogical account of the associated concepts and methods. It is aimed at graduate students and researchers in the fields of mathematical physics and mathematics.

The introduction (Chap. 1) provides a summary of the material (with some details and problems) with which the reader is assumed to be familiar, with the necessary references to bridge gaps which are sure to be left over. It also explains some rudiments of the physical ideas behind the models whose understanding is aimed at: particles in random or quasi-periodic potentials, including Anderson-like models and models of quasi-crystals, and atoms and molecules in external time-dependent fields. We do not touch (except for brief remarks) on the N-body problem (see [105, 170] and references given there, as well as [209, 210]). Systems with infinite number of degrees of freedom are left to the interlude.

We shall be concerned in Chaps. 2–6 with the decay (or nondecay) of the return probability (or related quantities associated to the so-called RAGE theorem) for one-particle systems, with the exception of part of Chap. 6, which is devoted to the approach of equilibrium in classical mechanics and some special systems of infinite number of degrees of freedom (quantum spin systems and quantum field theory). Chapter 2 deals with the pointwise decay associated to the absolutely continuous (a.c.) spectrum, specially by the method of stationary phase. The a.c. spectrum is, of course, deeply connected to scattering theory, a beautiful subject which we only touch in this book, viz. in precisely this section, in order to relate the decay of certain quantities essential to prove existence of the wave-operators to the rate of (space) fall-off of the potential. This is illustrated in the simplest possible way of how to obtain decay in the L^p-sense from pointwise decay, essentially the converse problem as discussed in Sec. 3.2. There are other instances in this book in which the a.c. spectrum is the central element of discussion: in the time–energy uncertainty relation of Sec. 3.2, in the discussion of stability in Sec. 3.4, in the Anderson-like transition in dimension $n \geq 2$ in Sec. 4.2 and in the approach to resonances in Chap. 5.

Our treatment of the stationary phase method in Chap. 2 will also serve as an introduction to the (modified) stationary phase method used in Sec. 4.3, to study pointwise decay for a class of models (with "sparse" potentials) whose spectrum is purely singular continuous (s.c.). This is no accident: we are particularly interested in some aspects of exotic spectra (such as s.c. or dense pure point (p.p.)) which have not been covered in the classic textbook references [54, 207–210]. Connected to the latter are several types of decay — decay in the average, decay in the L^p-sense, and pointwise decay — which will be seen to appear naturally in different mathematical and physical settings. In particular, decay in the L^p-sense is related both to pointwise decay (Chap. 2) and to decay in the average (Secs. 3.1 and 3.2) and, from a physical standpoint, relates to a rigorous form of the time–energy uncertainty relation (Sec. 3.2). The study of pointwise decay for s.c. measures (Rajchman measures, Sec. 3.3) provides a bridge between ergodic theory, number theory and analysis, and we hope that our treatment clarifies the mathematical unity of these topics (in this context, otherwise see [138]), and conveys some of their beauty and subtlety.

Further application of these results is reserved to a class of sparse models in dimensions higher than two in Sec. 4.2, for which we prove the existence of an Anderson-like transition following our paper [179]. The existence or not of the Anderson transition — posed in Chap. 3 — is one of the outstanding open problems in mathematical physics. For this reason, an important core of this book is the sparse model considered in Chap. 4. In Sec. 4.1 we endeavor to provide a completely self-contained new treatment of this model, based on [180] and on unpublished material [65], the latter emphasizing the connection with dynamical systems and ergodic theory. Appendix B also contains some unpublished material from [62], which should be helpful to render the subject matter specially comprehensive.

The basic emphasis on one-dimensional sparse models has an important technical reason: the profound approach of Gilbert and Pearson [92] connecting the space asymptotics of eigenfunctions of the restrictions of a large class of Sturm–Liouville operators to finite regions to the spectral theory of these same operators in infinite space through the concept of subordinacy (B.14) is only available in one-dimension. Using this theory in an important transfer matrix version of Last and Simon [163], the surprise is that, even in a regime of strong sparsity, a spectral transition from s.c. to p.p. spectrum (first shown by Zlatoš in [279]) may be proved: this is Theorem 4.1 in Chap. 4. An important unifying feature of our proof of the p.p. spectrum is the valuable role played by the Weyl theorems on uniform distribution (of the so-called Prüfer angles, see Sec. 4.2 and Theorem 4.3) discussed in Sec. 3.3 in connection with the Rajchman measures mentioned above. Finally, one may pass to higher dimensions by forming the separable models (4.28): their spectra are shown to exhibit an a.c. part around the center of the band (Theorem 4.6): the theory of decay on the average plays a central role in this proof.

The natural appearance of the time–energy uncertainty relation in the above-mentioned context of decay in the L^p-sense brings once more to our attention the double role of time in quantum physics [38], as an external parameter and as an observable. Indeed, it is interesting that the last view — time of occurrence of an effect as an observable — has been generally defined only recently by Brunetti and Fredenhagen [34, 36], this having also led to a time–energy uncertainty relation different from the one in Sec. 3.2, as will be remarked there.

An important topic treated in our book is (dynamical) stability. Section 3.4 is devoted to some of the important stability concepts in quantum mechanics: they are related to the various spectral types as well as dynamical regimes, viz. the localization region, subdiffusive and normal, as well as ballistic regimes. These notions are enrichned by the interplay, in models, between two antipodal phenomena: resonance and randomness, leading to the subtle instability of tunneling, of which a striking example is the existence of p.p. spectrum in a regime of strong sparsity of the model of Chap. 4 (see Remark 4.5), see also (5.57a) et seq. for a dynamical example. A further example of instability occurs in the model of Chap. 5, in connection to multiphoton ionization. Further notions of stability are studied in Chap. 6 for systems with infinite number of degrees of freedom.

Chapters 5 and 6 will be devoted to two special and important types of decays: quasi-exponential decay and exponential decay.

Chapter 5 deals with quasi-exponential decay, i.e., exponential decay for times is neither too short nor very long: the problem of existence of resonances. We discuss briefly the several concepts of resonance, but concentrate on the quite natural approach which appeared in recent years, related to the direct physical/mathematical explanation of Gamow decay

[76, 91], and the so-called Gamow vectors [50] arising in the Borel sum (shown to exist) of the asymptotic series for the wave-function. As an application, the theory of multiphoton ionization in a one-dimensional model of a delta-function atom in an external sinusoidal electric field [121] is sketched. In spite of the one-dimensional nature of the model, it has been applied with success to the ion H^- in an external electric field, for reasons reviewed there.

We start Chap. 6 by reviewing the folklore result that strict exponential decay is not possible for systems with Hamiltonians bounded from below (semibounded). This is the case for all physical systems. The approximation of neglecting back-reaction, i.e., replacing the physical Hamiltonian by an effective time-dependent Hamiltonian, is, however, frequently a very good approximation, widely used in quantum optics and atomic physics. One example is the external field Hamiltonians introduced in Chap. 3, which includes, as special case, the constant-field Stark effect. Some of these models are prototypes of the so-called *quantum chaos*, for instance, the kicked rotor (3.137a)–(3.137c). These models may exhibit exponential decay, and an elegant mathematical framework for their description is provided by the so-called quantum Anosov systems introduced by Emch, Narnhofer, Sewell and Thirring [80], which are seen to include, besides the Stark effect, the models of atoms and oscillators in external periodic and quasi-periodic fields introduced in Chap. 3. As discussed there, the latter are relevant to describe real physical systems, e.g., Paul-Penning traps [39]. We describe this theory in Sec. 6.2.

Our description in Sec. 6.2 follows, mostly, [130, 228]. The latter treatment emphasizes the connection with classical mechanics, leading naturally to the subject of Sec. 6.3: some aspects of the connection between quantum mechanics and classical mechanics (classical dynamical systems). We speak here not of the classical limit of quantum mechanics [108], which is well-known to be very singular, but of results, ideas and methods in classical mechanics which have suggested analogous, albeit necessarily different in some essential features, in quantum mechanics. Some of these also appear in Chap. 3 in relation to the KAM perturbation theory and stability problems, and Howland's method in classical mechanics.

Section 6.3 starts with a brief introductory motivation in Sec. 6.3.1: there, we remark (and justify) that we shall restrict our discussion to the approach to equilibrium in *closed* systems. We then go on to Sec. 6.3.2, which is the first part of the study of the approach to equilibrium in classical (statistical) mechanics, dealing with ergodicity, mixing and connecting it with the Anosov property. We introduce the Gibbs entropy and the second law, with "coarse graining" formulated along the ergodic approach of Penrose [202]. In Sec. 6.3.3, we further motivate some of the ideas through a presentation of the Ehrenfest model, following a very instructive set of lecture notes [89]. Section 6.3.4 is concerned with the initial state (in particular, of the Universe, see below) and macroscopic states: Boltzmann versus Gibbs entropy. Our treatment unifies the approaches of [161] with that of [72], providing a general discussion in the framework of evolution of densities. Unlike the evolution of points (trajectories), the evolution of densities, through the Ruelle–Perron–Frobenius operator introduced in App. A (which contains a, hopefully, rather complete survey of classical ergodic theory) presents deep analogies with the study of approach to equilibrium in classical statistical mechanics: on passing from points to, e.g., characteristic functions of intervals, one already observes "macroscopic behavior", since an interval already needs a macroscopic (in the limit infinite) number of points to build a density. These analogies only became apparent in the last century due to the seminal work of Borel, Renyi and Ulam and von Neumann. Section 6.3.5 then summarizes the mostly open problems concerning the approach to equilibrium in quantum mechanics. For this chapter, as a whole, the discussion by Lebowitz [167] has been invaluable; we further refer to it as a complementary overview.

Sections 6.3.3 and 6.3.4 discuss what is part of the more profound "arrow of time" problem. As remarked by Penrose [202], the laws and phenomena usually considered include "the second law of thermodynamics, the expansion of the Universe, the use of retarded potentials in electrodynamics, the decay of the K^0 meson, and our subjective experience that time passes". From the point of view of the average time spent by the system near, and far from, equilibrium, the Ehrenfest model in Sec. 6.3.3 shows a possible way out of the "paradox of irreversibility" well formulated by Schrödinger [232] in his remark quoted there, but this point of view does not account for the actual process of entropy growth mentioned by Schrödinger. It is believed by many (but still somewhat controversial, see the discussion in [167]) that a full explanation has to do with the initial state of the Universe: as remarked by Feynman (see [84], quoted in [167]), "it is necessary to add to the physical laws the hypothesis that in the past the Universe was more ordered, in the technical sense, than it is today...". This technical sense is low Boltzmann entropy, and this lends further interest to the subject matter of Sec. 6.3.4.

It turns out that the concept of mixing in classical mechanics is naturally related to an analogous concept in quantum mechanics, but now for systems with an infinite number of degrees of freedom, due to the existence in the latter of an invariant state. We discuss some of the fundamentals of the theory which we shall need in the interlude, in Sec. 6.4, drawing on several excellent existing textbook references [122, 237] for greater detail. We start the study of the aforementioned systems in Sec. 6.5, which is divided, for clarity, into three parts, Secs. 6.5.1, 6.5.2 and 6.5.3.

In Sec. 6.5.1, we define the quantum analogues of the mixing property, and the related properties of return to equilibrium and weak asymptotic abelianness. We also briefly discuss dynamical stability. The remaining two sections are devoted to illustrative examples.

In Secs. 6.5.2, we study mixing and weak asymptotic abelianness for the vacuum and thermal states in relativistic quantum field theory (rqft). Following [124], we explore the fact that in rqft, time-like and space-like decay are related by (Poincaré) covariance, and that space-like cluster properties are a consequence of microscopic causality [87].

In Sec. 6.5.3, approach to equilibrium is studied in one of the few models of quantum spin systems for which the rates of decay may be computed and/or estimated by the Emch–Radin model [79, 206]. A recent extension [274] to random systems includes exponential decay, and permits a unified discussion of rates of decay in connection to stability, which is provided there.

Appendix C aims to be a rather complete and comprehensive exposition of the theory of symmetric Cantor sets, essential to complement the reading of parts of Chap. 3.

Open problems are discussed throughout the book in the appropriate chapters/sections.

Acknowledgments

W.F.W. would like to thank several people for discussions and/or collaborations and correspondence on specific chapters in the book: O. Costin, K. Fredenhagen, M. Huang, C. Jäkel, H.-R. Jauslin, J. L. Lebowitz, H. Narnhofer, G. L. Sewell, and W. Thirring. He is particularly grateful to O. Costin and M. Huang for clarification of points in their joint work reported in Chap. 5, as well as for sending him the original figures.

Together, we would like to thank Silas Luis de Carvalho for his contributions in joint work and Ph.D. thesis. Special thanks are due to Joel L. Lebowitz. His scientific influence in both our academic trajectories, independently, has been very significant. Some of this is apparent in Chaps. 3, 5 and 6, but we would like to leave here a token of our gratitude to the very high scientific and ethical standards set by him in the statistical mechanics community throughout several decades, in particular, through the *Journal of Statistical Physics*, of which he is the chief editor.

Contents

Chapter 1

Introduction: A Summary of Mathematical and Physical Background for One-Particle Quantum Mechanics

In this chapter, we pose some problems which will be discussed throughout the book, at the same time, fixing some important notations and levels of knowledge assumed from the reader.

As remarked in the preface, we shall be concerned with the decay of various quantities in time such as the return or survival probability which lie at the basis of various phenomena in quantum physics. We shall be dealing with one-particle systems subject to certain potentials which may or may not depend on time (in the positive case, we speak of external potentials). The basic objects of the theory are a self-adjoint Hamiltonian H on a (separable) Hilbert space $\mathcal{H} = L^2(\mathbb{R}^n)$ or $\mathcal{H} = l^2(\mathbb{Z}^n)$ (spin will not occur in this context), $x \equiv (x_1, \ldots, x_n)$ denoting a point in $\mathbb{R}^n$, n the space dimension and

$$H = H_0 + V \tag{1.1}$$

with

$$H_0 = -\frac{\hbar^2}{2m}\triangle \tag{1.2a}$$

or

$$H_{0l} = -\triangle_l \tag{1.2b}$$

denoting the free Hamiltonian, where the Laplacian $\triangle$ is given by

$$\triangle = \sum_{i=1}^{n} \frac{\partial^2}{\partial x_i^2} \tag{1.2c}$$

and the difference Laplacian

$$(-\triangle_l u)(n) = \sum_{i=1}^{n}[2u(n) - u(n + \delta_i) - u(n - \delta_i)] \qquad (1.2d)$$

with $n \in \mathbb{Z}^n$ and $u \in l^2(\mathbb{Z}^n)$, and

$$\delta_i \equiv (0, \ldots, 1, \ldots, 0) \qquad (1.2e)$$

are unit vectors along the n coordinate axes, and V is a multiplication operator (possibly time-dependent)

$$(Vu)(\vec{x}) = V(\vec{x})u(\vec{x}) \quad \text{with } u \in L^2(\mathbb{R}^n) \text{ and } \vec{x} \in \mathbb{R}^n \qquad (1.3a)$$

or

$$(Vu)(n) = V(n)u(n) \quad \text{with } n \in \mathbb{Z}^n \text{ and } u \in l^2(\mathbb{Z}^n). \qquad (1.3b)$$

A short introduction to operators in Hilbert spaces may be found, for instance, in [105, pp. 6–10]; self-adjointness in [105, pp. 17–19]. We shall also assume as well-known the properties of the Fourier transform [105, pp. 20–23]. The Hamiltonian described by (1.2b) and (1.2d) is the tight-binding Hamiltonian which describes tunneling and bands for a particle (e.g., electron) in an n-dimensional hypercubic lattice [85, Chap. 13], and V, given by (1.3b), might describe an imperfection or impurity (see, again, [85, Chap. 13]. As such, it is, for suitable V, depending on a random parameter, of paradigmatic importance in the study of the Anderson model (see Chap. 3), which describes delocalized electrons in a conduction band in semiconductors, as well as localized (impurity) states. For an elementary but lucid introduction to the physics, see [85, Chaps. 13 and 14]. It is, however, easy to see what happens in a very rough way. Assume that we have a one-dimensional chain (for simplicity with unit spacing) of hydrogen atoms, each in its ground state, and that we add an extra electron to the chain, producing a slightly bound negative ion of energy E_0, with a fixed probability amplitude A_0 to tunnel to its nearest neighbors at sites $n - 1$ and $n + 1$: this is expressed by the two first terms in (1.2d) with $A_0 = 1$. Assume that the diagonal term in (1.3b) is such that

$$V(n) - 2 = \frac{E_0}{A_0}. \qquad (1.3c)$$

We have then (up to an irrelevant constant multiplier) the Schrödinger equation

$$i\hbar\frac{du(n)}{dt} = E_0 u(n) - A_0(u(n+1) - u(n-1)) \qquad (1.3d)$$

which, with the ansatz

$$u(n) = \exp\left(i\frac{(kn - Et)}{\hbar}\right), \qquad (1.3e)$$

leads to the energy eigenvalues

$$E = E(k) = E_0 - 2A_0 \cos k. \qquad (1.3f)$$

If k runs through $[-\pi, \pi)$ (the one-dimensional Brillouin zone), then (1.3f) describes a band of energies — the *conduction band*. The approximation of nearest neighbors is justified by the rapid (exponential) fall-off of the eigenfunctions, e.g., in a double-well [105], characteristic of tunneling, a fundamental and quite subtle quantum-mechanical phenomenon (see Chap. 3 about the instability of tunneling and references). The band of energies in (1.3f) is the interval $[E_0 - 2A_0, E_0 + 2A_0]$; excited energies in each hydrogen atom would yield different bands $[E_i - 2A_i, E_i + 2A_i]$, which may not overlap depending on the relations between the various A_i: gaps may occur, a well-known purely quantum-mechanical phenomenon basic to the theory of the transistor (see [85, Chap. 14]). Finally, if $V(n)$ is not given by (1.3c) but changes at some sites, one has a description of imperfections in the lattice — this V may even be a random variable (see further), in which case we have a model of random impurities — the Anderson model! There, as we shall see, a new phenomenon arises, that of localized states (in contrast to the delocalized states (1.3e)), which are believed to have their physical origin in the subtle phenomenon of tunneling instability (see also Chap. 3, Sec. 3.4). We close this short physical description with a sentence in [85]: "We have now explained a remarkable mystery — how an electron in a crystal (like an extra electron put into germanium) can ride through the crystal and flow perfectly freely even though it has to hit all the atoms. It does so by having its amplitudes going pip-pip-pip from one atom to the next, working its way through the crystal. That is how a solid can conduct electricity."

Recall that the evolution of the wave-function u for a particle described by a self-adjoint Hamiltonian H (i.e., $H = H^\dagger$, where $A^\dagger$ denotes the

Hermitian conjugate, see [105, pp. 17–20] for the subtleties associated to this concept) is determined by the Schrödinger equation

$$ i\hbar \frac{\partial u}{\partial t} = Hu, \tag{1.4a} $$

with the initial condition

$$ u(t = 0) = u_0, \tag{1.4b} $$

where $u_0 \in \mathcal{H}$. Equations (1.4a) and (1.4b) are solved by

$$ u(t) = U(t)u_0, \tag{1.4c} $$

where $t \longrightarrow U(t)$ is a one-parameter family of unitary operators, i.e., $U(t)U^\dagger(t) = U^\dagger(t)U(t) = \mathbb{I}$ (see [105, Definition 2.7]), where the dagger denotes Hermitian conjugate and $\mathbb{I}$ will always denote the identity operator. Now $U(t)$ is given by

$$ U(t) = \exp\left(-itH/\hbar\right) \tag{1.5} $$

and enjoys the important group property

$$ U(t)U(s) = U(t+s) \quad \text{with } t \text{ and } s \in \mathbb{R} \tag{1.6a} $$

together with

$$ U(0) = \mathbb{I}. \tag{1.6b} $$

If H is not a bounded operator (see [105, p. 9]), then (1.5) is defined by the forthcoming functional calculus, from which also (1.6a) follows. For the free continuum case, with $H = H_0$ given by (1.2a) and (1.2c) (see [105, Chap. 2]):

$$ (\exp(-iH_0 t/\hbar)u)(x) = (2i\pi\hbar t/m)^{-n/2} \int_{\mathbb{R}^n} \exp\left(\frac{im|x-y|^2}{2\hbar t}\right) u(y)d^n y. \tag{1.7} $$

Above, and in the rest of the book, $d^n x$ denotes the n-dimensional (Lebesgue) measure (volume measure), and $|x|^2 \equiv \sum_{i=1}^n x_i^2$ denotes the square of the Euclidean distance. An immediate consequence of (1.7) is the bound

$$ \| \exp(-iH_0 t/\hbar)u \|_\infty \leq (2\pi\hbar t/m)^{-n/2} \|u\|_1. \tag{1.8} $$

We now explain the notation above. A function u which is measurable on a nonempty open set Ω of $\mathbb{R}^n$ (which may be the whole of $\mathbb{R}^n$) is said to be essentially bounded on Ω iff there is a constant K such that $|u(x)| \leq K$ almost everywhere (a.e.) on Ω, i.e., except for a subset of Ω with zero measure. The greatest lower bound of such constants K is called the essential supremum of $|u|$ on Ω, and is denoted by $\operatorname{esssup}_{x \in \Omega}|u(x)|$. We denote by $L^\infty(\Omega)$ the vector space of all functions u that are essentially bounded on Ω, functions being identified if they are equal a.e. on Ω. The functional $\|.\|_\infty$, defined by

$$\|u\|_\infty = \operatorname{esssup}_{x \in \Omega}|u(x)|, \tag{1.9a}$$

is a norm on $L^\infty(\Omega)$. For p a positive real number $p \geq 1$, we denote by $L^p(\Omega)$ the class of all measurable functions u defined on Ω for which

$$\|u\|_p^p \equiv \int_\Omega |u(x)|^p d^n x \tag{1.9b}$$

is finite, and identify in $L^p(\Omega)$ functions that are equal a.e. on Ω. The so-identified spaces play an essential role in analysis (see [169, Chap. 2], also for concepts such as measurability with which the reader may not be familiar). In particular, we have an important theorem.

Theorem 1.1 (Hölder inequality). *Let* $1 \leq p \leq \infty$ *and* p' *denote the conjugate exponent defined by*

$$\frac{1}{p} + \frac{1}{p'} = 1$$

which also satisfies $1 \leq p' \leq \infty$. *If* $u \in L^p(\Omega)$, $v \in L^{p'}(\Omega)$, *then* $uv \in L^1(\Omega)$ *and*

$$\int_\Omega |u(x)v(x)| d^n x \leq \|u\|_p \|v\|_{p'}. \tag{1.10}$$

Equality in (1.10) only holds if $|u(x)|^p$ *and* $|v(x)|^{p'}$ *are proportional a.e. on* Ω, *and if* $1 < p < \infty$ *with special cases* $p = 1$ *and* $p = \infty$.

For a proof, see [169, Theorem 2.3 and special equality cases on p. 46]. The special case $p = 2$ in (1.10) is the *Schwarz inequality*.

We now recall some further standard material (see, e.g., [105, Chap. 2, pp. 11–23]). Let us denote by $D(A)$ the domain of an operator A on the

Hilbert space $\mathcal{H}$. Recall that A is symmetric iff, for all u and v in $D(A)$,

$$(Au, v) = (u, Av), \tag{1.11}$$

where $(\cdot, \cdot)$ denotes the scalar or inner product on $\mathcal{H}$, antilinear in the first argument. Since $|u(x, t)|^2$ is, for each instant of time t, a probability distribution, by the physical interpretation of quantum theory [85; 105, Chap. 1], it is required that

$$(u, u) \equiv \int_{\mathbb{R}^3} |u(x, t)|^2 d^n x = 1 \tag{1.12}$$

at all times t: (1.12) is the mathematical expression of conservation of probability. For u, a solution of (1.4a) and (1.4b), with u in $D(H)$,

$$\begin{aligned}
\frac{d(u, u)}{dt} &= \left(\frac{\partial u}{\partial t}, u\right) + \left(u, \frac{\partial u}{\partial t}\right) \\
&= -\frac{(Hu, u) - (u, Hu)}{i\hbar} = 0,
\end{aligned}$$

iff H is symmetric, i.e., satisfies (1.11), as follows from the polarization identity

$$(u, v) = 1/4(\|u + v\|^2 - \|u - v\|^2 - i\|u + iv\|^2 + i\|u - iv\|^2). \tag{1.13}$$

Problem 1.1. *Prove* (1.13).

Denote by

$$p_j = -i\hbar \frac{\partial}{\partial x_j} \quad \text{with } j = 1, \ldots, n \tag{1.14a}$$

the momentum operator, which satisfies with the coordinate x_j (considered as multiplication operator) the *formal* canonical commutation relations (CCR)

$$[x_j, p_j] = i\hbar. \tag{1.14b}$$

One way to give mathematical sense to (1.14b) is to consider matrix elements of the l.h.s. between states u and v, where both u and v belong to $D_j \equiv D(x_j) \cap D(p_j)$, but this is not very useful, and we refer to [270] for a lucid discussion of the CCR and Weyl's form of the CCR, in connection to von Neumann's uniqueness theorem. If, however, $u \in D_j$, then

$$\triangle x_j \triangle p_j \geq \hbar/2, \tag{1.14c}$$

where

$$(\triangle A)^2 \equiv ((A - (A)_u)^2)_u \qquad (1.14d)$$

denotes the dispersion of the operator in the state u, and $(A)_u \equiv (u, Au)$. Equations (1.14c) and (1.14d) are the *Heisenberg uncertainty principle* (see [105, Theorem 4.1, p. 30]). For the z-component of the angular momentum $L_z = -i\hbar\frac{\partial}{\partial\phi}$, and the azimuthal angle ϕ ranging in $[0, 2\pi)$, defined as operators on $\mathcal{H} = L^2(0, 2\pi)$, the formal relation $[\phi, L_z] = i\hbar$ analogous to (1.14b) is also satisfied. Here, however, domain questions have to be considered carefully, and it is not enough to require that $u \in D(\phi) \cap D(L_z)$ to have the Heisenberg uncertainty relation (1.14b) and (1.14c), i.e.,

$$(\triangle\phi)_u(\triangle L_z)_u \geq \hbar/2 \qquad (1.15)$$

satisfied. Consider, e.g., the domain

$$D_\gamma = \{f \in L^2(0, 2\pi) : f \text{ absolutely continuous, } f' \in L^2(0, 2\pi),$$
$$f(2\pi) = \exp(i\gamma)f(0) \text{ with } \gamma \in \mathbb{R}\}. \qquad (1.16)$$

Above, f' denotes the derivative of f with respect to ϕ, and it is said to be *absolutely continuous* iff there is a function $g \in L^1(0, 2\pi)$ such that for all $0 \leq \phi_0 < \phi \leq 2\pi$ one has $f(\phi) - f(\phi_0) = \int_{\phi_0}^{\phi} g(\phi')d\phi'$. It follows that f has a derivative g almost everywhere. It is easy to prove (see, e.g., [26, Chap. 19, p. 261]) that both ϕ and L_z are self-adjoint on D_γ, for any $\gamma \in \mathbb{R}$. This is a good occasion for the reader familiarize himself with the distinction between symmetric and self-adjoint operator. However:

Problem 1.2. *Show that (1.15) is false for $u \in D_\gamma$, except for one value of the parameter γ.*

We shall come back to this example in Chap. 3, but note that the above problem has been analyzed in full generality in [203].

We shall also need some of the basic elements of spectral theory. The spectrum $\sigma(A)$ of a self-adjoint operator A on $\mathcal{H}$ is the subset of $\mathbb{C}$ given by

$$\sigma(A) \equiv \{\lambda \in \mathbb{C} : A - \lambda\mathbb{I} \text{ does not have a bounded inverse}\}. \qquad (1.17)$$

The complement $\rho(A) \equiv \mathbb{C}\backslash\sigma(A)$ is called the resolvent set of A, which means, by (1.17), that for $\lambda \in \rho(A)$, the resolvent

$$r_\lambda(A) \equiv (A - \lambda\mathbb{I})^{-1} \qquad (1.18)$$

is a bounded operator. The resolvent is an analytic (operator-valued) function of $\lambda \in \rho(A)$, as follows from the first resolvent identity

$$r_z(A) = r_{z_0}(A) - (z_0 - z)r_{z_0}(A)r_z(A) \tag{1.19}$$

for z and $z_0 \in \rho(A)$.

Problem 1.3. *Prove* (1.19).

By (1.19),

$$r_z(A) = [\mathbb{I} - (z - z_0)r_{z_0}(A)]^{-1}r_{z_0}(A) \tag{1.20}$$

and by (1.20), if $|z - z_0| < \|r_{z_0}(A)\|^{-1}$, we may expand the first inverse on the r.h.s. of (1.20) in a Neumann series

$$r_z(A) = \sum_{j=0}^{\infty}(z - z_0)^j r_{z_0}(A)^{j+1},$$

and hence $r_z(A)$ is analytic in the neighborhood of any $z_0 \in \rho(A)$. It follows that $\rho(A)$ is an open set, and, consequently, that the spectrum $\sigma(A)$ of an operator is always a closed set. A straightforward reason for λ being in $\sigma(A)$, i.e., for the noninvertibility of $A - \lambda\mathbf{I}$, is $(A - \lambda\mathbb{I})\Psi = 0$ having a nonzero solution $\Psi \in \mathcal{H}$. If that is the case, λ is called an *eigenvalue* of A, and Ψ a corresponding *eigenvector*. The *multiplicity* of the eigenvalue λ is the dimension of the subspace

$$\{v \in \mathcal{H} : (A - \lambda\mathbb{I})v = 0\} \tag{1.21a}$$

(it may be infinite). A first important splitting of $\sigma(A)$ into disjoint subsets is:

$$\sigma(A) = \sigma_{\mathrm{d}}(A) \cup \sigma_{\mathrm{ess}}(A), \tag{1.21b}$$

where

$$\sigma_{\mathrm{d}}(A) \equiv \{\lambda \in \mathbb{C} : \lambda \text{ does not have a bounded inverse}\} \tag{1.21c}$$

is the *discrete* spectrum. Its complement

$$\sigma_{\mathrm{ess}}(A) \equiv \sigma(A)\backslash\sigma_{\mathrm{d}}(A) \tag{1.21d}$$

is called the *essential* spectrum, to which we come back in Chap. 3. We shall see there that, by a fundamental theorem of Weyl, it is characterized (in the case of a self-adjoint operator) by the equation $(A - \lambda\mathbb{I})\Psi = 0$ "almost" having a nonzero solution, in a sense to be made precise there.

A different splitting of $\sigma(A)$ into disjoint parts is provided by the famous *spectral theorem*. We shall provide two versions of this theorem: the spectral theorem for operators with so-called simple spectrum (a suitable generalization for infinite dimension of matrices with no eigenvalue degeneracy) and generalizations thereof, and the spectral theorem in terms of a resolution of unity as a generalization of the spectral theorem for the so-called compact operators. Both versions will be employed in this monograph, as they throw a different light into the functional analytic properties and, therefore, the use of one or the other may turn out to be more natural or simpler in different situations. As a preliminary, we need some essentials of measure theory. An excellent reference for the following is [127]; see also [207]. We follow [127] in this sketch and refer to this reference for further details.

Let M be a set and $\mathcal{F}$ a sigma-algebra in M. The pair $(M, \mathcal{F})$ is called a measure space. If M is a metric space, the minimal sigma-algebra in M containing all open sets is called the Borel sigma-algebra, and a measure μ on the Borel sigma-algebra is called a Borel measure. If μ is Borel, the complement of the largest open set $\mathcal{O}$ such that $\mu(\mathcal{O}) = 0$ is called the *support* of μ (denoted supp μ).

If M is a locally compact metric space (see [207] for the topological notions), a Borel measure μ is said to be *regular* iff, $\forall E \in \mathcal{F}$, $\mu(E) = \inf\{\mu(V) : E \subset V$ and V is open$\} = \sup\{\mu(K) : E \supset K$ and K is compact$\}$. If $(M, \mathcal{F})$ is a measure space, and $E \in \mathcal{F}$, a countable collection of sets $\{E_i\}$ in $\mathcal{F}$ is called a partition of E if $E_i \cap E_j = \emptyset$ for $i \neq j$, and $E = \bigcup_j E_j$. A complex measure on $(M, \mathcal{F})$ is a function $\mu : \mathcal{F} \to \mathbb{C}$ such that

$$\mu(E) = \sum_{i=1}^{\infty} \mu(E_i) \qquad (1.22)$$

for every $E \in \mathcal{F}$ and *every* partition $\{E_i\}$ of E. It follows that (1.22) is absolutely convergent. From the definition, a complex measure takes only finite values, in contrast to the usual *positive* measures which take values in $[0, \infty]$. The set function $|\mu|$ on $\mathcal{F}$, defined by

$$|\mu|(E) = \sup \sum_i |\mu(E_i)| \qquad (1.23)$$

where the supremum is taken over all partitions $\{E_i\}$ of E, is called the *total variation* of μ: if the latter is bounded, μ is said to be of *bounded variation*. Let, now, M be a locally compact metric space. A continuous function $f : M \to \mathbb{C}$ vanishes at infinity if $\forall \epsilon > 0$ there exists a compact set K_ϵ such that $|f(x)| < \epsilon$ for $x \notin K_\epsilon$. Let $C_0(M)$ denote the vector space of all continuous functions vanishing at infinity, endowed with the supremum norm $\|f\| = \sup_{x \in M} |f(x)|$. $C_0(M)$ is a Banach space, and we denote by $C_0^\star(M)$ its dual. The following theorem is of central importance (see [220, Theorem 6.19]):

Theorem 1.2 (Riesz representation theorem). *Let $\phi \in C_0^\star(M)$. Then, there exists a unique regular complex Borel measure μ such that*

$$\phi(f) = \int_M f d\mu \tag{1.24}$$

for all $f \in C_0(M)$. Moreover $\|\phi\| = |\mu|(M)$.

Let $(M, \mathcal{F})$ be a measure space and ν_1 and ν_2 complex measures concentrated on disjoint sets (a measure μ is concentrated on the set $A \in \mathcal{F}$ if $\mu(E) = \mu(E \cap A)$, $\forall E \in \mathcal{F}$). Then, we say that ν_1 and ν_2 are *mutually singular* or *orthogonal* and write $\nu_1 \perp \nu_2$. If ν is a complex measure and μ a positive measure, we say that ν is *absolutely continuous* (a.c.) with respect to μ and write $\nu \ll \mu$ if $\mu(E) = 0 \Rightarrow \nu(E) = 0$. The following theorem is also fundamental (see [220, Theorem 6.9]):

Theorem 1.3 (Lebesgue–Radon–Nikodym theorem). *Let ν be a complex measure and μ a positive sigma-finite measure on $(M, \mathcal{F})$ (μ is sigma-finite iff there exists a countable family $\{A_i\}_{i=1}^\infty$ of subsets of $\mathcal{F}$ s.t. $M = \bigcup_{i=1}^\infty A_i$ and $\mu(A_i) < \infty$, $\forall i \in \{1, \ldots, \infty\}$). Then, there exists a unique pair of complex measures ν_a and ν_s s.t. $\nu_a \perp \nu_s$, $\nu_a \ll \mu$, $\nu_s \perp \mu$, and*

$$\nu = \nu_a + \nu_s. \tag{1.25a}$$

Moreover, there exists a unique $f \in L^1(\mathbb{R}, d\mu)$ s.t. $\forall E \in \mathcal{F}$,

$$\nu_a(E) = \int_E f d\mu. \tag{1.25b}$$

If $M = \mathbb{R}$ and μ is Lebesgue measure, ν_{ac} denotes the part of ν a.c. w.r.t. Lebesgue measure, and ν_{sing} the part which is singular to Lebesgue measure. A point $\lambda \in \mathbb{R}$ is an *atom* or *singleton* of ν if $\nu(\{\lambda\}) \neq 0$. Let A_ν denote the set of all atoms of ν. The set A_ν is countable and $\sum_{\lambda \in A_\nu} |\nu(\{\lambda\})| < \infty$. The pure point (p.p.) part of ν is defined by

$$\nu_{pp}(E) = \sum_{\lambda \in E \cap A_\nu} \nu(\{\lambda\}). \tag{1.26}$$

The measure

$$\nu_{\text{sc}} \equiv \nu_{\text{sing}} - \nu_{\text{pp}} \tag{1.27}$$

is called the singular continuous part of ν. We shall see examples of such measures in Chap. 3 and App. C.

We need a final definition in the above sketch of measure theory. Let μ be a complex Borel measure on $\mathbb{R}$. Its *Fourier–Stieltjes transform* $\hat{\mu}$ (F.S. transform) is defined by

$$\hat{\mu}(t) = \int_{\mathbb{R}^n} \exp(-itx) d\mu(x). \tag{1.28}$$

Here $\hat{\mu}$ is also called the *characteristic function* of the measure μ. Since

$$|\hat{\mu}(t+h) - \hat{\mu}(t)| \leq \int_{\mathbb{R}^n} |\exp(-ihx) - 1| d|\mu|, \tag{1.29}$$

the function $t \in \mathbb{R} \to \hat{\mu}(t) \in \mathbb{C}$ is uniformly continuous. The same follows if μ is a finite positive Borel measure.

The concept of multiplicity $n_\lambda \leq n$ of an eigenvalue λ of a finite matrix M which appears is, of course, the same as given above, only in this case the Hilbert space is a vector space V of finite dimension, say n. If $n_\lambda = 1$, λ is said to be a simple eigenvalue; similarly, the spectrum of M is *simple* if any $\Psi \in V$ may be written

$$\Psi = \alpha_1 \Psi_1 + \cdots + \alpha_n \Psi_n \tag{1.30a}$$

with

$$(M - \lambda_k)\Psi_k = 0 \tag{1.30b}$$

for $k \in \{1, \ldots, n\}$ and

$$\lambda_i \neq \lambda_j \quad \text{if } i \neq j. \tag{1.30c}$$

A general functional $\Phi(M)$ of the matrix M is defined in a straightforward manner by

$$\Phi(M)\Psi = \Phi(\lambda_1)\alpha_1 \Psi_1 + \Phi(\lambda_2)\alpha_2 \Psi_2 + \cdots + \Phi(\lambda_n)\alpha_n \Psi_n. \tag{1.30d}$$

Now, if $\lambda_1 = \lambda_2$, for instance, Ψ_1 and Ψ_2 enter into (1.30d) in the same combination $\alpha_1 \Psi_1 + \alpha_2 \Psi_2$ as in the initial vector Ψ given by (1.30a), and thus, no matter how many vectors of type $\Phi(M)\Psi$ we choose, there will not be among them a full system of linearly independent vectors in V. This reformulation allows the extension of the concept of simple spectrum to any self-adjoint operator A on a separable Hilbert space $\mathcal{H}$. We start with a definition: a collection of vectors $\mathcal{C} = \{\Psi_n\}_{n \in \Gamma}$, where Γ is a countable set, is called *cyclic* for A, if the closure of the linear span of the set of vectors

$$\{r_z(A)\Psi_n : n \in \Gamma, \ z \in \mathbb{C}\backslash\mathbb{R}\}, \tag{1.31}$$

where $r_z(A)$ is defined by (1.18), is equal to $\mathcal{H}$. The choice of the function Φ in (1.30d) as the resolvent of A, which is bounded if $z \in \mathbb{C}\backslash\mathbb{R}$, is a convenient one. A cyclic set for A always exists: we may take $\{\Psi_n\}_{n \in \Gamma}$ to be an orthonormal basis of $\mathcal{H}$. If $\mathcal{C} = \{\Psi\}$, then Ψ is called a *cyclic vector* for A, and A is said to have *simple spectrum* [2, Chap. VI, p. 242] or to be *multiplicity free* [207, VII, p. 232]. Simple spectrum has been shown recently to characterize the singular spectrum of the Anderson model [125].

The spectral theorem becomes particularly simple in the cyclic case which will appear several times in the Jacobi matrix examples, in particular, in Chaps. 3 and 4, and App. B:

Theorem 1.4 ([127, Theorem 3.11]). *Let A be a self-adjoint operator on $\mathcal{H}$, and $\Psi \in \mathcal{H}$. There exists a unique finite positive Borel measure μ_Ψ on $\mathbb{R}$ such that $\mu_\Psi(\mathbb{R}) = \|\Psi\|^2$ and*

$$(\Psi, (A - z)^{-1}\Psi) = \int_{\mathbb{R}^n} \frac{d\mu_\Psi(t)}{t - z} \tag{1.32}$$

for any $z \in \mathbb{C}\backslash\mathbb{R}$.

The measure μ_Ψ is called the *spectral measure* for A and Ψ.

Problem 1.4. *Show from (1.32) that, if Φ and Ψ are in $\mathcal{H}$, there exists a unique complex measure $\mu_{\phi,\Psi}$ on $\mathbb{R}$ such that*

$$(\Phi, (A - z)^{-1}\Psi) = \int_{\mathbb{R}^n} \frac{d\mu_{\Phi,\Psi}(t)}{t - z} \tag{1.33}$$

for $z \in \mathbb{C}\backslash\mathbb{R}$.

Hint: use the polarization identity (1.13) for existence, and the fact that the set of functions $\{(x - z)^{-1} : z \in \mathbb{C}\backslash\mathbb{R}\}$ is dense in $C_0(\mathbb{R})$ (together with the Riesz representation theorem, Theorem 1.2); prove the latter fact completing the steps of [127, Problem [2], p. 29].

We have, now

Theorem 1.5 ([127, Theorem 3.13]) (Spectral theorem for self-adjoint operators with simple spectrum). *Assume that Ψ is a cyclic vector for the self-adjoint operator A. Then, A is unitarily equivalent* to the operator of multiplication by λ on $L^2(\mathbb{R}, d\mu_\Psi(\lambda))$. In particular, $\sigma(A) = \operatorname{supp} \mu_\Psi$.

Brief sketch of proof (for the rest, see the reference given above). W.l.o.g. assume $\Psi \neq 0$. We have $(A - z)^{-1}\Psi = (A - w)^{-1}\Psi$ iff $z = w$. For $z \in \mathbb{C}\backslash\mathbb{R}$, set $r_z(\lambda) = (\lambda - z)^{-1}$. We have that $r_z \in L^2(\mathbb{R}, d\mu_\Psi)$ and the linear span of $\{r_z\}_{z \in \mathbb{C}\backslash\mathbb{R}}$ is dense in $L^2(\mathbb{R}, d\mu_\Psi)$ (see Problem 1.3). Set

$$U(A - z)^{-1}\Psi = r_z. \tag{1.34}$$

If $\bar{z} \neq w$, then

$$
\begin{aligned}
(r_z, r_w)_{L^2(\mathbb{R}, d\mu_\Psi)} &= \int_{\mathbb{R}} r_{\bar{z}} r_w d\mu_\Psi \\
&= \int_{\mathbb{R}} \frac{(r_{\bar{z}} - r_w)}{\bar{z} - w} d\mu_\Psi \\
&= \frac{(\Psi, (A - \bar{z})^{-1}\Psi) - (\Psi, (A - w)^{-1}\Psi)}{\bar{z} - w} \\
&= ((A - \bar{z})^{-1}\Psi, (A - w)^{-1}\Psi)
\end{aligned}
\tag{1.35}
$$

by the resolvent identity (1.19) and Theorem 1.4. Both sides of (1.35) are continuous as $\bar{z} \to w$, therefore, (1.35) holds for all z and w both in $\mathbb{C}\backslash\mathbb{R}$. Hence, the map (1.34) extends to a unitary $U : \mathcal{H} \to L^2(\mathbb{R}, d\mu_\Psi)$. From this, it is easy to prove the final assertion. $\square$

Several extensions of the above theorem, i.e., to arbitrary self-adjoint operators, as well as in different wordings, are available [127; 207, Chap. VII]. We follow a different route because we shall explicitly use only the above version in Chaps. 3 and 4, and for the other results — the functional calculus, the decomposition in spectral subspaces — a second important alternative way of formulating the spectral theorem, in terms of the spectral family or resolution of the identity, is often more useful in general situations, in which the multiplicity is hard to establish.

A *projection operator* P is a bounded symmetric operator on $\mathcal{H}$ such that $P^2 = P$. For a pair of bounded operators A, B on $\mathcal{H}$, we shall write $A \leq B$ whenever $(\Psi, A\Psi) \leq (\Psi, B\Psi)$, $\forall \Psi \in \mathcal{H}$. Consider, now, a pair of projection operators P_1, P_2. A necessary and sufficient condition for $P_1 \leq P_2$ is that

$$P_1 P_2 = P_2 P_1 = P_1. \tag{1.36}$$

Problem 1.5. *Prove the above statement.*

Hint: $P_1 \leq P_2$ means that P_i project onto subspaces $\mathcal{H}_i$, $i = 1, 2$, of $\mathcal{H}$ s.t. $\mathcal{H}_1 \subseteq \mathcal{H}_2$.

In the finite-dimensional case (1.30a), the Hilbert space may be written as

$$\mathcal{H} = \bigoplus_{k=1}^{m} \mathcal{H}_k, \tag{1.37}$$

with each $k = 1, \ldots, m$, $n_k = \dim \mathcal{H}_k$ being the dimension of eigenspace $\mathcal{H}_k$ corresponding to the eigenvalue λ_k ((1.30c) is not assumed here), i.e., multiplicities n_k are allowed with $\sum_{k=1}^{m} n_k = n = \dim V$. Let P_k denote the projector onto $\mathcal{H}_k$. The matrix M can then be written as

$$M = \sum_{k=1}^{m} \lambda_k P_k. \tag{1.38}$$

We need some preliminary notions. A subset E of a normed space X is said to be compact if any sequence of points in E has a subsequence converging to a point in E in the topology of X. The Bolzano–Weierstrass theorem [218] asserts that a closed (i.e., containing all sequential limits) bounded set is compact.

The structure (1.37) and (1.38) generalizes naturally to *compact operators*:

Definition 1.1 ([2, II.30] or [207, VI.5]). A bounded operator A on $\mathcal{H}$ is said to be *compact* iff for every bounded sequence $\{\Psi_n\} \subset \mathcal{H}$, $\{A\Psi_n\}$ has a convergent subsequence.

In other words, a compact operator maps a bounded set into a compact set. Important examples of compact operators are Hilbert–Schmidt and trace-class operators, with which we assume the reader is acquainted: see [105, p. 122], for a quick overview.

We shall say that a sequence $\{\Psi_n\}_{n=1,2,\ldots}$ tends to $\Psi \in \mathcal{H}$ in the norm topology iff $\lim_{n\to\infty} \|\Psi_n - \Psi\| = 0$: this is the usual concept of convergence in $\mathcal{H}$. The same sequence is said to tend *weakly* to Ψ iff $\forall \Phi \in \mathcal{H}, \lim_{n\to\infty}(\Phi, \Psi_n) = (\Phi, \Psi)$. Clearly, norm convergence implies weak convergence. In correspondence to these two topologies, one also speaks of (norm or strong) compactness and weak compactness. A sequence of bounded operators $\{T_n\}_{n=1,2,\ldots}$ is said to converge *in norm* (denoted n-$\lim_{n\to\infty} T_n = T$) to a bounded operator T iff $\lim_{n\to\infty} \|T_n - T\| = 0$, where $\|T\| = \sup_{\|\Psi\|=1} \|T\Psi\|$ is the operator norm; it is said to converge *strongly* (denoted by s-$\lim_{n\to\infty} T_n = T$) iff $\lim_{n\to\infty} \|T_n\Psi - T\Psi\| = 0$, $\forall \Psi \in \mathcal{H}$ and *weakly* (denoted by w-$\lim_{n\to\infty} T_n = T$) iff $\lim_{n\to\infty}(\Phi, T_n\Psi) = (\Phi, T\Psi)$, $\forall \Phi$ and $\Psi \in \mathcal{H}$.

An important property of compact operators is that they are, in a sense, "closest possible" to matrices in finite-dimensional spaces as (1.38) above, in the following precise sense.

Theorem 1.6 ([207, Theorem VI.13]). *Any compact operator T on a separable Hilbert space $\mathcal{H}$ is a norm limit of a sequence of operators of finite rank, i.e., there exist orthonormal sequences $\phi_j, j = 1, 2, \ldots,$ such that*

$$T = \text{n-}\lim_{n \to \infty} \sum_{j=1}^{n} (\phi_j, \cdot)\phi_j. \tag{1.39}$$

A significant distinction between the strong and weak topologies in Hilbert spaces is that the unit ball $\{\Psi \in \mathcal{H} : \|\Psi\| \leq 1\}$ is never compact if the dimension of $\mathcal{H}$ is infinite: indeed any orthonormal basis $\{\phi_j\}_{j=1,2,\ldots}$ lies in the unit ball but has no convergent subsequence because $\|\phi_i - \phi_j\| = 2$ if $i \neq j$. In this connection, it is remarkable that:

Theorem 1.7 ([2, p. 66]). *Every bounded set in $\mathcal{H}$ is weakly compact.*

A compact self-adjoint operator has discrete spectrum (see (1.21b)) which may accumulate at most at zero ([2, V. 64] or [207, Theorem VI.16, p. 203]). This shows that the discrete spectrum may not be closed (although its complement, the essential spectrum, is always closed, see Chap. 3). An example from physics is the discrete spectrum of the (reduced) Hamiltonian of the hydrogen atom (see, e.g., [105, p. 76]), which accumulates at zero, the lower bound of the essential (a.c.) spectrum.

From the above-mentioned cited results, for a compact self-adjoint operator A, we may write

$$\mathcal{H} = \mathcal{H}_0 \oplus \mathcal{H}_1 \oplus \mathcal{H}_2 \oplus \cdots \tag{1.40}$$

with each $\mathcal{H}_k$ corresponding to a real number λ_k, where $\lambda_0 = 0$ and $\lambda_k \neq \lambda_i$ for $k \neq i$. In each subspace $\mathcal{H}_k$, the application of the operator A is reduced to multiplication by λ_k:

$$A\Psi = \lambda_k \Psi \quad \text{if } \Psi \in \mathcal{H}_k. \tag{1.41}$$

Denoting by P_k the projector onto $\mathcal{H}_k$, we have:

$$\mathbb{I} = P_0 + P_1 + P_2 + \cdots \tag{1.42}$$

and

$$A = \lambda_1 P_1 + \lambda_2 P_2 + \cdots. \tag{1.43}$$

Equation (1.43) may be considered as the spectral theorem for a compact self-adjoint operator. How can it be generalized? This generalization joins smoothly to both the spectral theorem for compact operators and to the functional calculus alluded to before, and goes as follows (see [2, VI] or [207, pp. 234–235]). Let G_λ denote the subspace of $\mathcal{H}$ given by

$$G_\lambda \equiv \bigoplus_{0 \le k < \lambda} \mathcal{H}_k \tag{1.44}$$

in correspondence with (1.40), and let E_λ denote the projector onto G_λ. We see from (1.44) that $E_{\lambda-0}$ and $E_{\lambda+0}$ both exist, and $E_{\lambda-0} = E_\lambda$, i.e., E is left-continuous, and

$$E_{\lambda_k+0} - E_{\lambda_k} = P_k. \tag{1.45}$$

We may thus write (1.42) and (1.43) in the form (for any $\Psi \in \mathcal{H}$):

$$\Psi = \mathbb{I}\Psi = \int_\alpha^\beta dE_\lambda \Psi \tag{1.46}$$

and

$$A\Psi = \int_\alpha^\beta \lambda dE_\lambda \Psi, \tag{1.47}$$

where $[\alpha, \beta]$ is assumed to enclose all the eigenvalues of A. Clearly (1.47) is just an alternative way of writing the discrete sum (1.43), but in this form, a generalization to all self-adjoint operators (not necessarily compact) will be seen to hold. For this purpose, we pose the following definition.

Definition 1.2. As *resolution of unity* or *spectral family*, we denote a one-parameter family of projectors E_λ, defined on an interval $[\alpha, \beta]$ which may be finite or infinite (in the latter case, by definition, $E_{-\infty} \equiv$ s-$\lim_{\lambda \to -\infty} E_\lambda$ and $E_\infty \equiv$ s-$\lim_{\lambda \to \infty} E_\lambda$), such that:

(a) $E_u E_v = E_s$ if $s = \min\{u, v\}$;
(b) $E_{\lambda-0} = E_\lambda, \forall \alpha < \lambda < \beta$ in the strong operator sense (i.e., when acting on any $\Psi \in \mathcal{H}$);
(c) $E_\alpha = 0$ and $E_\beta = \mathbb{I}$.

From Definition 1.2, $\forall \Psi \in \mathcal{H}$, the expression

$$(E_\lambda \Psi, \Psi) \equiv \mu_\Psi(\lambda) \tag{1.48}$$

is a left-continuous monotonously increasing function of bounded variation (see definition after (1.23)), for which

$$\mu_\Psi(\alpha) = 0 \tag{1.49a}$$

and

$$\mu_\Psi(\beta) = (\Psi, \Psi), \tag{1.49b}$$

which is precisely the spectral measure previously defined in the case of operators with simple spectrum. For any interval $\Delta = [\lambda', \lambda''] \subset [\alpha, \beta]$, denote $E_{\lambda''} - E_{\lambda'} \equiv E(\Delta)$. By (c)

$$E(\Delta_1)E(\Delta_2) = E(\Delta) \tag{1.50a}$$

where

$$\Delta = \Delta_1 \cap \Delta_2 \tag{1.50b}$$

It is not difficult to show (see, e.g., [2, VI, pp. 204–207]) the usual form of the spectral theorem.

Theorem 1.8. *To each resolution of unity $E_\lambda(-\infty \leq \lambda \leq \infty)$, there corresponds a self-adjoint operator*

$$A = \int_{-\infty}^{\infty} \lambda dE_\lambda \tag{1.51a}$$

on the domain D_A:

$$D_A = \left\{ \Psi \in \mathcal{H} : \int_{-\infty}^{\infty} \lambda^2 d(E_\lambda \Psi, \Psi) < \infty \right\} \tag{1.51b}$$

and

$$\|A\Psi\|^2 = \int_{-\infty}^{\infty} \lambda^2 d(E_\lambda \Psi, \Psi) \quad \forall \Psi \in D_A. \tag{1.51c}$$

Theorem 1.8 allows the definition of the so-called *functional calculus*, i.e., of general functions (or functionals) of the self-adjoint operator A by the formula

$$f(A)\Psi = \int_{-\infty}^{\infty} f(\lambda)dE_\lambda \Psi, \tag{1.51d}$$

where f is a.e. defined and bounded measurable w.r.t. the operator-valued measure E_λ, and Ψ are such that

$$\int_{-\infty}^{\infty} |f(\lambda)|^2 d(E_\lambda \Psi, \Psi) < \infty \qquad (1.51e)$$

is satisfied: we assume that the set of Ψ satisfying (1.51e) builds a dense set in $\mathcal{H}$. We refer to [2, p. 257] for further details but remark that the case $f(\lambda) = \exp(it\lambda)$ defines the unitary operator $U(t) = \exp(itA)$ alluded to before, as a bounded operator.

In terms of the spectral measure defined by (1.48), we may define

$$\mathcal{H}_{ac} \equiv \{\Psi \in \mathcal{H} : \mu_\Psi \text{ is a.c. w.r.t. Lebesgue measure}\} \qquad (1.52a)$$

and

$$\mathcal{H}_s \equiv \{\Psi \in \mathcal{H} : \mu_\Psi \text{ is singular w.r.t. Lebesgue measure}\}. \qquad (1.52b)$$

The following important decomposition follows from Theorem 1.8:

$$\mathcal{H} = \mathcal{H}_{ac} \oplus \mathcal{H}_s. \qquad (1.52c)$$

Indeed, if $\Psi \in \mathcal{H}_{ac}$, $\Phi \in \mathcal{H}_s$, and S is any Borel subset of $\mathbb{R}$,

$$\left| \int_S d(E_\lambda \Psi, \Phi) \right|^2 \leq \int_S d(E_\lambda \Psi, \Psi) \int_S d(E_\lambda \Phi, \Phi). \qquad (1.53a)$$

Problem 1.6. *Prove* (1.53a).

Hint: Define $\int_S d(E_\lambda \Psi, \Phi) \equiv (\Psi, E_S \Phi)$ for any Borel set S and prove that $|\int_S d(\Psi, E_\lambda \Phi)| \leq C\|\Psi\|$ for a constant C which defines, for each Φ, a bounded linear functional on $\mathcal{H}$. By the Riesz lemma [207, II, p. 143], there exists a vector, which we denote by $E_S \Phi$, s.t. $(\Psi, E_S \Phi) = \int_S d(\Psi, E_\lambda \Phi)$. We have

$$|(\Psi, E_S \Phi)|^2 \leq |(E_S \Psi, E_S \Phi)|^2 \leq (\Psi, E_S \Psi)(\Phi, E_S \Phi) \qquad (1.53b)$$

using the Schwarz inequality, the symmetry of E_S and the projection property for E_S:

$$E_S^2 = E_S.$$

The above property may be proved by successive applications of (a) of Definition 1.2, e.g., $(\Psi, E_S E_\mu \Phi) = \int_{-\infty}^{\infty} \chi_S(\lambda) d(\Psi, E_\mu E_\lambda \Phi)$.

If $\Phi \in \mathcal{H}_s$, there exists a Borel set B of Lebesgue measure zero s.t.

$$\int_{\mathbb{R}\setminus B} d(E_\lambda \Phi, \Phi) = 0 \qquad (1.54a)$$

and, by definition of $\mathcal{H}_{ac}$, if $\Psi \in \mathcal{H}_{ac}$,

$$\int_B d(E_\lambda \Psi, \Psi) = 0. \qquad (1.54b)$$

Inserting (1.54a) and (1.54b) into (1.53), we obtain $(\Psi, E_S \Phi) = 0$ for both $S = \mathbb{R}\setminus B$ and $S = B$, thus

$$(\Psi, \Phi) = \int_{\mathbb{R}} d(E_\lambda \Psi, \Phi) = \int_{\mathbb{R}\setminus B} d(E_\lambda \Psi, \Phi) + \int_B d(E_\lambda \Psi, \Phi) = 0,$$

and thus $\mathcal{H}_{ac} \perp \mathcal{H}_s$. On the other hand, we may write, for any $\theta \in \mathcal{H}$ and any Borel set B, $\theta = \int_{\mathbb{R}\setminus B} dE_\lambda \theta + \int_B dE_\lambda \theta$, which concludes the proof of (1.52c). By further splitting the singular spectral measure according to (1.26) and (1.27) and defining

$$\mathcal{H}_{sc} \equiv \{\Psi \in \mathcal{H} : \mu_\Psi \text{ is s.c. w.r.t. Lebesgue measure}\} \qquad (1.55a)$$

and

$$\mathcal{H}_{pp} \equiv \{\Psi \in \mathcal{H} : \mu_\Psi \text{ is p.p. w.r.t. Lebesgue measure}\}, \qquad (1.55b)$$

we may write (1.52c) in the form

$$\mathcal{H} = \mathcal{H}_{ac} \oplus \mathcal{H}_{sc} \oplus \mathcal{H}_{pp}. \qquad (1.55c)$$

Under the above decomposition, the pure point spectrum $\sigma_{pp}(A)$ of a self-adjoint operator A, which is the spectrum of A restricted to $\mathcal{H}_{pp}$, is the closure of the set of eigenvalues of A (compare with the discussion of [207, p. 230], who define $\sigma_{pp}(A)$ as the actual set of eigenvalues). The spectra $\sigma_{sc}(A)$ and $\sigma_{ac}(A)$ are defined analogously, as the spectra of A restricted to the subspaces $\mathcal{H}_{sc}$ and $\mathcal{H}_{ac}$, respectively: these spectra need not be disjoint and, as we shall see in Chap. 3, the s.c. spectrum may have nonzero Lebesgue measure.

We conclude this introduction with a last important remark: (1.51a) is a symbolic expression whose meaning is given by

$$\int_\alpha^\beta \lambda dE_\lambda \equiv \operatorname*{s-lim}_{|Z| \to 0} \sum_{j=1}^n \lambda_j' E_{[\lambda_{j-1}, \lambda_j)},$$

where the strong limit of the r.h.s., when the partition Z of the interval $[\alpha.\beta]$ gets finer and finer, i.e., the diameter $|Z|$ of the partition tends to zero, may be shown to exist; above λ'_j is any point inside the interval $[\lambda_{j-1}, \lambda_j)$. Finally, this is true for finite α and β: if any (or both) of the limits is infinite, there is a further strong limit to be considered. For the complete argument, see [2, VI, p. 218] or [26, pp. 344–346].

Chapter 2

Spreading and Asymptotic Decay of Free Wave Packets: The Method of Stationary Phase and van der Corput's Approach

In this chapter, we shall be almost exclusively concerned with the propagation of free wave packets and their decay in time (pointwise or in L^p, see Chap. 3 for much more about this distinction). This means that for most of the time our Hamiltonian will be $H_0 = \frac{-\hbar^2}{2m}\triangle$, where $\triangle \equiv \sum_{i=1}^{n}\frac{\partial^2}{\partial x_i^2}$ is the Laplacian on $\mathcal{H} = L^2(\mathbb{R}^n)$, or $H_{0l} = -\triangle_l$ which is the difference Laplacian (1.2d) and (1.2e) on $\mathcal{H} = l^2(\mathbb{Z}^n)$, with n denoting the dimension as before. The difference Laplacian describes the tight-binding model whose physical significance was discussed in Chap. 1, for $E_0 = 2A$ in (1.3f); in this case, the energy $E = E(k)$ is, for sufficiently small modulus of the wave vector $|k|$, of the form $2Ak^2$. This means that the group velocity

$$v_{\mathrm{g}} = \frac{1}{\hbar}\frac{\partial E}{\partial k}(k) \tag{2.1a}$$

equals

$$v_{\mathrm{g}} = \frac{2A}{\hbar}k, \tag{2.1b}$$

meaning that the electron acts like a classical particle with effective mass $m_{\mathrm{eff}} = \frac{\hbar^2}{2A}$ for large wavelengths.

The spectra of both H_0 and H_{0l} are absolutely continuous. We may, therefore, view this chapter as an exposition of the most elementary examples of decay of states in $\mathcal{H}_{\mathrm{ac}}$. A proof of absolute continuity is rather easy: consider the self-adjoint multiplication operator H defined by $(Hu)(x) = f(x)u(x)$ on $\mathcal{H} = L^2(\Omega)$, where f is a real-valued measurable function on a region Ω of $\mathbb{R}^n$; let E_λ denote its spectral family (definition 1.2). By definition, the range of E_λ is the set of $u \in \mathcal{H}$ s.t. $u(x) \equiv 0$ if

$f(x) > \lambda$ and, therefore,

$$\|E(\Delta)u\|^2 = \int_{f^{-1}(\Delta)} |u(x)|^2 d^n x. \tag{2.2}$$

If f is continuously differentiable and $(\nabla f)(x) \neq 0$ a.e. on Ω, it follows from (2.2) that H has purely a.c. spectrum. We have

$$(\widetilde{H_0 u})(k) = \frac{\hbar^2 k^2}{2m} \tilde{u}(k) \tag{2.3}$$

and

$$(\widetilde{H_{0l} u})(\varphi) = 2 \sum_{i=1}^{n} (1 - \cos \varphi_i) \tilde{u}(\varphi) \tag{2.4}$$

with $\varphi \equiv (\varphi_i)_{i=1}^n$, on $\mathcal{H} = L^2(\mathbb{R}^n, d^n k)$ and $\mathcal{H} = L^2(0, 2\pi)$, respectively. Above,

$$\tilde{u}(k) = \frac{1}{(2\pi)^n} \int d^n x \exp(-ik \cdot x) u(x) \tag{2.5a}$$

and

$$\tilde{u}(\varphi) = \frac{1}{(2\pi)^n} \sum_{n \in \mathbb{Z}^n} \exp(-in \cdot \varphi) u(n) \tag{2.5b}$$

with

$$k \cdot x \equiv \sum_{i=1}^{n} k_i x_i \tag{2.5c}$$

and

$$n \cdot \varphi \equiv \sum_{i=1}^{n} n_i \varphi_i , \tag{2.5d}$$

a notation which will be used consistently throughout the book. By (2.2) et ff., (2.3) and (2.4), both H_0 and H_{0l} have purely a.c. spectrum, with

$$\sigma(H_0) = [0, \infty) \tag{2.6}$$

and

$$\sigma(H_{0l}) = [0, 4n], \tag{2.7a}$$

respectively. Frequently we shall consider, instead of H_{0l}, the operator

$$-H_{0l} + 2\mathbb{I} \tag{2.7b}$$

upon absorbing the term proportional to the identity in the potential. This operator has spectrum symmetric w.r.t. the origin (as follows from a unitary transformation); we shall denote it by the same symbol H_{0l}, as long as no confusion arises. Its spectrum is

$$\sigma(-H_{0l} + 2\mathbb{I}) = [-2n, 2n]. \tag{2.7c}$$

From (2.3), we have the evolution in momentum space

$$\tilde{u}_t(k) = \exp(-i\omega(k)t)\tilde{u}_0(k), \tag{2.8a}$$

where

$$\omega(k) = \frac{E(k)}{\hbar} = \frac{\hbar k^2}{2m}. \tag{2.8b}$$

The expectation values of the coordinates may be computed immediately from (2.8a):

$$\langle x_i \rangle_t = \int d^n x x_i |u_t(x)|^2 = (u_t, x_i u_t) = \left(\tilde{u}_t, i\frac{\partial \tilde{u}_t}{\partial k_i} \right) \tag{2.9}$$

since $i\frac{\partial \tilde{u}_t}{\partial k_i}$ is the Fourier transform of $x_i u_t$; by (2.8a) it follows that

$$i\frac{\partial \tilde{u}_t}{\partial k_i}(k) = t\frac{\partial \omega}{\partial k_i}\tilde{u}_t(k) + \exp(-i\omega t)i\frac{\partial \tilde{u}_0}{\partial k_i}. \tag{2.10}$$

From $|\tilde{u}_t(k)|^2 = |\tilde{u}_0(k)|^2$, we obtain from (2.9) and (2.10),

$$\langle x_i \rangle_t = \langle x_i \rangle_0 + t \int d^n k \frac{\partial \omega}{\partial k_i} |\tilde{u}_0(k)|^2 , \tag{2.11}$$

where, by (2.1a), the coefficient of t in the second term on the r.h.s. of (2.11) may be interpreted as a *group velocity* of the wave packet.

Problem 2.1. *Show that* (2.11) *implies*

$$\langle x_{i_1} \cdots x_{i_n} \rangle_t = \left(\frac{t\hbar}{m}\right)^n \int d^n k k_{i_1} \cdots k_{i_n} |\tilde{u}_0|^2 + O(t^{n-1}) \tag{2.12a}$$

as $t \to \infty$ *with* $i_1, \ldots, i_n \in \{1, \ldots, n\}$ *and thus*

$$(\Delta x_i)^2 = \langle (x_i - \langle x_i \rangle_t)^2 \rangle_t = \left(\frac{t}{\hbar}\right)^2 (\Delta p_i)^2 + O(t) \tag{2.12b}$$

as $t \to \infty$, *for any* $i = 1, \ldots, n$.

The meaning of (2.12b) is that the wave packet *spreads linearly in time*, since there are no states (in $\mathcal{H}$) with $(\Delta p_i)^2 = 0$. By (2.11), this phenomenon is common to quantum mechanics and classical wave theory, as long as the group velocity (2.1a) is nonzero, i.e., there is *dispersion*. The main difference between these theories lies in the fact that only in the former the uncertainty relation (1.14c) holds — a fact, which, incidentally, implies that $(\Delta p_i)^2 \neq 0$ if $u \in D(x_i) \cap D(p_i)$. Equation (2.12a) also shows explicitly that states in the a.c. spectrum of H_0 are prototypes of the ballistic regime — the forthcoming (3.138b).

The linear spreading of free wave packets has consequences also for their pointwise decay. Suppose the initial wave packet u_0 is a Gaussian of width σ, i.e., $u_0(x) = A \exp(-x^2/\sigma^2)$. By passing to momentum space or using (1.7) directly, it is easy to see that u_t is a Gaussian with width $\sigma_t = O(t)$ and amplitude $A_t = O(t^{-n/2})$: this is, of course, consistent with (2.12b) (with linear term equal to zero). The fact that $A_t = O(t^{-n/2})$ may be seen as a direct consequence of linear spreading and the conservation of probability (1.12): in the present case, this follows by a simple change of variable.

The above-mentioned pointwise decay is basic to *scattering theory*, a beautiful theory which we do not touch in this book, except for the development in the present chapter: see [105] for a quick overview, and [7, 209] for deep treatments. Our discussion would be, however, too incomplete if we did not at least point out the bare essentials of certain types of decay to scattering, and we do this in the next two paragraphs, partly following [239]. The idea is that if the potential decays sufficiently rapidly at infinity, a decay of the type (1.8) persists for the quantities of importance in scattering theory, viz. (1.8).

Let us assume that we are given a one-particle system in $\mathcal{H} = L^2(\mathbb{R}^3)$, described by a self-adjoint Hamiltonian (1.1), $H = H_0 + V$, where V is a multiplication operator (1.3a), with $V(\cdot)$ satisfying certain conditions (see, e.g., [105, Chap. 6, Theorems 6.2 and 6.3]). It is assumed that H is already a reduced description, with V denoting the interaction between the particle and a scattering center, choosing the time origin $t = 0$ to be the "moment of collision", in which the state of the system is given by a wave-function Ψ. If the potential decays sufficiently rapidly at infinity (see later), we expect $\exp(-itH/\hbar)\Psi$ to be asymptotically ($t \ll 0$) described by a free evolution $u_t = \exp(-itH_0/\hbar)\Phi_i$, which represents the motion of the particle long before the collision (Φ_i would be specified by, e.g., a momentum distribution peaked around a certain value), i.e.,

$$\| \exp(-itH_0/\hbar)\Phi_i - \exp(-itH/\hbar)\Psi\| \to 0, \quad \text{as } t \to -\infty \tag{2.13a}$$

and, since $\exp(-itH/\hbar)$ is unitary:

$$\Psi = \lim_{t \to -\infty} \exp(itH/\hbar)\exp(-itH_0/\hbar)\Phi_i. \tag{2.13b}$$

Long after the collision, the particle becomes free again, and a state $\exp(-itH/\hbar)\Phi$ becomes asymptotically, i.e., for $t \gg 0$, equal to $\exp(-itH_0/\hbar)\Phi_f)$, where Φ_f characterizes the momentum distribution of the outgoing state:

$$\| \exp(-itH_0/\hbar)\Phi_f - \exp(-itH/\hbar)\Phi\| \to 0, \quad \text{as } t \to \infty \tag{2.14a}$$

or, again by unitarity of $\exp(-itH/\hbar)$,

$$\Phi_f = \lim_{t \to \infty} \exp(itH_0/\hbar)\exp(-itH/\hbar)\Phi. \tag{2.14b}$$

We expect that the limits (2.13b) and (2.14b) exist for all $\Phi_i, \Phi_f \in \mathcal{H}$ i.e., in the strong operator topology: in other words, the so-called *wave-operators*

$$\Omega^{\pm}(H, H_0) \equiv \underset{t \to \mp\infty}{\text{s-lim}} \exp(itH/\hbar)\exp(-itH_0/\hbar) \tag{2.15}$$

exist. The *S matrix* or *scattering matrix* gives the asymptotic behavior in the future in terms of the asymptotic behavior in the past, i.e.,

$$|(\Phi_f, S\Phi_i)|^2 \tag{2.16a}$$

is the probability that a state which has the form $\exp(-itH_0/\hbar)\Phi_i$ in the distant past looks like $\exp(-itH_0/\hbar)\Phi_f$ in the distant future, that is to say, $(\Phi_f, S\Phi_i)$ is the overlap of $\Omega^+\Phi_i$ and $\Omega^-\Phi_f$, which leads to

$$S = (\Omega^-)^\dagger \Omega^+. \tag{2.16b}$$

The quantities (2.16a) are precisely those observed in scattering experiments.

Both $\Omega^{\pm}$ are isometric, i.e., norm preserving, by definition (2.15): $\|\Omega^{\pm}\Phi\| = \|\Phi\|$. Further, it is immediate (check!) that $(\Omega^+)^\dagger$ is zero on $(\text{Ran}\,\Omega^+)^\perp$. Thus, $(\Omega^+)^\dagger$ is the inverse of Ω^+ on $\text{Ran}\,\Omega^+$, and zero on the complement, and a similar assertion is true for $(\Omega^-)^\dagger$. Thus, S is isometric iff $\operatorname{Ran}\Omega^+ \subset \operatorname{Ran}\Omega^-$, and similarly $S^\dagger$ is isometric iff $\operatorname{Ran}\Omega^- \subset \operatorname{Ran}\Omega^+$. Thus, $S^\dagger S = SS^\dagger = \mathbb{I}$, i.e., S is unitary iff

$$\operatorname{Ran}\Omega^- = \operatorname{Ran}\Omega^+. \tag{2.17}$$

Equation (2.17) is the property of *weak asymptotic completeness*: see [208] for further elaboration on this essential property, but we remark that the construction by Pearson [199], which provides an example of a physically reasonable potential violating (2.17), will play a role in Sec. 3.3.

Now, let (we omit now $\hbar$ for notational economy):

$$W(t) \equiv \exp(itH)\exp(-itH_0). \tag{2.18a}$$

Formally,

$$\frac{dW(t)}{dt}\Phi = i\exp(itH)V\exp(-itH_0)\Phi. \tag{2.18b}$$

We have the following theorem.

Theorem 2.1 (Cook's method). *Suppose that $H\Phi - H_0\Phi = V\Phi$ for $\Phi \in D(H) \cap D(H_0)$. Suppose moreover that there is a total set $\mathcal{D} \subset D(H_0)$ such that $\exp(-itH_0\Phi) \in D(H)$ if $|t| \geq 1$ and that*

$$\int_1^\infty \|V\exp(\pm itH_0)\Phi\|dt < \infty \tag{2.19}$$

for all $\Phi \in \mathcal{D}$. Then, $\Omega^\pm(H, H_0)$ exist.

Proof. (See [239]). Since $W(t)$ is linear, and norm-bounded (by one), proof of existence of $\Omega^\pm(H, H_0)$ need only be done for a total set, i.e., a set whose finite linear combinations are dense (e.g., Gaussians). We have

$$\|[W(t) - W(s)]\Phi\| \leq \int_s^t \left\|\frac{dW(u)}{du}\Psi\right\| du,$$

and the limit $t, s \to \pm\infty$ of the l.h.s. of the inequality above exists by (2.18b), which is justified by the domain hypothesis, and (2.19). $\square$

We now show (again following [239]) that (2.19) may be shown in an elementary and very instructive way as a consequence of pointwise decay (1.8), which we write as

$$\|\exp(-itH_0)\Psi\|_\infty \leq Ct^{-3/2}\|\Phi\|_1, \tag{2.20}$$

where C is a constant. Hölder's inequality (1.10) implies that, if $f \in L^s(\mathbb{R}^n)$ $\cap L^t(\mathbb{R}^n)$, and $q^{-1} = \theta s^{-1} + (1-\theta)t^{-1}$, with $0 < \theta < 1$, and $1 \le s < q < t \le \infty$, then $f \in L^q(\mathbb{R}^n)$ and

$$\|f\|_q \le \|f\|_s^\theta \|f\|_t^{1-\theta}. \tag{2.21}$$

Problem 2.2. *Prove* (2.21) *from Hölder's inequality* (1.10).

Hint: Consider first $t < \infty$, treating the case $t = \infty$ separately, and define $v = \frac{p}{\theta q}$, then $v \ge 1$. Write $\|u\|_q^q = \int_{\mathbb{R}^n} |u(x)|^{\theta q} |u(x)|^{(1-\theta)q} dx$ and apply (1.10) to this expression using the exponent v.

Hölder's inequality also implies

$$\|V\eta\|_p \le \|V\|_r \|\eta\|_q \quad \text{with } p^{-1} = r^{-1} + q^{-1}. \tag{2.22}$$

This is seen by considering r/p and q/p as conjugate exponents and applying (1.10) to the integral of $|V|^p |\eta|^p$. Now, let $n = 3$. From (2.22), the unitarity of $\exp(-itH_0)$, which implies $\| \exp(-itH_0\Phi)\|_2 = \|\Phi\|_2$, (2.22), with $p = 2$, and (2.21), with $s = 2$ and $t = \infty$, it follows that (check!):

$$\|V \exp(-itH_0\Phi)\|_2 \le Ct^{-3/r} \|V\|_r, \tag{2.23}$$

where $r \ge 2$. By (2.23), (2.19) holds as long as

$$V(x) = V_1(x) + V_2(x) \quad \text{for a.e. } x \in \mathbb{R}^3 \tag{2.24a}$$

with

$$V_1 \in L^2(\mathbb{R}^3) \tag{2.24b}$$

and

$$V_2 \in L^r(\mathbb{R}^3) \quad \text{with } r < 3. \tag{2.24c}$$

The potential

$$V(x) = |x|^{-1-\varepsilon} \quad \text{with } \varepsilon > 0 \tag{2.25}$$

satisfies (2.24a), with $V_1(x) \equiv V(x)\chi(|x| \le 1)$, and $V_2(x) \equiv V(x)(1 - \chi(|x| \le 1))$, where $\chi(A)$ denotes the characteristic function of a region A. This shows that pointwise decay (2.20) of the free wave packet, which is, physically, a consequence of linear spreading and conservation of probability, implies the necessary decay (2.19) for the existence of scattering states, as long as the space rate of fall-off of the potential is, roughly, faster than Coulomb (in the latter case a modification of the definition (2.15) of the wave-operators is necessary, which takes into account the long range of the Coulomb potential, see [71]).

We have presented this result in this chapter for two reasons: first, it is the most economical [239] way of estimating the effect of the interaction potential in proving the decay of the r.h.s. of (2.23), which is necessary in Theorem 2.1, i.e., not requiring anything beyond Hölder's inequality; second, it illustrates — again, in the simplest form, but in a physically relevant example — how pointwise decay (2.20) may be used to obtain L^2-decay (2.23); the inverse problem, obtaining pointwise decay from L^2-decay, is treated at greater length in Chap. 3. More advanced methods, namely, the Riesz–Thorin convexity theorem (see [208, Theorem IX.17, p. 17] or [193]), may be used to obtain stronger estimates, see [239] or [208].

The last topic of this chapter is the *method of stationary phase*, which was advocated by Rudolf Haag in his seminal paper with Brenig [33]. This method, and variations thereof, will be important in Chap. 4, which deals with a lattice model with a potential, i.e., with a Hamiltonian H of the form (1.1), with $H_0 = -\triangle_l$ (more precisely, the forthcoming (3.84a), with V a sparse potential of the form (3.84c)). Accordingly, we shall use it as an introduction to prove the pointwise decay of free lattice wave packets in this chapter. Both our method in Chap. 4 and the free lattice case rely on the forthcoming (Proposition 2.2) van der Corput's approach, which explains the special emphasis in the title of this chapter.

Our exposition of the stationary phase method follows the profound, and at the same time eminently pedagogical, exposé of Stein in [250], together with some excerpts from Zygmund's classic [280], but it is evidently no substitute to (rather an elementary introduction to a tiny part of) the treatment by these masters! An excellent, but more advanced treatment is the recent book by Costin [49]. From the perspective of microlocal analysis (which is not needed in this book), the extensive treatment in [117] is invaluable; an excellent introduction to the latter is given in [100].

We start with the behavior for large positive λ of the integral

$$I(\lambda) = \int_a^b \exp(i\lambda\Phi(x))\Psi(x)dx, \qquad (2.26)$$

where Φ is a real-valued smooth function (the *phase*) and Ψ is complex-valued and smooth (often with compact support in (a, b)). According to [250], the basic facts of $I(\lambda)$ depend on three principles: *localization*, *scaling* and *asymptotics*.

Localization means that, if Φ has compact support in (a, b), the asymptotic behavior of $I(\lambda)$ is determined by the points where $\Phi'(x) = 0$, i.e., the points of *stationary phase*. We have:

Proposition 2.1. *Let* Φ *and* Ψ *be smooth functions such that* Ψ *has compact support in* (a, b), *and* $\Phi'(x) \neq 0$ *for all* $x \in [a, b]$. *Then,*

$$I(\lambda) = O(\lambda^{-N}), \quad \text{as } \lambda \to \infty \tag{2.27}$$

for all $N \geq 0$.

Proof. Let D denote the differential operator

$$(Df)(x) = (i\lambda\Phi'(x))^{-1}\frac{df}{dx} \tag{2.28a}$$

and ^tD its transpose

$$(^tDf)(x) = -\frac{d}{dx}\left(\frac{f}{i\lambda\Phi'(x)}\right). \tag{2.28b}$$

Then, $D^N(\exp(i\lambda\Phi)) = \exp(i\lambda\Phi)$ for all N, and by partial integration

$$\int_a^b \exp(i\lambda\Phi(x))\Psi dx = \int_a^b (D^N \exp(i\lambda\Phi))\Psi dx$$

$$= \int_a^b \exp(i\lambda\Phi)(^tD^N\Psi)dx. \tag{2.28c}$$

From (2.28b) and (2.28c), $|I(\lambda)| \leq A_N\lambda^{-N}$. $\qquad\square$

If Ψ is not assumed to vanish near the endpoints of the interval $[a.b]$, (2.28c) must be modified by including the extra boundary terms. The example $\Phi(x) = x$ with $\Psi(x) = 1$ shows that $O(\lambda^{-1})$ is the best general decay possible in this case, but if Φ and Ψ are *periodic*, i.e., $\Phi^{(k)}(a) = \Phi^{(k)}(b)$ and $\Psi^{(k)}(a) = \Psi^{(k)}(b)$, for all $k \geq 0$, the boundary terms cancel, and (2.27) is again seen to hold.

Scaling: Suppose, we know only that

$$|\Phi^{(k)}(x)| \geq \rho > 0 \tag{2.29}$$

for some *fixed* k, and we wish to obtain an estimate of $I(\lambda)$ for $\Psi(x) \equiv 1$ which is *independent* of a and b. The change of variable (scaling) $x \to (\lambda)^{-1/k}x'$ shows that the only possible estimate for the integral is $O(\lambda^{-1/k})$. This estimate actually does hold, by the following important result.

Proposition 2.2 (van der Corput). *Suppose Φ is real-valued and smooth in (a,b) and (2.29) holds for all $x \in (a,b)$. Then, for c_k independent of Φ and λ,*

$$|I(\lambda)| = \left| \int_a^b \exp(i\lambda\Phi(x))dx \right| \leq c_k(\lambda\rho)^{-1/k} \qquad (2.30)$$

holds when: (i) $k \geq 2$ *or* (ii) $k = 1$ *and* $\Phi(x)$, $a \leq x \leq b$, *has a monotonic derivative* $\Phi'(x)$, *and there exists* $\mu > 0$ *such that* $\Phi' \geq \mu$ *or* $\Phi' \leq -\mu$ *in* (a,b).

Proof. Consider first (ii). By hypothesis,

$$I(\lambda) = (i\lambda)^{-1} \int_a^b \frac{d\exp(i\lambda\Phi(x))}{\Phi'(x)}. \qquad (2.31)$$

Applying the second mean-value theorem [218, Theorem 6.32, p. 124] to the real and imaginary parts of (2.31), we obtain (2.30). We show (i) only for $k = 2$ (see [250] for the general case). Corresponding to (2.29), let

$$\Phi^{(2)}(x) \geq \rho, \qquad (2.32)$$

otherwise, we may replace Φ by $-\Phi$ and I by its complex conjugate. Then, Φ' is increasing. Suppose initially that Φ' is of constant sign in (a,b), say positive. Then, if $a < \gamma < b$, it follows from (2.32) that $\Phi' \geq (\gamma - a)\rho$ in (γ, b). Therefore,

$$|I(\lambda)| \leq \left| \int_a^\gamma \exp(i\lambda\Phi(x))dx \right| + \left| \int_\gamma^b \exp(i\lambda\Phi(x))dx \right|$$

$$\leq \gamma - a + \frac{2}{(\gamma - a)\rho\lambda}. \qquad (2.33)$$

Choosing γ in (2.33) in order to render the r.h.s. of (2.33) a minimum, we find $|I(\lambda)| \leq 2\sqrt{2}/\sqrt{\rho\lambda}$. In the general case, (a,b) is the union of two intervals in each of which Φ' is of constant sign and (2.30) (with $c_2 = 4\sqrt{2}$) follows from adding the inequalities for these two intervals. $\qquad \square$

Under the same assumptions on Φ of Proposition 2.2, it follows that:

Proposition 2.3.

$$\left| \int_a^b \exp(i\lambda\Phi(x))\Psi(x)dx \right| \leq c_k(\lambda\rho)^{-1/k} \left(|\Psi(b)| + \int_a^b |\Psi'(x)|dx \right). \qquad (2.34)$$

Proof. Write the l.h.s. of (2.34) as $|\int_a^b F'(x)\Psi(x)dx|$, with $F(x) = \int_a^x \exp$ $(i\lambda\Phi(t))dt$, and integrate by parts using (2.30). $\qquad\qquad\square$

Asymptotics: The third principle describes the full asymptotic development of $I(\lambda)$ using both localization and scaling. We know that when the support of Ψ is a compact subset of (a, b), the behavior of $I(\lambda)$ is determined by the critical points of Φ, i.e., those points x_0 such that $\Phi'(x_0) = 0$. Assuming now that the support of Ψ is so small that it contains only one critical point of Φ, the character of the asymptotic expansion depends on the smallest $k \geq 2$ for which

$$\Phi^{(k)}(x_0) \neq 0, \tag{2.35}$$

and is given in terms of powers of λ in a way consistent with Proposition 2.2:

Proposition 2.4 (Method of stationary phase). *Assume* (2.35) *and*

$$\Phi(x_0) = \Phi'(x_0) = \cdots = \Phi^{(k-1)}(x_0) = 0. \tag{2.36}$$

If Ψ is supported in a sufficiently small neighborhood of x_0, then

$$I(\lambda) = \int \exp(i\lambda\Phi(x))\Psi(x)dx \sim \lambda^{-1/k}\sum_{j=0}^{\infty} a_j\lambda^{-j/k} \tag{2.37a}$$

in the sense that, for all non-negative integers N and r,

$$\left(\frac{d}{d\lambda}\right)^r\left(I(\lambda) - \lambda^{-1/k}\sum_{j=0}^{N} a_j\lambda^{-j/k}\right) = O(\lambda^{-r-(N+1)/k}) \tag{2.37b}$$

as $\lambda \to \infty$.

Proof. We follow [250] quite literally: these arguments are, however, important in toto, and for this reason we reproduce them here. It is convenient to divide the proof into four steps:

Step 1. We observe that

$$\int_{-\infty}^{\infty} \exp(i\lambda x^2)x^l \exp(-x^2)dx \sim \lambda^{-(l+1)/2}\sum_{j=0}^{\infty} c_j^l\lambda^{-j} \tag{2.38a}$$

for any non-negative even integer l (if l is odd the integral vanishes). The l.h.s. of (2.38a) is $\int_{-\infty}^{\infty} \exp[-(1-i\lambda)x^2]x^l dx$. Set $z = (1-i\lambda)^{1/2}x$. Note that the rapid decay of $\exp(-x^2)$ allows replacement of the contour $(1 - i\lambda)^{1/2}\mathbb{R}$ by $\mathbb{R}$, arriving at

$$(1 - i\lambda)^{-1/2-l/2} \int_{-\infty}^{\infty} \exp(-x^2)x^l dx.$$

The principal branch of $z^{-(l+1)/2}$ in the complex plane has been fixed cutting the plane along the negative real axis; with this determination

$$(1 - i\lambda)^{-(l+1)/2} = \lambda^{-(l+1)/2}(\lambda^{-1} - i)^{-(l+1)/2}$$

if $\lambda > 0$. The power series expansion of $(w - i)^{-(l+1)/2}$ on the disc $|w| < 1$ yields (2.38a) with $w = \lambda^{-1} \to 0$.

Step 2. Observe now that if $\eta \in C_0^\infty$ and l is a non-negative integer,

$$\left| \int_{-\infty}^{\infty} \exp(i\lambda x^2)x^l \eta(x)dx \right| \leq A\lambda^{-1/2-l/2}. \tag{2.38b}$$

To prove (2.38b), let α be a C^∞ function with the property that $\alpha(x) = 1$ for $|x| \leq 1$, and $\alpha(x) = 0$ for $|x| \geq 2$, and write

$$\int \exp(i\lambda x^2)x^l \eta(x)dx = \int \exp(i\lambda x^2)x^l \eta(x)\alpha(x/\varepsilon)dx$$

$$+ \int \exp(i\lambda x^2)x^l \eta(x)[1 - \alpha(x/\varepsilon)]dx, \tag{2.38c}$$

where $\varepsilon > 0$ will be chosen shortly. The first integral is bounded by $C\varepsilon^{l+1}$, on the second we use $D\exp(i\lambda x^2) = \exp(i\lambda x^2)$, where D is the operator (2.28a) with $\Phi(x) = x^2$, and integrate by parts, obtaining the bound

$$\frac{C_N}{\lambda^N} \int_{|x| \geq \varepsilon} |x|^{l-2N} = C_N' \lambda^{-N} \varepsilon^{l-2N+1}$$

if $l - 2N < -1$. The two previous bounds in (2.38c) yield the global bound $C_N[\varepsilon^{l+1} + \lambda^{-N}\varepsilon^{l-2N+1}]$, and choosing $\varepsilon = \lambda^{-1/2}$ (with $N > (l+1)/2$) we obtain (2.38b).

Problem 2.3. *By a similar, but simpler, argument of integration by parts, show that*

$$\int \exp(i\lambda x^2)g(x)dx = O(\lambda^{-N}), \quad \forall N \geq 0 \tag{2.38d}$$

whenever $g \in \mathcal{S}(\mathbb{R})$, *where* $\mathcal{S}$ *denotes the Schwartz space of infinitely differentiable functions such that*

$$\sup_{x \in \mathbb{R}} \left| x^a \left(\frac{dg}{dx} \right)^b \right| \leq C_{a,b}, \quad \forall \text{ integers } a \text{ and } b,$$

where $C_{a,b}$ *are finite constants, and in addition* $g(0) = 0$.

The Schwartz space, fundamental in distribution theory, is economically and elegantly described in [193] in connection with Fourier transformation.

Step 3. We now prove the proposition in the special case $\Phi(x) = x^2$. In order to do this, write

$$\int \exp(i\lambda x^2) \Psi(x) dx = \int \exp(i\lambda x^2) \exp(-x^2) \left(\exp(x^2) \Psi(x) \right) \tilde{\Psi}(x) dx,$$

$$(2.38e)$$

where $\tilde{\Psi} \in C_0^\infty$ is equal to one on the support of Ψ. For each N, write the Taylor expansion

$$\exp(x^2) \Psi(x) = \sum_{j=0}^{N} b_j x^j + x^{N+1} R_N(x) = P(x) + x^{N+1} R_N(x).$$

Substituting this expansion in (2.38e) yields three terms:

$$\sum_{j=0}^{N} b_j \int_{-\infty}^{\infty} \exp(i\lambda x^2) \exp(-x^2) x^j \, dx, \qquad (2.38f)$$

$$\int_{-\infty}^{\infty} \exp(i\lambda x^2) x^{N+1} R_N(x) \exp(-x^2) \tilde{\Psi}(x) dx, \qquad (2.38g)$$

and

$$\int_{-\infty}^{\infty} \exp(i\lambda x^2) P(x) \exp(-x^2) [\tilde{\Psi}(x) - 1] dx. \qquad (2.38h)$$

For (2.38f), use (2.38a), for (2.38g) use (2.38b), and for (2.38h) use (2.38d); thus we arrive (check!) at the combination (2.37a) for the special case $\Phi(x) = x^2$.

Step 4. For the general case with $k = 2$, write $\Phi(x) = c(x - x_0)^2 + O(|x - x_0|^3)$ with $c \neq 0$ and set $\Phi(x) = c(x - x_0)^2 [1 + \varepsilon(x)]$, where $\varepsilon(\cdot)$ is a smooth $O(|x - x_0|)$ function, hence $|\varepsilon(x)| < 1$ when x is sufficiently close to x_0. Moreover, $\Phi'(x) \neq 0$ when $x \neq x_0$ is sufficiently close to x_0.

Fix a neighborhood U of x_0 such that both these conditions hold on U, and let $y \equiv (x - x_0)[1 + \varepsilon(x)]^{1/2}$. Then the mapping $x \to y$ is a diffeomorphism (a differentiable one-to-one mapping with a differentiable inverse) from U to a neighborhood of $y = 0$, and of course $cy^2 = \Phi(x)$. Thus $\int \exp(i\lambda\Phi(x))\Psi(x)dx = \int \exp(i\lambda cy^2)\tilde{\Psi}(y)dy$ with $\tilde{\Psi} \in C_0^\infty$ if the support of Ψ lies in U. Equation (2.37a) follows thus for general Φ satisfying (2.35b), for $k = 2$. The proof for higher k is based on the identity $\int_0^\infty \exp(i\lambda x^k) \exp(-x^k)x^l dx = c_{k,l}(1 - i\lambda)^{(l+1)/k}$ along similar lines. $\square$

Equation (2.37a) shows that in the case $k = 2$ the main contribution to $I(\lambda)$ is

$$a_0 \lambda^{-1/2} \tag{2.39}$$

with

$$a_0 = \exp(i\lambda\phi(x_0)) \left(\frac{2\pi}{-i\Phi''(x_0)} \right)^{1/2} \Psi(x_0)$$

if we drop in (2.36) the restriction $\Phi(x_0) = 0$.

Sometimes, albeit rarely, it is possible to find the expansion (2.37a) directly in $\mathbb{R}^n$:

Problem 2.4. *Use* (1.7) *to find the asymptotic expansion of* $(\exp(-iH_0 t/\hbar)u)(x)$ *for* $u \in C_0^\infty(\mathbb{R}^n)$.

Hint: Expand $\exp(i\frac{m|x-y|^2}{2\hbar t})$.

In the lattice, however, the situation is not so simple, and, as an application of Proposition 2.4, we consider the return or survival probability (the forthcoming (3.1a)) for a dense set of initial states Ψ_0:

$$|(\Psi_0, \Psi(t))|^2 = |(\Psi_0, \exp(-itH/\hbar)\Psi_0)|^2, \tag{2.40}$$

where $H = H_{0,l} = -\triangle_l$ is the difference Laplacian (1.2d) and (1.2e) responsible for the dynamics of free spin waves. Since we shall be concerned with the one-dimensional case in Sec. 4.2 (see, however, Sec. 4.1 and part of Sec. 4.2 for the separable multidimensional case), we shall, by comparison, restrict ourselves to dimension $d = 1$. By (1.2d) and (1.2e), the finite difference Laplacian may be written ($\hbar = 1$) as:

$$\triangle_l = T + T^{-1} - 2\mathbb{I}, \tag{2.41a}$$

where T is the operator of translation to the right by one lattice step:

$$(Tu)(n) = u(n + 1), \quad n \in \mathbb{Z} \quad \text{and} \quad u \in l^2(\mathbb{Z}). \tag{2.41b}$$

By (2.41a),

$$\exp(-itH_{0l}) = \exp(-2it)\exp(t[(iT) - (iT)^{-1}]). \tag{2.41c}$$

The generating function for Bessel functions of integer order is (see [1, 9.1.41]):

$$\exp[(z/2(\rho - \rho^{-1})] = \sum_{n=-\infty}^{\infty} \rho^n J_n(z) \tag{2.41d}$$

and hence, from (2.41c),

$$\exp(-itH_{0l}) = \exp(-2it) \sum_{n=-\infty}^{\infty} i^n J_n(2t)T^n. \tag{2.41e}$$

It follows from (2.41e) that, for $\Psi_0 \in l^2(\mathbb{Z})$ and for all $n' \in \mathbb{Z}$,

$$(\exp(-itH_{0l}\Psi_0))(n') = \exp(-2it) \sum_{m \in \mathbf{Z}} i^{m-n'} J_{m-n'}(2t)\Psi_0(m) \tag{2.42}$$

(see [159] for this quick derivation). Equation (2.42) is the analogue of (1.7) for $d = 1$. Alternatively, we may, by (2.4), define H_{0l} as an operator on the (Fourier) space $L^2(0, 2\pi)$ by

$$(H_{0l}\tilde{\Psi}_0)(\varphi) = 2(1 - \cos\varphi)\tilde{\Psi}_0(\varphi) \tag{2.43}$$

for all $\varphi \in [0, 2\pi]$ and for all $\tilde{\Psi}_0 \in L^2(0, 2\pi)$, and thus, in correspondence to (2.42) (by the convolution theorem)

$$(\exp(-iH_{0l}t)\Psi_0)(n') = \frac{1}{2\pi} \int_0^{2\pi} d\varphi \exp(-2it + 2it\cos\varphi - in'\varphi)\tilde{\Psi}_0(\varphi). \tag{2.44}$$

Equation (2.44) is of the form (2.26), with $\lambda = 2t$, $\Phi(x) = \cos x$, and $\Psi(x) = \exp(-in'x)$. Noticing that Φ' vanishes only at $x_0 = 0$, $x_0 = \pi$ and $x_0 = 2\pi$ in $[0, 2\pi]$, and there $\Phi''(x_0) = \pm 1$, write $1 = \Phi_1 + \Phi_2 + \Phi_3$, where Φ_1 has small support near π and equals one near π, Φ_2 has small support near 0 and 2π, and equals one near both points. We now use Proposition 2.4 for the first two terms, and take into account that both Φ and $\Psi_3(x) \equiv \exp(-in'x)\Phi_3(x)$ are periodic of period 2π, implying that the boundary terms vanish as in Proposition 2.1. Notice, however, that the operator ${}^tD$ also acts on $\exp(-in'x)$, yielding a factor in' each time! We thus find

$$|(\exp(-itH_{0l}\Psi_0)(n')| \le C_{n'}|t|^{-1/2} \quad \text{for a constant } 0 < C_{n'} < \infty \tag{2.45}$$

for all $|t| \geq 1$. One should compare (2.45) with (2.20) (for $d = 1$, the $t^{-3/2}$ being replaced by $t^{-1/2}$), which is uniform in the coordinate (for $\Phi \in L^1$). In the lattice case, the uniformity does not hold; indeed, alternatively, from (2.42), one obtains instead

$$|(\exp(-itH_{0l}\Psi_0)(n')| \leq \sup_{m \in \mathbb{Z}} |J_m(2t)| \, \|\Psi_0\|_1, \qquad (2.46)$$

where $\|\Psi_0\|_1 = \sum_{m \in \mathbb{Z}} |\Psi_0(m)|$ is the $l^1(\mathbb{Z})$-norm of Ψ_0, assumed finite. We shall come back to (2.46) shortly.

We now reconsider (2.44) for general n', i.e., n' may be, for instance, $O(t)$. Let

$$\Phi(\varphi) \equiv 2\cos\varphi - \frac{n'\varphi}{t}. \qquad (2.47a)$$

Then,

$$\Phi'(\varphi) = -2\sin\varphi - \frac{n'}{t}. \qquad (2.47b)$$

The critical points $\Phi'(\varphi) = 0$ correspond by (2.47b) to those wave vectors $\varphi = k$ such that the group velocity (2.1a) $v_g = \frac{\partial E}{\partial k} = -2\sin k = \frac{n'}{t}$, the latter being the points in the velocity space of a classical particle. Outside this "classically allowed region", the wave packet decays faster than any power of $|t|$ by Proposition 2.1 (for an n-dimensional generalization of this picture, see [209, App. 1 to XI-3: stationary phase methods]).

Problem 2.5. *Show that*

$$\sup_{n' \in \mathbb{Z}} |(\exp(-itH_{0l}\Psi_0)(n')| \leq A|t|^{-1/3} \qquad (2.48)$$

for some constant $A > 0$.

Hint: In (2.47a), note that $|\Phi''(\varphi)| + |\Phi'''(\varphi)| \geq c > 0$ and apply Proposition 2.2.

That the estimate in (2.48) is best possible may be seen by the asymptotic expansion.

Problem 2.6. *Show that*

$$\exp(-itH_{0l}\Psi_0)(n' = 2t) = c(2t)^{-1/3} + O((2t)^{-2/3}), \quad \textit{as } t \to \infty,$$

$$(2.49a)$$

where, for $\tilde{\Psi}_0(\varphi) = 1$,

$$c = \frac{\Gamma(1/3)}{\pi 2^{2/3} 3^{1/6}}. \tag{2.49b}$$

Hint: Use Proposition 2.4 and note that, when $2t = n'$, $\Phi'(3\pi/2) = \Phi''(3\pi/2) = 0$, but $\Phi'''(3\pi/2) \neq 0$.

For $\tilde{\Psi}_0(\varphi) = 1$, or $\Psi_0 = \delta_0$ (the Kronecker delta at the origin),

$$(\exp(-itH_{0l}\Psi_0))(n') = \exp(-2it)i^{n'} J_{n'}(2t) \tag{2.50a}$$

by [1, (9.1.21)]; the corresponding return probability amplitude is

$$(\Psi_0, \Psi_0(t)) = \exp(-2it)i^{n'} J_0(2t) \tag{2.50b}$$

which is $O(|t|^{-1/2})$ by [1, (9.2.1)]. Since the latter bound holds for $J_n(2t)$, for any *fixed* n, i.e., independent of t, we obtain that there exists a dense set $\mathcal{D}$ consisting of finite linear combinations of the vectors δ_i, $i = 1, \ldots, M$, such that

$$|(\Psi_0, \Psi_0(t))| \leq A_M |t|^{-1/2} \tag{2.50c}$$

which is seen to be equivalent to (2.45). In this special case $\Psi_0 = \delta_0$, Problem 2.6 corresponds to the asymptotic expansion

$$J_m(m) = cm^{-1/3} + O(m^{-2/3}) \tag{2.51}$$

with c as in (2.49b), which is related with the best value $A = 0.7857\ldots$ in the inequality

$$\sup_{m \in \mathbb{Z}} |J_m(t)| \leq A|t|^{-1/3} \tag{2.52}$$

(see [159]). Inequality (2.52), with some $A > 0$ independent of t, follows from Problem 2.5, (2.42) and (2.50a) and yields a corresponding bound in (2.46) equal to the r.h.s. of (2.52): this is the bound which replaces (1.8) in the lattice case!

We have seen that van der Corput's proposition is very powerful: this will be further developed in our Lemma 4.1 in Sec. 4.3. A further method of van der Corput, combining the Poisson summation formula (see [280, Vol. 1, Chap. II, (13), p. 68]), with the stationary phase method has also proved very powerful in the study of exponential (Gaussian) sums, see [98].

The method of stationary phase is also important in the theory of decay of smooth solutions of the Klein–Gordon equation [208, Theorem XI.17, p. 43], which is the basic ingredient of the Haag–Ruelle theory of scattering of fields and particles (see [135], also see [13] and the excellent summary in [158]).

We urge the reader to complement this introduction with the reading of [250, Chap. VIII], in particular regarding the situation in $\mathbb{R}^n$ with $n > 1$, which is complicated by the multiplicity and complexity of the critical points for $n > 1$, and the decay estimates for the Fourier transforms of measures carried on surfaces, where the conditions required on the phase Φ come about because of "curvature" conditions on the surfaces: an elementary example of the latter will occur in Sec. 3.3.

Chapter 3

The Relation Between Time-Like Decay and Spectral Properties

In this chapter, we investigate an important topic, viz. the relation between time-like decay and spectral properties. A first question to be posed is: what decays? In one-particle quantum mechanics, a prototypic quantity is the *return or survival probability*

$$|(\Psi_0, \Psi(t))|^2, \tag{3.1a}$$

where

$$\Psi(t) = \exp(-itH)\Psi_0, \tag{3.1b}$$

see (1.4a)–(1.5): we again choose units such that $\hbar = 1$ and above Ψ_0 is any (initial) state in $\mathcal{H}$. More generally, one might be interested in the probability of finding the particle, initially in the state Ψ_0, in the (arbitrary) state Φ after a time t, i.e.,

$$|(\Phi, \Psi(t))|^2. \tag{3.1c}$$

In the RAGE theorem, we shall be concerned with the decay in the average (Cesàro) sense of quantities of type

$$\|K \exp(-itH)\Psi_0\|^2 \tag{3.1d}$$

under certain conditions on the operator K, see (1) and (2) of Theorem 3.2.

In this study, it will be useful to distinguish between different types of decay, which will be labeled as sections:

3.1. Decay in the average (Cesàro) sense
3.2. Decay in the L^p — sense
3.3. Pointwise decay.

We shall see that there is a marked difference between the results and the methods employed to study these three cases. Such differences — particularly regarding Secs. 3.1 and 3.3 — will reappear in ergodic theory (App. A). Section 3.4 will be devoted to quantum stability.

3.1. Decay on the Average Sense

3.1.1. *Preliminaries: Wiener's, RAGE and Weyl theorems*

In Chap. 1, we introduced the Lebesgue–Radon–Nikodym decomposition of a complex Borel measure μ on $\mathbb{R}$ (see (1.25a)–(1.27)):

$$\mu = \mu_{\text{ac}} + \mu_{\text{sc}} + \mu_{\text{pp}}. \tag{3.2a}$$

Let $\hat{\mu}$ denote as in (1.28) the F.S. transform of μ. A fundamental theorem of Wiener (see [127, Theorem 1.6]) relates decay on the mean of $\hat{\mu}$ to the nonexistence of the p.p. part μ_{pp}. Let A_μ (see (1.26)) denote the set of atoms of μ (counting multiplicities). We have the following theorem:

Theorem 3.1 (Wiener's theorem). *Let μ be a signed Borel measure. Then*

$$\lim_{T \to \infty} \frac{1}{T} \int_0^T |\hat{\mu}(t)|^2 dt = \sum_{x \in A_\mu} \mu(\{x\})^2. \tag{3.2b}$$

Proof (see [127, Theorem 1.6]). Note that

$$|\hat{\mu}(t)|^2 = \hat{\mu}(t)\overline{\hat{\mu}}(t) = \int_{\mathbb{R}^2} \exp(-it(x-y)) d\mu(x) d\mu(y).$$

Let

$$K_T(x,y) = \frac{1}{T} \int_0^T \exp(-it(x-y)) dt$$

$$= \begin{cases} (1 - \exp(-iT(x-y)))/(iT(x-y)) & \text{if } x \neq y, \\ 1 & \text{if } x = y. \end{cases}$$

Then,

$$\frac{1}{T} \int_0^T |\hat{\mu}(t)|^2 dt = \int_{\mathbb{R}^2} K_T(x,y) d\mu(x) d\mu(y).$$

Since

$$\lim_{T \to \infty} K_T(x,y) = \begin{cases} 0 & \text{if } x \neq y, \\ 1 & \text{if } x = y, \end{cases}$$

and $|K_T(x, y)| \leq 1$, it follows from the dominated convergence theorem and Fubini's theorem that

$$\lim_{T \to \infty} \frac{1}{T} \int_0^T |\hat{\mu}(t)|^2 dt = \int_{\mathbb{R}} \mu(\{x\}) d\mu(x) = \sum_{x \in A_\mu} \mu(\{x\})^2. \qquad \Box$$

We now return to the setting of Chap. 1, which describes one-particle quantum dynamics (in the continuum or on the lattice). For any function f of time t, we denote its Cesàro time average by

$$\langle f \rangle_T \equiv \langle f(t) \rangle \equiv \frac{1}{T} \int_0^T f(t) dt. \qquad (3.3)$$

Let us denote the continuous spectral subspace of H by $\mathcal{H}_c \equiv \mathcal{H}_{ac} \oplus \mathcal{H}_{sc}$, where the latter have been defined in (1.52a) and (1.55a), respectively. We begin with the following important theorem.

Theorem 3.2 (RAGE) (see [6, 81, 221]).

(1) *Let K be a compact operator (Definition 1.1),*

$$f(t) \equiv \|K \exp(-itH)\Psi\|^2 \qquad (3.4a)$$

with $\Psi \in \mathcal{H}_c$. Then,

$$\lim_{T \to \infty} \langle f \rangle_T = 0. \qquad (3.4b)$$

(2) *The same result holds if K is bounded and $K(H + i)^{-1}$ is compact.*

Proof (see [127]). The proof of (1) may be reduced to Theorem 3.1 by writing K as a norm limit of finite-rank operators (1.39) (in the general case K need not be self-adjoint and in (1.39) one would have $(\Phi_n, \cdot)\Theta_n$ instead). Use of the triangle inequality and induction leads to further reduction to the rank-one operator $K = (\Phi, \cdot)\Theta$.

Problem 3.1. *Prove this.*

It follows that it suffices to show that for $\Phi \in \mathcal{H}$ and $\Psi \in \mathcal{H}_c$,

$$\lim_{T \to \infty} \langle |(\Phi, \exp(-itH)\Psi)|^2 \rangle_T = 0. \qquad (3.4c)$$

Let $\chi_c(H)$ denote the characteristic function of the continuous spectrum, defined by the functional calculus (1.51a)–(1.51e). Then,

$$(\Phi, \exp(-itH)\Psi) = (\Phi, \exp(-itH)\chi_c(H)\Psi) = (\chi_c(H)\Phi, \exp(-itH)\Psi),$$

so that w.l.o.g. we may assume $\Phi \in \mathcal{H}_c$. Finally, by the polarization identity, we may assume $\Phi = \Psi$. Since $d\mu_\Psi$ is continuous, the result follows from Theorem 3.1.

(2) $\overline{\mathcal{D}} \equiv D(H) \cap \mathcal{H}_c$ is dense in $\mathcal{H}_c$, and it thus suffices to prove the statement for $\Psi \in \overline{\mathcal{D}}$. We write

$$\|K \exp(-itH)\Psi\| = \|K(H+i)^{-1} \exp(-itH)(H+i)\Psi\|$$

and thus the result follows from (1).   □

The above ingenious, economical proof is a simplification due to Jaksic [127] of the proof in [54].

Now, let a one-particle Schrödinger operator $H = H_0 + V$, with $H_0 = -\triangle/(2m)$, $\triangle$ being the Laplacian, and V a multiplication operator on $\mathcal{H} = L^2(\mathbb{R}^3)$ be given, and choose

$$K = P_S, \tag{3.5a}$$

where S is a bounded region of $\mathbb{R}^3$, and

$$(P_S f)(x) = \begin{cases} f(x) & \text{if } x \in S, \\ 0 & \text{if } x \notin S. \end{cases} \tag{3.5b}$$

We first observe that if A is a multiplication operator on $\mathcal{H}$ such that $A \in L^2(\mathbb{R}^3)$,

$$(p|A(H_0+i)^{-1}|q) = \hat{A}(p-q)\left(i - \frac{|q|^2}{2m}\right)^{-1} \tag{3.6}$$

is a Hilbert–Schmidt (HS) kernel (see [207, Chap. VI]), and

$$\|A(H_0+i)^{-1}\|_{HS}^2 = \int d^3p\, d^3q\, |(p|A(H_0+i)^{-1}|q)|^2$$

$$= \|A\|^2 \int_0^\infty 4\pi q^2 dq \frac{1}{|i - |q|^2/(2m)|^2} < \infty. \tag{3.7}$$

Thus $A(H_0+i)^{-1}$ is a HS operator, which implies compactness (see [207, Chap. VI]). By (3.5a) and (3.5b), $K \in L^2(\mathbb{R}^3)$, and thus $K(H_0+i)^{-1}$ is compact. On the other hand,

$$K(H+i)^{-1} = K(H_0+i)^{-1} + K((H+i)^{-1} - (H_0+i)^{-1}) \tag{3.8}$$

and, by the resolvent equation (1.17),

$$(H_0+i)^{-1} - (H+i)^{-1} = (H_0+i)^{-1}V(H+i)^{-1}. \tag{3.9}$$

If $V \in L^2$, by the same argument (3.6) and the fact that $K(H_0 + i)^{-1}$ is compact, it follows from (3.7) that $K(H + i)^{-1}$ is compact. This may be proved, however, under much less stringent assumptions on V, which includes virtually all potentials occurring in physical applications (see [248, Theorem 4, p. 268]).

By (3.3)–(3.5b), we see that Theorem 3.2 has the physical interpretation that states in the continuous spectral subspace have zero mean "sojourn time" in any finite region of space, while, for states in the p.p. component $\mathcal{H}_{pp}$ this mean sojourn time is nonzero by Theorem 3.1. On the other hand, the theorem treats the a.c. component and the s.c. component on the same footing. We shall come back to this question from the point of view of the present section (RAGE theorems for exotic spectra) in Theorem 3.9. We wish, however, to briefly point out that in several important cases (recall (1.21a)–(1.21d)) $\sigma_{ess}(H) = \sigma_{ess}(H_0) = \sigma(-\triangle) = [0, \infty)$, and that the eigenvalues of H in the complement of $\sigma(H_0)$ consist of discrete eigenvalues of finite multiplicity which can accumulate at most at the bottom 0 of the essential spectrum, see [207, Theorem VI-16; 210]; under certain conditions, $\sigma_{ess}(H) = \sigma_{ac}(H)$, and the RAGE theorem provides a complete geometric distinction between the (absolutely) continuous and point spectrum. These facts are beautifully presented for the general N-body problem in [209, 210].

It is, nevertheless, important for future developments that the reader knows what is the main idea behind the proofs of assertions such as $\sigma_{ess}(A) = \sigma_{ess}(B)$, for two self-adjoint operators A and B. Definitions (1.21a)–(1.21d) imply that, for a self-adjoint operator A, a point $\lambda_0 \in \sigma_{ess}(A)$ iff, for all $\varepsilon > 0$, $\delta \equiv (\lambda_0 - \varepsilon, \lambda_0 + \varepsilon)$ is such that E_δ has infinite rank (i.e., the dimension of its range is infinite), where $\{E_\lambda\}$ is the spectral family associated to A (Definition 1.2). Points of $\sigma_{ess}(A)$ are thus either isolated points of infinite multiplicity or nonisolated points. The latter may lie in $\sigma_c(A)$, the continuous spectrum of A (including the case of point eigenvalues embedded in the continuum), or belong to the dense p.p. spectrum. We have the fundamental lemma.

Lemma 3.1 (Weyl). *Let A be a self-adjoint operator. Then, $\lambda_0 \in \sigma_{ess}(A)$ iff there exists an orthonormal sequence $\{\Psi_n\}_{n=1,2,...}$ such that $\Psi_n \in D(A)$ for all $n = 1, 2, \ldots$ and*

$$\|(A - \lambda_0 I)\Psi_n\| \to 0 \quad as \ n \to \infty. \tag{3.10}$$

For the proof, see [127, Theorem 3.25]. One of the theorems of the previously mentioned type, which covers a large number of applications, is the following theorem.

Theorem 3.3 (Weyl–Kato). *Let H_0 and H be self-adjoint and bounded from below, $H_0 \geq -M_1\mathbb{I}$, $H \geq -M_2\mathbb{I}$, with M_1 and M_2 non-negative constants and $\alpha < \min(-M_1, -M_2)$. If*

$$(H + \alpha)^{-1} - (H_0 + \alpha)^{-1} \equiv B \tag{3.11}$$

is a compact operator, then $\sigma_{\text{ess}}(H) = \sigma_{\text{ess}}(H_0)$.

Proof. Let $A = (H_0 + \alpha)^{-1}$, $C = (H + \alpha)^{-1}$, we need to prove that $\sigma_{\text{ess}}(C) = \sigma_{\text{ess}}(A + B) = \sigma_{\text{ess}}(A)$. By symmetry, it suffices to prove that $\sigma_{\text{ess}}(A + B) \subset \sigma_{\text{ess}}(A)$. Let $\lambda_0 \in \sigma_{\text{ess}}(A + B)$, and Ψ_n be the orthonormal sequence of Lemma 3.1 for the operator $A + B$. By Bessel's inequality [207, Chap. II, p. 38], Ψ_n converges weakly to zero, and, since B is compact, it follows that $B\Psi_n \to 0$, i.e., the convergence is strong (see [207, Theorem VI]). Thus, by (3.10), $\|(A - \lambda_0\mathbf{I})\Psi_n\| \to 0$, which implies that $\lambda_0 \in \sigma_{\text{ess}}(A)$ by Lemma 3.1. $\qquad\square$

3.1.2. *Models of exotic spectra, quantum KAM theorems and Howland's theorem*

We now turn to RAGE theorems for *exotic spectra*, which, following [164], we define as either of the three types: singular continuous (s.c.) spectrum, dense (or thick) pure point (p.p.) spectrum and recurrent absolutely continuous (a.c.) spectrum. The latter (together with the complementary transient a.c. spectrum) was introduced by Avron and Simon in their seminal paper [16], and is defined in App. C. It should be mentioned at this point that Theorem 3.1 is also applicable to dense p.p. spectra, and that Theorem 3.2 does not distinguish between recurrent and transient a.c. spectra. In fact, as remarked before, it treats even the a.c. and s.c. spectra on the same footing. We shall be concerned, in this section, primarily with a RAGE theorem for s.c. spectra (Theorem 3.9), and in the next section both the recurrent a.c. spectrum (Theorem 3.14) and the s.c. versus a.c. spectrum from a different point of view (Theorem 3.16 and Corollary 3.1) will make their appearance.

We start by motivating the study of exotic spectra by means of examples. As remarked by Last [164] in his beautiful review of exotic spectra, the s.c. spectrum was, before 1978 (which marked the appearance of [210]), "a non-occurring phenomenon, which complicated life by requiring some effort to prove it did not occur". In the mean time, the situation changed dramatically: s.c. spectrum has appeared in quantum mechanics in a variety of situations, starting with Pearson's seminal paper [200] of 1978, which constructs a one-dimensional Schrödinger operator

$$H = \frac{-d^2}{dx^2} + V \tag{3.12a}$$

on $L^2(\mathbb{R})$, with V multiplication by an even function on $\mathbb{R}$ such that, for $x \geq 0$,

$$V(x) = \sum_{n=0}^{\infty} g_n f(x - a_n). \tag{3.12b}$$

Above, f is a C^∞ function with support in $(-1/2, 1/2)$, $f \geq 0$, and

$$\sum_{n=0}^{\infty} g_n^2 = \infty, \tag{3.12c}$$

but in such a way that

$$g_n \to 0 \quad \text{as } n \to \infty. \tag{3.12d}$$

The positions a_n of the "bumps" in V in (3.12b) satisfy:

$$a_{n+1} - a_n = \exp(\exp(n)). \tag{3.12e}$$

By (3.12b) and (3.12d), $V(x) \to 0$ as $x \to \infty$. In [200] it was proved that H, given by (3.12a)–(3.12e), has purely s.c. spectrum. The above potential is an example of a *sparse potential*, which has been studied since for a much wider class (see [133, 152, 188]). Possible physical significance of random sparse models for the Anderson transition along a program set up by Molchanov (see [188], also [68, Chap. 5]), will be reviewed in Chap. 4 and some aspects of the physical interpretation of the localization phenomenon in the Anderson model, related to the instability of tunneling, will be reviewed in Sec. 3.4 of the present chapter.

A different important class of models in which the s.c. spectrum made an appearance was almost-periodic models of almost Mathieu type, defined by the Hamiltonian

$$(H_{\lambda,\alpha,\theta}\Psi)(n) = \Psi(n+1) + \Psi(n-1) + \lambda\cos(2\pi\alpha n + \theta)\Psi(n), \tag{3.13}$$

where λ, α, θ are real parameters, on $\mathcal{H} = l^2(\mathbb{Z})$. Jitomirskaya and Simon [134] proved that, for $|\lambda| > 2$, and any irrational α, there is a dense G_δ (a countable intersection of open sets) set of θ for which $H_{\lambda,\alpha,\theta}$ has purely s.c. spectrum. There is, in addition, a huge number of results on this model, which practically exhaust its spectral analysis, see the review of Jitomirskaya in [132]. Interest in almost-periodic models such as (3.13) is connected both with the explanation of charge transport in the integer quantum Hall effect, as well as with the theory of quasicrystals, and certain alloys which exhibit an almost-periodic structure, discovered by Schechtman *et al.* [230]; we refer to [164] for further comments and references, but shall come back to the problem of quasicrystals briefly from another point of view in Sec. 3.4.

Another very important example of exotic spectrum is the thick or dense p.p. spectrum. Define (see [54, Chap. 9, pp. 164–166]) the random operator H_ω on $\mathcal{H} = l^2(\mathbb{Z}^n)$ by (1.2b), with V given, instead of (1.3b), by

$$(V_\omega u)(n) = \omega(n)u(n), \quad \text{with } n \in \mathbb{Z}^n \text{ and } u \in \mathcal{H}, \qquad (3.14)$$

where ω is a random variable (r.v.) on a probability space $(\Omega, \mathcal{F}, P)$ in the following way. Now $\mathcal{F}$ is a sigma algebra on Ω, P a probability measure on $(\Omega, \mathcal{F})$, and we may take w.l.o.g.

$$\Omega = \times_{\mathbb{Z}^n} S, \qquad (3.15)$$

where S is a Borel subset of $\mathbb{R}$, and $\mathcal{F}$ is the sigma algebra generated by the cylinder sets, i.e., sets of the form

$$\{\omega : \omega_{i_1} \in A_1, \ldots, \omega_{in} \in A_n\} \qquad (3.16)$$

for $i_1, \ldots, i_n \in \mathbb{Z}^n$ and $A_1, \ldots, A_n$ Borel sets in $\mathbb{R}$. Define the shift operator T_i on Ω by

$$(T_i\omega)(j) = \omega(j - i). \qquad (3.17)$$

We assume that $\{\omega(i)\}$ is a family of independent, identically distributed (i.i.d.) r.v. with common distribution P_0: in this case the measure P is the product measure

$$dP = \times_{i \in \mathbb{Z}^n} dP_0. \qquad (3.18)$$

The Hamiltonian defined by (1.2b), (3.14) with the above specifications is called the *Anderson model*, originally proposed by Anderson [8] to describe the localization (absence of diffusion, see Sec. 3.4) in heavily doped semiconductors (for instance, Si doped with a neighboring element which contributes excess electrons, e.g., P). We refer to [266] for the ergodicity of P, i.e., the property that either $P(A) = 0$ or $P(A) = 1$ for any shift-invariant set A, i.e., a set A such that $T_i^{-1}A = A$ for all $i \in \mathbb{Z}^n$, and [42, Chap. 2], for the background in probability theory.

Define, for $u \in l^2(\mathbb{Z}^n)$,

$$(U_i u)(n) = u(n - i). \tag{3.19}$$

A stochastic operator $H_\omega = \triangle_l + V_\omega$ satisfying

$$H_{T_i\omega} = U_i H_\omega U_i^\dagger \tag{3.20}$$

is called an *ergodic operator*.

Problem 3.2. *Prove that* (3.14) *and* (3.17) *imply* (3.20).

Equation (3.20) is the key property behind the basic Pastur's theorem, which asserts, among other things, that $\sigma_{\text{dis}}(H_\omega) = \emptyset$ P-a.s. (see [54, Theorem 9] for a proof and references). Thus, by (1.21), $\sigma(H_\omega) = \sigma_{\text{ess}}(H_\omega)$ P-a.s. A very important consequence of this fact is that the p.p. spectrum which is found to cover a nonempty interval in the Anderson model is everywhere dense or "thick" in the language of Avron and Simon [16]. The first paper to prove that, for $n = 1$, $\sigma(H_\omega) = \sigma_{\text{pp}}(H_\omega)$ P-a.s. was [95], see also [157]. Later, several results proving the existence of p.p. spectrum, i.e., localized states for $n \geq 2$ for large disorder or low energy, that is, at the edges of the band, appeared: we refer to [132] for a very nice review and references. The original prediction of Anderson [9] of a sharp spectral transition to delocalized states filling a segment symmetric with respect to the center of the band and consisting of a.c. spectrum (for more about this, see Sec. 4.1), at two symmetrically (w.r.t. the band's center) placed points — which build the so-called mobility edge — has never been proved. For the Anderson model on the Bethe lattice, Klein [153] succeeded to prove existence of a.c. spectrum on an interval around the center of the band: see [132] for comments and references. This problem remains as one of the great open problems in mathematical physics.

A final important context in which s.c. — or dense p.p. — spectra occurs (not mentioned in [164]) is that of "quantum chaos"; see [129, 272] for reviews. The systems studied under this general nomenclature are such that the dynamics of their classical counterparts is chaotic. There is, however, no real quantum chaos, for reasons reviewed in Chap. 6. Paradigms of such systems are atoms and oscillators in external periodic or almost-periodic fields (see also Sec. 3.3 for corresponding dynamical stability aspects):

$$H_1(t) = \beta\sigma_z + \varepsilon f(t)\sigma_x \qquad (3.21)$$

or

$$H_2(t) = \omega_0 a^\dagger a + \lambda f(t)(a + a^\dagger). \qquad (3.22)$$

Above, σ_z and σ_x are Pauli matrices, and a and $a^\dagger$ are standard Boson creation and annihilation operators, satisfying $[a, a^\dagger] = \mathbb{I}$. Here β, λ, ε and ω_0 are real parameters, and f is either a periodic or almost-periodic function (see Definition 3.6) of time t. The standard example of an almost-periodic function is the sum of two periodic functions whose periods are incommensurate, i.e., their ratio is an irrational number. Equation (3.21) describes an atom or molecule in the two-level approximation, that is, assuming that only two levels of energies $\pm\beta$ are involved (a common and good approximation in laser physics, when the external radiation has precisely the frequency 2β, i.e., at resonance), with the second term representing the interaction with the (supposedly classical) radiation field. For the semiclassical approximation of the radiation field using coherent states see [182, Chap. 8.2], and for an excellent pedagogical description of the two-level approximation, we refer to [194, Chap. 6.3]. Equation (3.22) is, for suitable f, a model of a quadrupole radio-frequency trap (Paul-Penning trap), to which we return in Chap. 6.

In order to pose the problems more precisely, consider the periodic case first, and let us assume that we are given a Hamiltonian

$$H(t) = H_0 + V(t), \qquad (3.23a)$$

where H_0 is a self-adjoint operator on a Hilbert space $\mathcal{H}$ with discrete spectrum $\{E_n\}_{n=1,2,...}$ and V is, for instance, a bounded periodic operator, with

$$V(t + T) = V(t). \qquad (3.23b)$$

The idea of how to treat (3.23a) and (3.23b) stems from Howland's method in classical mechanics (see [208, Chap. X, p. 290]): starting from a time-dependent Hamiltonian $H(p, q, t)$, one chooses an equivalent Hamiltonian description which is conservative with respect to a new, fictitious time variable η. The phase space is extended in order to include ordinary time as a canonical variable, with the "energy" as conjugate momentum. The new

Hamiltonian is $K(p, q; E, t) = H(p, q, t) + E$, and the Hamilton equations of motion for t and E are

$$\frac{dt}{d\eta} = \frac{\partial K}{\partial E} = 1$$

$$\frac{dE}{d\eta} = -\frac{\partial K}{\partial t} = -\frac{\partial H}{\partial t},$$

showing that $\eta = t + $ const., and that the variation of E equals minus the variation of H, leading to the identification of E as the energy of an external field responsible for the variation of the energy $H(t)$ of the system.

This method of substituting time by a new dynamical variable corresponds to the introduction, in quantum mechanics, of the *Floquet or quasienergy operator* (see [119, 128, 129])

$$K(t) = K_0(t) - V(t), \tag{3.24a}$$

where

$$K_0 = i\frac{\partial}{\partial t} - H_0, \tag{3.24b}$$

on the enlarged Hilbert space

$$\mathcal{K} = \mathcal{H} \otimes L^2[0, T]. \tag{3.24c}$$

By (3.24b), we see that $i\frac{\partial}{\partial t}$ should be defined on a domain such as D_γ in (1.16), with 2π replaced by T, in order to define a self-adjoint operator; due to (3.23b), we take $\gamma = 0$, i.e., the condition

$$u(T) = u(0). \tag{3.24d}$$

Let $U(t, s)$ denote the propagator defined as in (1.4c), but now, for t and s both in $[0, T]$ and satisfying the evolution equation

$$i\frac{\partial U(t, s)}{\partial t} = (H_0 + V(t))U(t, s) \tag{3.25a}$$

with

$$U(t, t) = \mathbb{I}, \tag{3.25b}$$

and

$$U(t, s) = U(t, r)U(r, s), \quad \text{for } T \geq t \geq r \geq s \geq 0, \tag{3.25c}$$

we assume such a family exists. For bounded V, this follows from the Dyson expansion (see [208, Chap. X, p. 282]).

As we shall see presently in a more general context, the so-called monodromy operator $U(T, 0)$ (evolution along one period) is such that $U(T, 0) \otimes \mathbf{I}$ is unitarily equivalent to $\exp(-iKT)$, with K the Floquet operator defined by (3.24a)–(3.24d). It suffices thus to study K. The spectrum of K_0 given by (3.24b) is, by (3.24d),

$$E_{n,m} = \omega m + E_n, \quad \text{with } m \in \mathbb{Z} \text{ and } n = 1, 2, \ldots, \tag{3.26}$$

where $\omega \equiv \frac{2\pi}{T}$. Unless there are some commensurability conditions between ω and E_n this is, in general, dense p.p.. We shall see that, for suitable perturbations $V(t)$, the essential spectrum of K_0 is preserved in K and that, at the same time, K has no a.c. spectrum (Theorem 3.6): we have, thus here, the simplest example of exotic spectrum!

Following [129], we now consider the general situation described by the Hamiltonian

$$H(t) = H_0(x) + V(x, \theta(t)), \tag{3.27a}$$

where x denote the internal dynamical variables of a system, which act on the Hilbert space $\mathcal{H}$, and

$$\theta(t) = g_t(\theta), \tag{3.27b}$$

where g_t is an invertible flow corresponding to the trajectory of a *classical* dynamical system on a manifold Ω (in general multidimensional), having an ergodic measure μ (see App. A). Thus, $\theta(t)$ is a classical variable whose time dependence is *independent* of the state of the system evolving according to $H(t)$ — in the manner of an "external bath". Let $U(t, s; \theta)$ be the unitary propagator associated to (3.24a)–(3.24d), strongly continuous in t and s, and such that for all $a \in \mathbb{R}$,

$$U(t + a, s + a; \theta) = U(t, s; g_a\theta). \tag{3.27c}$$

In analogy to (3.24a)–(3.24d), define the family of operators on

$$\mathcal{K} = \mathcal{H} \otimes L^2(\Omega, d\mu) \tag{3.27d}$$

given by, for $\Psi \in \mathcal{K}$,

$$[W(t)\Psi](\theta) = U(0, -t; \theta)\tau_{-t}\Psi(\theta) = \tau_{-t}U(t, 0; \theta)\Psi(\theta), \tag{3.27e}$$

where

$$(\tau_t\Psi)(\theta) \equiv \Psi(g_{-t}\theta). \tag{3.27f}$$

Then [129] W is a strongly continuous family of unitary operators with

$$W(t) = \exp(-iKt), \tag{3.27g}$$

and

$$(K\Psi)(\theta) \equiv -i\frac{d}{dt}\Psi(g_t\theta)\Big|_{t=0} + H(\theta)\Psi \tag{3.27h}$$

is the *generalized quasienergy operator* [129]. If the flow has a generator G,

$$\frac{d\theta}{dt} = G(\theta(t)), \tag{3.27i}$$

then

$$K = -iG(\theta) \cdot \frac{\partial}{\partial\theta} + H(\theta), \tag{3.27j}$$

where the dot product has been defined in (2.5c) (for an n-dimensional manifold).

Problem 3.3. *Show that K, defined by (3.27h), is given by (3.27j) and (3.27i).*

We list a few examples:

(1) Periodic force:

$$H = H_0 + f(x)\cos(\omega t + \theta), \tag{3.28a}$$

$\Omega = S^1$ (the circle), $\theta(t) = \theta + \omega t$, $d\mu = d\theta$, and

$$K = -i\omega\frac{\partial}{\partial\theta} + H(\theta) \tag{3.28b}$$

with

$$H(\theta) = H_0 + f(x)\cos\theta. \tag{3.28c}$$

(2) Quasi-periodic force with two frequencies ω_1 and ω_2:

$$H = H_0 + f(x)[\cos(\omega_1 t + \theta_1) + \cos(\omega_2 t + \theta_2)] \tag{3.29a}$$

with $\Omega = S^1 \times S^1$, $g_t(\theta_1, \theta_2) = (\theta_1 + \omega_1 t, \theta_2 + \omega_2 t)$, $d\mu = d\theta_1 d\theta_2$, and

$$K = -i\omega_1\frac{\partial}{\partial\theta_1} - i\omega_2\frac{\partial}{\partial\theta_2} + H(\theta_1, \theta_2) \tag{3.29b}$$

with

$$H(\theta) = (\theta_1, \theta_2)) = H_0 + f(x)(\cos\theta_1 + \cos\theta_2). \tag{3.29c}$$

We now turn to possible methods of study of the spectrum of K, given by (3.24a)–(3.24d) or (3.27j) (in special cases (3.28b) and (3.29b)). As remarked, the spectrum (3.26) is dense if there are no commensurability conditions between ω and E_n. In order to explain this better, consider, for simplicity, the special case (3.22), for which $E_n = \omega_0(n + 1/2)$: if ω and ω_0 are not commensurate, the point spectrum $\{E_{n,m}\}_{n=0,1,\ldots;m\in\mathbb{Z}}$ is nondegenerate and dense in the real line (this is proved along the same lines as (3.30e) et seq.)). A perturbation (3.24a)

$$K = K_0 + \varepsilon Y \tag{3.30a}$$

(where we set $Y = V(t)$) is such that the lower-order terms in the formal perturbation expansion of the eigenvalues are

$$E^{\varepsilon}_{n,m} = E_{n,m} + \varepsilon Y_{n,m;n,m} + \varepsilon^2 \sum_{(n_1,m_1)\neq(n,m)} \frac{|Y_{n_1,m_1;n,m}|^2}{E_{n,m} - E_{n_1,m_1}} + \cdots . \tag{3.30b}$$

In case (3.29b), for instance,

$$E_{n,m} = \omega_1 n + \omega_2 m, \tag{3.30c}$$

and thus, for $(n,m) \neq (n_1,m_1)$,

$$E_{n,m} - E_{n_1,m_1} = \omega_1(n - n_1) + \omega_2(m - m_1). \tag{3.30d}$$

For $(n - n_1, m - m_1) \neq (0,0)$, the denominators (3.30d) in (3.30b) become, in general, arbitrarily small for large $(n - n_1)^2 + (m - m_1)^2$. Consider, e.g., $0 < \frac{\omega_1}{\omega_2} < 1$, and $n - n_1 \neq 0$. We may write

$$E_{n,m} - E_{n_1,m_1} = \omega_2(n - n_1) \left(\frac{\omega_1}{\omega_2} + \frac{m - m_1}{n - n_1} \right). \tag{3.30e}$$

It turns out that approximants $(\frac{p_k}{q_k})_{k=1,2,\ldots}$ of a given irrational number $0 < \alpha = \frac{\omega_1}{\omega_2} < 1$ are such that, for $k = 1, 2, \ldots$,

$$\left| \alpha - \frac{p_k}{q_k} \right| < \frac{1}{q_k^2} \tag{3.30f}$$

(see [150, Theorem 9, p. 9]). Putting (3.30f) into (3.30e), we see that there exist certain sequences $m - m_1 \to -\infty$, $n - n_1 \to \infty$ (or vice versa) in (3.30b) such that the denominators $E_{n,m} - E_{(n_1,m_1)} = O((n - n_1)^{-1})$, i.e., become arbitrarily small for large $(n - n_1)^2 + (m - m_1)^2$. This problem of "small denominators" is mathematically similar to the one encountered in classical mechanics; there, one manages to control the perturbation by a KAM (for Kolmogorov, Arnold and Moser) iteration, see, e.g., [257] for a simple model. A similar type of KAM perturbation has been developed in quantum mechanics; It was Belissard [21] who started this program, which had wide impact on several areas of mathematics and physics related to "quantum chaos". Considering, in particular, (3.21), (3.29b), we assume that in $H(\theta)$ in (3.29c), $f(x) = \sigma_x$ and

$$H_0 = \beta \sigma_z. \tag{3.30g}$$

We have then

$$K = K_0 + \epsilon \sigma_x V(\theta), \tag{3.30h}$$

where

$$K_0 = -i\omega_1 \frac{\partial}{\partial \theta_1} - i\omega_2 \frac{\partial}{\partial \theta_2} + H_0. \tag{3.30i}$$

Above, $V(\theta)$ will be a general function of $\theta = (\theta_1, \theta_2)$ satisfying certain hypotheses, including the special case of the sum of two cosines in (3.29c). By (3.30i), K_0 has a p.p. spectrum with eigenvalues

$$\lambda_{n,m} \equiv \beta m + n \cdot \omega = \beta m + n_1 \omega_1 + n_2 \omega_2 \tag{3.30j}$$

with $m \in \{-1, 1\}$, $n \in \mathbb{Z}^2$, which form a dense subset of $\mathbb{R}$.

Theorem 3.4 (KAM theorem of Bleher, Jauslin and Lebowitz [27]). *Let V be analytic in the strip $\{\theta : \Im \theta_j < r_0 \text{ for } j = 1, 2\}$. Assume, w.l.o.g $\alpha \equiv \frac{\omega_2}{\omega_1} > 1$ (satisfying diophantine condition stated below after (3.30t)) and*

$$(2\beta/\omega_1)_{\mathrm{mod}\ 1} > 0. \tag{3.30k}$$

Then, for any given $\eta > 0$ and fixed ω_1 there is a set of α, $S_\eta \subset (1, \infty)$ of Lebesgue measure $|S_\eta| < \eta$ and a value $\epsilon_c(\eta)$ s.t. if $\alpha \in (1, \infty) \backslash S_\eta$ and $\epsilon < \epsilon_c$, the spectrum of K is pure point.

Brief sketch of proof. One constructs a unitary operator $U(\overline{\alpha}, \epsilon)$ s.t.

$$UKU^{-1} = K_0 + \delta g,$$

where $\delta g = \begin{pmatrix} g_+(\alpha, \epsilon) & 0 \\ 0 & g_-(\alpha, \epsilon) \end{pmatrix}$ where $g_\pm$ are independent of θ. Stated equivalently, U transforms K into an operator which is diagonal in the basis of eigenfunctions of K_0, which are of the form

$$\Psi_{n,m} = \begin{cases} \exp(in \cdot \theta)\chi_+ & \text{if } m = 1, \\ \exp(in \cdot \theta)\chi_- & \text{if } m = -1, \end{cases}$$

where $\chi_\pm$ are the eigenfunctions of σ_z corresponding to the eigenvalues ± 1. They correspond to the eigenvalues $\omega \cdot n \pm \beta$, where $\omega = (\omega_1, \omega_2)$. The transformation U is constructed by iteration $U = \cdots U_j \cdots U_2 U_1$: at the jth step the order of the θ-dependent perturbation is reduced from order ϵ^j to ϵ^{2j}. In this respect, the method is analogous to the KAM method in classical mechanics [257], in which each order uses as input the result obtained in the preceding one. The first $U_1 = \exp(i\epsilon W_1)$ satisfies

$$U_1(K_0 + \epsilon V)U_1^{-1} = K_0 + \epsilon V + i\epsilon[W_1, K_0] + O(\epsilon^2). \tag{3.30l}$$

If we choose W_1 such as to satisfy the so-called cohomological equation:

$$i[K_0, W_1] = V + \delta g, \tag{3.30m}$$

then $U_1 K U_1^{-1}$ is diagonal in the same basis as K_0 to order ϵ^2. The next step performs a similar operation on the $O(\epsilon^2)$ term, and so on until the product $U_N \cdots U_1$ yields $H_N + O(\epsilon^{2N})$, where H_N is diagonal. For $N = 1$, we obtain from (3.30m) for the nondiagonal elements of W_1,

$$W_1(m, m', n - n') = -i\frac{(\Psi_{n,m}, V\Psi_{n',m'})}{\lambda_{n,m} - \lambda_{n',m'}} \tag{3.30n}$$

while the diagonal elements furnish:

$$\delta g(m, m) = -(\Psi_{n,m}, V\Psi_{n,m}). \tag{3.30o}$$

The denominator of (3.30n) is, for

$$(n, m) \neq (n', m'), \tag{3.30p}$$

never zero, by the nonresonance condition (3.30k). Note that $(\omega_1, \omega_2, 2\beta)$ plays the role of the classical "frequency vector" and by assuming α irrational and (3.30k), none of the frequencies can be an integer multiple of each other. As in the classical theory, resonances are a source of instability and have to be excluded. If $m = m'$, however, by (3.30p) $n \neq n'$, but, in this case, the denominator in (3.30n) may become arbitrarily small as $|n - n'| \to \infty$, due to the phenomenon of "small denominators" previously discussed. Let $0 < \nu = 1/\alpha < 1$. We assume, as in the classical theory [257], that ν is a *diophantine number* of type σ (see App. A), i.e., there exists $\gamma > 0$ such that

$$\left|\nu - \frac{s}{r}\right| \geq \frac{\gamma}{r^\sigma}, \quad \text{for all } \frac{s}{r} \in \mathbb{Q}. \tag{3.30q}$$

Almost every number in $(0, 1)$ is diophantine of power $\sigma > 2$ (see App. A). Assumption (3.30q) does not, however, suffice for the proof to go through, for the following reason.

Even if the initial diagonal term (3.30o) is zero (this happens whenever V possesses no constant term in a Fourier expansion, e.g., in case (3.29c)), it must be remembered that the iteration uses the previous approximant as input, i.e., the $O(\epsilon^2)$ term in (3.30l) for the second iteration, which has, in general, a nonzero diagonal term g_2. In fact, it is these diagonal terms in the various iterations whose sum forms the final diagonal Hamiltonian. The cohomological equation of order $k + 1$ is thus of the form

$$i[W_{k+1}, D_k] = V_k + \delta g \tag{3.30r}$$

with

$$\delta g(m, m) = -(\Psi_{m,n}, V\Psi_{m,n}). \tag{3.30s}$$

Above,

$$D_k = K_0 + g_k, \tag{3.30t}$$

where g_k is the above-mentioned diagonal term coming from the iteration of order k. The diophantine condition is thus imposed on the denominator

$$D_k(m, n) - D_k(m', n') = \omega \cdot (n - n') + (m - m')\beta + g_k(m) - g_k(m').$$

Since the g_k depend on α, in order to obtain diophantine estimates, one has to control the size of g_k as well as its variation as a function of α: this has been achieved in [27] using the norm introduced by Combescure [47]. The diophantine condition on the D_k has on the r.h.s. a quantity of the type (3.30q), i.e., $\frac{\gamma}{1+|n-n'|^\sigma}$ with $\sigma > 2$. This leads to polynomial growth in $|n - n'|$, which must be matched by the exponential fall-off of the matrix elements of V due to the assumption of analyticity in the strip: this leads to a loss of analyticity, i.e., the analyticity-strip shrinks at each iteration. At the end of the iteration, it must be checked that the initial assumed class of Hamiltonians has been preserved, i.e., the final strip has nonzero width. This concludes our crude sketch, whose aim was to clarify the main ideas.

To the reader who is not familiar with the KAM method in classical mechanics, it is important to study one case in which the estimates can be carried out to the end with minimal effort. We strongly recommend the article by Chandre and Jauslin, which treats a model of interest in its own sake [40].

It is of interest in several areas (quantum optics, laser theory) to consider what happens for *large coupling* in (3.30g) and (3.30h), or, alternatively (by scaling), setting $\beta = \epsilon$ sufficiently small in (3.30g), and replacing in (3.30h) ϵ by one:

$$K = K_0 + (\cos\theta_1 + \cos\theta_2) + \nu)\sigma_x \qquad (3.30\text{u})$$

with

$$K_0 = -i\omega_1 \frac{\partial}{\partial\theta_1} - i\omega_2 \frac{\partial}{\partial\theta_2} + \epsilon\sigma_z. \qquad (3.30\text{v})$$

Theorem 3.5. *Under the assumptions of Theorem 3.4, with β replaced by ν in the resonance condition (3.30k), the assertions of that theorem remain true for the system defined by (3.30u) and (3.30v).*

For a proof, see [272, Proposition 1 and Theorem 3]. The method is to map the systems (3.30u) and (3.30v) to (3.30g) and (3.30h) by a suitable unitary transformation.

The above theorems may be viewed as examples of *quantum stability*, analogous in classical mechanics to the preservation of tori for small coupling and a frequency ratio which is "sufficiently irrational", that is, far removed from rational. As discussed above, the latter has the physical interpretation of being "sufficiently far" from a resonance condition: we return to these points in Sec. 3.4.

Concerning *nonperturbative* methods, we have the following remarkable result due to Howland [119]:

Theorem 3.6. *Let H_0 have a purely discrete nondegenerate spectrum $\{E_n\}_{n=0,1,\ldots}$ such that*

$$E_n - E_{n-1} \geq cn^\alpha, \qquad (3.31)$$

with c and α constants. Let V in (3.24a) be bounded, periodic with period T, with zero average

$$\int_0^T dt V(x, \omega t) = 0 \qquad (3.32)$$

and be r *times continuously differentiable, i.e., C^r with $r \geq [1/\alpha] + 1$, where* $[x]$ *means the integer part of x. Then, the quasienergy operator (3.24a), (3.29a)*

$$K = -i\omega \frac{\partial}{\partial \theta} + H_0 + V(x, \theta) \tag{3.33}$$

has no absolutely continuous spectrum.

The ingenious method of proof — the method of operator gauge transformations — consists in transforming K by a finite succession of transformations of type $\exp(iG(t))$; using (3.31) and (3.32), one obtains at the final step an operator $V(t)$ of trace class for each t. It follows then from scattering theory (essentially the Kato–Rosenblum theory, see [209, Theorem XI.7]) that the a.c. parts of K_0 and K are unitarily equivalent, but, since K_0 is p.p., K has no a.c. part. The above method shows that the essential spectrum of K is exotic, i.e., a possible mixture of dense p.p. and s.c. spectra. The fact that the p.p. spectrum is dense follows from Theorem 3.4, because the last $V(t)$, being of trace class, is compact, and therefore K has the same essential spectrum as K_0. Unlike the perturbative result, however, the p.p. spectrum could, in principle, be empty.

3.1.3. *UαH measures and decay on the average: Strichartz–Last theorem and Guarneri–Last–Combes theorem*

We now turn to a more specific and quite remarkable RAGE-like theorem for s.c. spectra due to Last [162], based on previous important work of Strichartz [252]. This theory generalized previous seminal work of Guarneri [101, 102] and Combes [45]. We closely follow [162].

Definition 3.1. Let S be a subset of $\mathbb{R}$, $\alpha \in [0, 1]$, and $\delta > 0$. Define

$$h_\delta^\alpha(S) \equiv \inf \left\{ \sum_{j=1}^{\infty} |C_j|^\alpha : S \subset \bigcup_{j=1}^{\infty} C_j \text{ with } |C_j| \leq \delta \right\}, \tag{3.34a}$$

where $|C|$ denotes the Lebesgue measure (length) of C, and

$$h^\alpha(S) = \lim_{\delta \to 0} h_\delta^\alpha(S) = \sup_{\delta > 0} h_\delta^\alpha(S). \tag{3.34b}$$

We call $h^\alpha$$\alpha$-dimensional Hausdorff measure on $\mathbb{R}$. h^1 agrees with Lebesgue measure, and h^0 is the counting measure, so that $\{h^\alpha : 0 \leq \alpha \leq 1\}$

is a family which interpolates continuously between the counting measure and Lebesgue measure. h^α is a regular Borel measure, see, e.g., [82, Theorem 1, p. 61] but is not a Radon measure. Recall that μ is a Radon measure if μ is Borel regular and $\mu(K) < \infty$ for each compact set $K \subset \mathbb{R}$: see [82, Chap. 1.1, p. 5]. The reason is that if $0 \le \alpha < 1$, $\mathbb{R}$ is not sigma-finite with respect to h^α; this follows from the elementary properties of h^α:

Lemma 3.2 ([82, Lemma 2, p. 65]). *Let $A \subset \mathbb{R}$, $0 \le \alpha < \beta \le 1$. We have:*

(i) $h^\alpha(A) < \infty$ *then* $h^\beta(A) = 0$;
(ii) $h^\beta(A) > 0$ *then* $h^\alpha(A) = +\infty$.

Proof. Let $h^\alpha(A) < \infty$ and $\delta > 0$. Then, there exist sets $\{C_j\}_{j=1}^\infty$ such that $|C_j| \le \delta$, $A \subset \bigcup_{j=1}^\infty C_j$ and

$$\sum_{j=1}^\infty |C_j|^\alpha \le h_\delta^\alpha(A) + 1 \le h^\alpha(A) + 1$$

whence

$$h_\delta^\beta(A) \le \sum_{j=1}^\infty |C_j|^\alpha |C_j|^{\beta-\alpha}$$

$$\le \delta^{\beta-\alpha}(h^\alpha(A) + 1)$$

and, taking $\delta \to 0$, it follows that $h^\beta(A) = 0$. This proves (i), and (ii) follows from (i). $\qquad\square$

Definition 3.2. The Hausdorff dimension of a set $A \subset \mathbb{R}$ is defined to be

$$\alpha(A) \equiv \inf\{\alpha, 0 \le \alpha \le 1 : h^\alpha(A) = 0\}.$$

We have $h^\beta(A) = 0$ for all $\beta > \alpha(A)$ and $h^\beta = \infty$ for all $\beta < \alpha(A)$. The actual value $h^{\alpha(A)}(A)$ may be any number between 0 and ∞, inclusive.

From Rogers–Taylor theory [217], one has the following basic notions.

Definition 3.3. Given $0 \le \alpha \le 1$, a measure μ is α-continuous (αc) if $\mu(S) = 0$, $\forall S$ such that $h^\alpha(S) = 0$. It is called α-singular (αs) if it is supported on some set S with $h^\alpha(S) = 0$. It is said to have exact dimension

α if, for all $\epsilon > 0$, it is both $\alpha - \epsilon$-continuous and $\alpha + \epsilon$-singular. One defines

$$D_\mu^\alpha(x) \equiv \limsup_{\epsilon \to 0} \frac{\mu(x - \epsilon, x + \epsilon)}{(2\epsilon)^\alpha} \tag{3.35a}$$

and

$$T_\infty \equiv \{x : D_\mu^\alpha(x) = \infty\}. \tag{3.35b}$$

Then $\mu(T_\infty \cap \cdot) \equiv \mu_{\alpha s}$ and $\mu((\mathbb{R}\backslash T_\infty) \cap \cdot) \equiv \mu_{\alpha c}$ are, respectively, α-singular and α-continuous, and it follows that each measure decomposes uniquely into an αc and an αs part:

$$\mu = \mu_{\alpha c} + \mu_{\alpha s} \tag{3.35c}$$

and, in analogy to the Lebesgue decomposition,

$$\mathcal{H} = \mathcal{H}_{\alpha c} \oplus \mathcal{H}_{\alpha s}, \tag{3.35d}$$

where $\mathcal{H}_{\alpha c} = \{\Psi : \mu_\Psi \text{ is } \alpha\text{-continuous }\}$, and similarly:

$$\mathcal{H}_{\alpha s} = \{\Psi : \mu_\Psi \text{ is } \alpha\text{-singular}\}$$

(see [162, Theorem 5.1] for (3.35d) et seq.). We also need the following important definition from [162, Definition 2.1]:

Definition 3.4. A Borel measure on $\mathbb{R}$ is *uniformly α-Hölder continuous* (UαH) if there exists C such that for every interval I with $|I| < 1$,

$$\mu(I) < C |I|^\alpha. \tag{3.36}$$

Note that a UαH measure is automatically a Radon measure. The importance of UαH measures is due to the following theorem.

Theorem 3.7 (Frostman's lemma). *Suppose A is compact. Then $h^\alpha(A) > 0$ holds iff A carries a probability measure μ satisfying (3.36).*

The proof may be found in [183, p. 112]. We shall also see that, by a theorem of Last [162, Theorem 5.2], UαH measures may be obtained by a process of closure. This fact is of even greater relevance to quantum mechanics than Frostman's lemma. In a very basic paper, Strichartz [252] proved (see also [162] for a slick proof):

Theorem 3.8 (Last [162], Strichartz [252]). *Let μ be a finite UαH measure and, for each $f \in L^2(\mathbb{R}, d\mu)$, denote*

$$\widehat{f\mu}(t) \equiv \int \exp(-ixt) f(x) d\mu(x). \tag{3.37}$$

Then, there exists C depending only on μ such that for all $f \in L^2(\mathbb{R}, d\mu)$ and $T > 0$,

$$\langle |\widehat{f\mu}|^2 \rangle_T < C \, \|f\|^2 \, T^{-\alpha}, \tag{3.38}$$

where $\|f\|$ denotes the L^2-norm of f.

Proof (see [162, Theorem 3.1(i)]). Suppose μ is UαH, and let $f \in L^2(\mathbb{R}, d\mu)$,

$$\langle |\widehat{f\mu}|^2 \rangle_T = \frac{1}{T} \int_0^T dt |\widehat{f\mu}|^2$$

$$\leq \frac{e}{T} \int_{-\infty}^{\infty} dt \exp(-t^2/T^2) |\widehat{f\mu}(t)|^2$$

$$= \frac{e}{T} \int_{-\infty}^{\infty} dt \exp(-t^2/T^2) \int d\mu(x) d\mu(y) f(x) \overline{f}(y) \exp[-i(x-y)t]$$

$$= \frac{e}{T} \int d\mu(x) d\mu(y) f(x) \overline{f}(y) \int_{-\infty}^{\infty} dt \exp(-t^2/T^2) \exp[-i(x-y)t]$$

$$= e\sqrt{\pi} \int d\mu(x) d\mu(y) f(x) \overline{f}(y) \exp[-(x-y)^2 T^2/4]$$

$$\leq e\sqrt{\pi} \int d\mu(x) d\mu(y) (|f(x)| \exp[-(x-y)^2 T^2/8])(|f(y)|$$

$$\times \exp[-(x-y)^2 T^2/8]). \tag{3.39}$$

Using the Schwartz inequality in (3.39),

$$\langle |\widehat{f\mu}|^2 \rangle_T \leq e\sqrt{(\pi)} \int d\mu(x) |f(x)|^2 \int d\mu(y) \exp[-(x-y)^2 T^2/4]. \tag{3.40}$$

Since μ is UαH, with $T > 1$,

$$\int d\mu(y) \exp[-(x-y)^2 T^2/4]$$

$$= \sum_{n=0}^{\infty} \int_{\{n/T \leq |x-y| \leq (n+1)/T\}} d\mu(y) \exp[-(x-y)^2 T^2/4]$$

$$\leq \sum_{n=0}^{\infty} 2CT^{-\alpha} \exp(-n^2/4). \tag{3.41}$$

Inserting (3.41) into (3.40) yields (3.38). $\qquad\square$

We now consider that we have a quantum dynamics, described by a self-adjoint operator H on a (separable) Hilbert space $\mathcal{H}$; μ_Ψ will denote the spectral measure (1.48). A corollary of Theorem 3.8 is (see [162, Lemma 3.2]) the following.

Lemma 3.3. *If μ_Ψ is $U\alpha H$, then there exists a constant c_Ψ such that, for all $\Psi \in \mathcal{H}$, for all $\Phi \in \mathcal{H}$ such that $\|\Phi\| \leq 1$,*

$$\langle |(\Phi, \Psi(t))|^2 \rangle_T < c_\Psi T^{-\alpha}, \tag{3.42}$$

where

$$\Psi(t) \equiv \exp(-itH)\Psi. \tag{3.43}$$

Proof. Let P denote the projection onto the cyclic subspace generated by H and Ψ. We have, by definition and (3.43),

$$(\Phi, \Psi(t)) = (P\Phi, \exp(-itH)\Psi). \tag{3.44}$$

By Theorem 1.5, $P\Phi$ is unitarily equivalent to a function $g \in L^2(\mathbb{R}, d\mu_\Psi)$, such that

$$\|P\Phi\|^2 = \int |g(\lambda)|^2 d\mu_\psi(\lambda) \leq \|\Phi\|^2, \tag{3.45}$$

and (3.42) of the lemma follows from (3.45) and Theorem 3.8. $\qquad\square$

Lemma 3.3 concerns the quantity on the l.h.s. of (3.42), which is the same as (3.1c), and has the physical interpretation mentioned there. Using it, it is not difficult to prove:

Theorem 3.9 (Last's RAGE theorem for s.c. spectra). *If μ_Ψ is $U\alpha H$ (Definition 3.4), there exists a constant c_Ψ such that, for any compact operator A, p any positive integer, and $T > 0$,*

$$\langle |(\Psi, A\Psi)| \rangle_T < c_\Psi^{1/p} \|A\|_p T^{-\alpha/p}, \tag{3.46}$$

where $\|A\|_p \equiv (\mathrm{Tr}\,|A|^p)^{1/p}$ and $|A| \equiv (A^\dagger A)^{1/2}$ where Tr denotes the trace.

Next, we shall be interested in the lattice case, with $\mathcal{H} = l^2(\mathbb{Z}^n)$, and define P_N to be the projection on a sphere of radius N, i.e.,

$$P_N \equiv \sum_{|n| \leq N} (e_n, \cdot)e_n, \tag{3.47a}$$

where e_n denotes the unit basis (1.2e). Note that n denotes both the dimension and the label of lattice sites, which should not lead to confusion!

Let us denote the moments of the position operator in $\mathcal{H}$ by

$$|X|^m \equiv \sum_{n \in \mathbb{Z}^n} |n|^m (e_n, \cdot) e_n. \tag{3.47b}$$

Choosing $A = P_N$ in Theorem 3.9, we have

$$\|P_N\|_1 \equiv \operatorname{Tr} P_N = \sum_{|n| \leq N} 1 \leq C_n N^n. \tag{3.48}$$

Further applications depend on the decomposition theorem of Rogers and Taylor [217] (see [162, Theorem 4.2]), which implies (see (3.35d)) that μ is α-continuous iff, for each $\epsilon > 0$, there are mutually singular Borel measures μ_1^ϵ and μ_2^ϵ such that $d\mu = d\mu_1^\epsilon + d\mu_2^\epsilon$, with μ_1^ϵ UαH and $\mu_2^\epsilon(\mathbb{R}) < \epsilon$. As a consequence, a powerful result [162, Theorem 5.2] implies that, if

$$\mathcal{H}_{\mathrm{uh}}(\alpha) = \{\Psi : \mu_\Psi \text{ is } U\alpha H\}, \tag{3.49a}$$

then Theorem 3.10 follows.

Theorem 3.10 ([162, Theorem 5.2]). *For every $\alpha \in [0,1]$, $\mathcal{H}_{\mathrm{uh}}(\alpha)$ is a vector space and*

$$\overline{\mathcal{H}_{\mathrm{uh}}(\alpha)} = \mathcal{H}_{\alpha c}, \tag{3.49b}$$

where the bar denotes norm closure in $\mathcal{H}$.

The importance of the above theorem is to guarantee that UαH measures may be generated by a process of closure. We now come back to (3.47a) and choose $\Psi \in \mathcal{H}$ (see (3.35d)), such that

$$P_{\alpha c} \Psi \neq 0, \tag{3.50a}$$

where $P_{\alpha c}$ is the orthogonal projection onto $\mathcal{H}_{\alpha c}$. By the above-mentioned decomposition theorem, one may choose a Borel set $S_1 \subseteq \mathbb{R}$ and

$$\Psi_1 = P_{S_1} P_{\alpha c} \Psi,$$
$$\Psi_2 = (1 - P_{S_1}) P_{\alpha c} \Psi, \tag{3.50b}$$

such that μ_{Ψ_1} is UαH (see [162, p. 429]), and then, by Theorem 3.9 and (3.48),

$$\langle \|P_N \Psi_1(t)\|^2 \rangle_T < c_{\Psi_1} C_n N^n T^{-\alpha}. \tag{3.50c}$$

Making, now, the choice

$$N^n T^{-\alpha} = \mathrm{const.}, \tag{3.50d}$$

call N_T the chosen N. We then find, from (3.47b), (3.50b)–(3.50d) that

$$\langle(\Psi(t),|X|^m\Psi(t))\rangle_T \geq \langle((1-P_{N_T})\Psi(t),|X|^m(1-P_{N_T})\Psi(t))\rangle_T$$
$$\geq c\|P_{\alpha c}\Psi\|^2 N_T^m > c_{\Psi,m}T^{m\alpha/n} \tag{3.50e}$$

with c a constant (see [162, p. 430] for the details), would lead to the remarkable Theorem 3.11.

Theorem 3.11 (Guarneri–Last–Combes). *If H is self-adjoint on $\mathcal{H}$ and (3.50a) holds, then, for each $m > 0$, there exists a constant $c_{\Psi,m}$ depending on Ψ and m such that, for all $T > 0$,*

$$\langle(\Psi(t),|X|^m\Psi(t))\rangle_T > c_{\Psi,m}T^{m\alpha/n}. \tag{3.50f}$$

The above theorem has important physical consequences, see Sec. 3.4.

We close this Section 3.1 with the statement of a version of Theorem 3.2 for time-periodic Hamiltonians (3.23a) and (3.23b) (or (3.28c) in the second version). Let the Floquet operator be written as

$$F = W(T) = \exp(-iKT), \tag{3.51}$$

which is unitarily equivalent to $U(T,0) \otimes \mathbb{I}$ on $\mathcal{H} \otimes L^2[0,T]$ as reviewed before (see the comments following (3.25c)). Let P_R be defined by (3.5b), with S a sphere of radius R centered at the origin.

Theorem 3.12 ([277]). *For any $R > 0$ and $\Psi_0 \in \mathcal{H}_c$, where $\mathcal{H}_c$ denotes the continuous spectral subspace corresponding to F,*

$$\lim_{N\to\infty} \frac{1}{N} \sum_{k=0}^{N-1} |(\Psi_k, P_R\Psi_k)|^2 = 0, \tag{3.52a}$$

where

$$\Psi_k \equiv F^k\Psi_0. \tag{3.52b}$$

Note the satisfying analogy to Theorem 3.2, which shows once more how natural is the notion of quasi-energy operator in this context. We refer to the original paper for details.

We shall shortly see that Theorem 3.8 implies the finiteness of the L^2-norm of $\hat{\mu}_\Psi$ — the Fourier–Stieltjes transform of μ_Ψ, defined in (1.28) — with respect to a certain measure. This will lead us naturally to the subject of the next section.

3.2. Decay in the L^p-Sense

3.2.1. *Relation between decay in the L^p-sense and decay on the average sense*

We now come back to Theorem 3.8 with $f \equiv 1$, which may be written as

$$\sup_T T^\alpha \langle |\hat{\mu}|^2 \rangle_T < \infty. \tag{3.53a}$$

For $\alpha \in (0,1)$, define the integral (where μ is a finite regular Borel measure, $\hat{\mu}$ its F.S. transform):

$$J_\alpha(\mu) \equiv \int_{-\infty}^{\infty} |\hat{\mu}(t)|^2 d\rho_\alpha(t), \tag{3.53b}$$

where ρ_α is the measure

$$d\rho_\alpha(t) = |t|^{\alpha-1} dt. \tag{3.53c}$$

Note that ρ_α is infinite, i.e., $\rho_\alpha(\mathbb{R}) = \infty$ if $\alpha \in (0,1)$ and that for every $T > 0$ the inequality $T^\alpha \langle |\hat{\mu}|^2 \rangle \leq J_\alpha(\mu)$, hence $J_\alpha(\mu) < \infty$ implies (3.53a). Thus, finiteness of the L^2-norm of $\hat{\mu}$ with respect to ρ_α implies (3.53a), which means that $|\hat{\mu}|^2$ decays in the average (Cesàro) sense: this justifies the title of the present section. Finally, one may consider *pointwise decay* of the type $|\hat{\mu}(t)| \leq C \|t\|^{-1/2-\delta}$ for all $t \in \mathbb{R}$, for some constant C and some $\delta > 0$ (many other variants may be considered, see Sec. 4.3). We have the implications:

pointwise decay $\rightarrow$ decay in the L^2 sense $\rightarrow$ decay on the average.
$$\tag{3.53d}$$

Accordingly, pointwise decay is hardest to prove and will be studied in Sec. 3.3.

"Almost" conversely to the first implication in (3.53d), we have:

Lemma 3.4 ([162, Lemma 5.2]). *If (3.53a) holds, then* $J_{\alpha-\epsilon}(\mu) < \infty$ *for any* $0 < \epsilon < \alpha$.

Proof. Since $|\hat{\mu}|^2 \leq 1$, $\int_0^1 |\hat{\mu}(t)|^2 t^{\alpha-\epsilon-1} dt < \infty$. By (3.53a),

$$\int_1^\infty |\hat{\mu}(t)|^2 t^{\alpha-\epsilon-1} dt = \sum_{n=0}^\infty \int_{2^n}^{2^{n+1}} |\hat{\mu}(t)|^2 t^{\alpha-\epsilon-1} dt$$

$$\leq \sum_{n=0}^\infty (2^n)^{\alpha-\epsilon-1} C(2^{n+1})^{1-\alpha}$$

$$\leq C 2^{1-\alpha} \sum_{n=0}^\infty 2^{-\epsilon n} < \infty. \qquad \square$$

We now turn to a different, but related topic: does finiteness of $J_\alpha(\mu)$ imply that μ is α-continuous? *In a certain sense*, this is a stronger property than (3.53a) as we shall see next. Let

$$E_\alpha(\mu) \equiv \int \int d\mu(x)d\mu(y)|x - y|^{-\alpha}. \qquad (3.54a)$$

E_α is sometimes called the α-energy of μ. For μ a probability measure supported on a bounded set $A \subset \mathbb{R}^n$, and $\alpha = n - 2$, where n is the dimension, E_α is related to the *capacity* of the set A, see [169, p. 289 et ff.]; for general α see [82]. It is worthwhile to consider the generalization of (3.53b)–(3.53d) to $\mathbb{R}^n$: let μ be a finite regular Borel measure on $\mathbb{R}^n$, and $\hat{\mu}$ denote its F.S. transform, defined as in (1.28)

$$\hat{\mu}(t) = \int d\mu(x) \exp(-it \cdot x) \qquad (3.54b)$$

with the notation (1.28). We define

$$J_\alpha^{(n)}(\mu) \equiv \int |\hat{\mu}(t)|^2 |t|^{\alpha-n} d^n t \qquad (3.54c)$$

which coincides with (3.53b) for $n = 1$. We also define the α-energy of μ

$$E_\alpha^{(n)}(\mu) = \int \int d\mu(x)d\mu(y)|x - y|^{-\alpha}. \qquad (3.54d)$$

The Fourier transform $\hat{g}$ of a function g is usually formally defined as in (3.54b) with $d\mu(x) = g(x)d^n x$, but we prefer to define $\hat{g}$ in the distributional sense by the requirement

$$\int_{\mathbb{R}^n} g(x)\overline{f}(x)d^n x = (2\pi)^{-n} \int_{\mathbb{R}^n} \hat{g}(k)\overline{\hat{f}}(k)d^n k \qquad (3.55)$$

with (3.55) holding for all $f \in C_0^\infty(\mathbb{R}^n)$. Let f be a function in $C_0^\infty(\mathbb{R}^n)$, and

$$0 < \alpha < n . \qquad (3.56)$$

Then, its Fourier transform $\hat{f}$ and all its derivatives decay faster than any power of $|k|$, as can easily be checked. Therefore, by (3.56), the function $h(k) \equiv |k|^{-\alpha}\hat{f}(k)$ belongs to $L_1(\mathbb{R}^n)$, and has, therefore, a Fourier transform in the sense of functions, which is, furthermore, a continuous function. It thus follows (see [169, Theorem 5.9(2)] for $x = 0$) that if

$$g \equiv g_\alpha(x) = |x|^{-\alpha} \qquad (3.57a)$$

is the so-called Riesz kernel, its Fourier transform in the sense of (3.55) is (cf. [183, (12.10), p. 161]):

$$\hat{g}_\alpha = c(\alpha, n) g_{n-\alpha} \tag{3.57b}$$

for a certain constant $0 < c(\alpha, n) < \infty$, known explicitly, but its actual value will be of no concern to us. From (3.57b), the following theorem may be proved.

Theorem 3.13.

$$J_\alpha^{(n)}(\mu) = C E_\alpha^{(n)}(\mu) \tag{3.58}$$

for a constant $0 < C < \infty$.

The proof of the above theorem is given in full detail in [183, Lemma 12.12, p. 162].

We now complete our argument:

Lemma 3.5 ([162, Lemma 5.1]). *If $J_\alpha(\mu) < \infty$, then μ is α-continuous.*

Proof. If $J_\alpha(\mu) < \infty$, then, by Theorem 3.13 (for $n = 1$), $E_\alpha(\mu) < \infty$, and thus $\int d\mu(y)|x - y|^{-\alpha} < \infty$ for a.e. x w.r.t. μ. But for every x and $\epsilon > 0$, $\mu(x - \epsilon, x + \epsilon) \le \epsilon^\alpha \int d\mu(y)|x - y|^{-\alpha}$, and hence $D_\mu^\alpha(x) < \infty$ for a.e. x w.r.t. μ. By (3.35a) et ff. μ is α-continuous. $\square$

On the other hand, μ need not be UαH (Definition 3.4)! If $\mu = \mu_\Psi$, i.e., we are in the context of quantum dynamics, and μ_Ψ denotes the spectral measure (1.49a) and (1.49b), then Theorem 3.10 assures that there exists a dense set of $\Psi \in \mathcal{H}$ such that μ_Ψ is UαH, and hence (3.38) holds (with $f \equiv 1$), i.e., $\hat{\mu}$ decays on the average as $T^{-\alpha/2}$, that is, for $n = 1$ we have just justified the assertion made after Lemma 3.4 that finiteness of $J_\alpha(\mu)$ implies that μ is α-continuous (Lemma 3.5), which is a stronger property than (3.53a) in the previously stated sense.

There is, also, a purely measure-theoretical analogue of Theorem 3.10, which is true in the more general context of $\mathbb{R}^n$. For $A \subset \mathbb{R}^n$ a Borel set, let

$$\mathcal{M}(A) = \{\mu : \mu \text{ is a compact supported Radon measure}$$
$$\operatorname{supp}\mu \subset A, 0 < \mu(\mathbb{R}^n) < \infty\}. \tag{3.59}$$

Recall that we have defined $\operatorname{supp}\mu$ in Chap. 1, and Radon measure in Sec. 3.1. By the definition of α-continuity (Definition 3.3) and (3.6), we see that Frostman's lemma in $\mathbb{R}^n$ (the $\mathbb{R}^n$-version of Theorem 3.7, see [183, Theorem 8.8, p. 112]) may be stated in the form: let ν be α-continuous and

$E = \operatorname{supp} \nu$. Then $\mathcal{M}(E)$ is not empty, and there exist a UαH measure $\tilde{\nu}$ and a constant $0 < c < \infty$ such that for all $x \in \mathbb{R}$, for all $0 < r < \infty$,

$$\tilde{\nu}(B(x,r)) \leq cr^{\alpha}, \tag{3.60}$$

where $B(x,r)$ is the open ball of center x and radius r (in $\mathbb{R}^n$). A stronger form of Lemma 3.4 also has an analogue in $\mathbb{R}^n$:

Lemma 3.6. *If $\tilde{\nu}$ satisfies* (3.60), *then* $J_{\alpha-\epsilon}^{(n)}(\tilde{\nu}) < \infty$ *for any* $0 < \epsilon < \alpha$.

Remark 3.1. The UαH property (3.60) for $n = 1$ implies (3.53a) by Theorem 3.8; therefore a stronger form of Lemma 3.4 may be stated: if μ is UαH, then $J_{\alpha-\epsilon}(\mu) < \infty$ for any $0 < \epsilon < \alpha$.

Proof. We have, by [169, Theorem 1.13] (layer cake representation), with $\Phi(t) = \nu([0,t)) = t$ and $f(x) = |x - y|^{-(\alpha-\epsilon)}$:

$$\int d\tilde{\nu}(y)|x - y|^{-(\alpha-\epsilon)} = \int_0^\infty \tilde{\nu}(\{y : |x - y|^{-(\alpha-\epsilon)} \geq u\})du$$

$$= \int_0^\infty \tilde{\nu}(B(x, u^{-\frac{1}{\alpha-\epsilon}}))du$$

$$= (\alpha - \epsilon) \int_0^\infty r^{-(\alpha-\epsilon)-1}\tilde{\nu}(B(x,r))dr$$

$$= (\alpha - \epsilon) \int_0^1 r^{-(\alpha-\epsilon)-1}\tilde{\nu}(B(x,r))dr$$

$$+ (\alpha - \epsilon) \int_1^\infty r^{-(\alpha-\epsilon)-1}\tilde{\nu}(B(x,r))dr. \tag{3.61}$$

The first integral in (3.61) is finite by $\epsilon > 0$ and (3.60), the second by $\epsilon - \alpha < 0$ and $\tilde{\nu}(\mathbb{R}^n) < \infty$. We have proved that the l.h.s. of (3.61) is finite for all $x \in \mathbb{R}^n$, thus $E_{\alpha-\epsilon}^{(n)} < \infty$ by definition (3.54a). By Theorem 3.13, $J_{\alpha-\epsilon}^{(n)}(\tilde{\nu}) < \infty$. $\square$

Our interest in the versions in $\mathbb{R}^n$ was two-fold: the crucial role played by properties such as (3.58) and (3.60) becomes clearer in a generalized context, and the connection to Sec. 3.3. (pointwise decay) also becomes clearer by a counterexample to the converse of the third implication in (3.53d), i.e., L^2 decay implying pointwise decay, in dimension $n \geq 2$, which will be presented at the end of this section.

3.2.2. Decay on the L^p-sense and absolute continuity

Condition (3.56), for $n = 1$, excludes the case $\alpha = 1$. For $\alpha = 1$, $J_\alpha(\mu)$ becomes

$$I(\mu) \equiv J_1(\mu) = \int_{-\infty}^{\infty} dt |\hat{\mu}(t)|^2, \tag{3.62}$$

which is exactly the L^2-norm of $\hat{\mu}$. In this case, an important folklore theorem states the result analogous to Lemma 3.4, namely, that μ is a.c. with respect to Lebesgue measure. Our proof given next emphasizes the role of the measure as a functional on the dual space:

Theorem 3.14. *Let $\mu \in M(\mathbb{R})$ be the space of all finite regular Borel measures on $\mathbb{R}$. If its F.S. transform $\hat{\mu} \in L^p(\mathbb{R}, \mathcal{L})$ for some*

$$1 \leq p \leq 2,$$

then μ is absolutely continuous with respect to Lebesgue measure $\mathcal{L}$.

Proof. Since $\mu \in M(\mathbb{R})$, $|\hat{\mu}(t)| \leq$ const. for all $t \in \mathbb{R}$ by definition of the F.S. transform, and thus, if $\hat{\mu} \in L^p$ for $1 \leq p \leq 2$, it follows that $\hat{\mu} \in L^2$. It thus suffices to prove the assertion for $\hat{\mu} \in L^2$. Since μ is a linear combination of at most four non-negative measures, we may assume μ to be non-negative.

Now $M(\mathbb{R})$ being the dual space of $C_0(\mathbb{R})$, the Banach space of all continuous functions on $\mathbb{R}$ which vanish at infinity (with the supremum norm), is uniquely determined by the functional

$$f \in C_0(\mathbb{R}) \rightarrow \int_{\mathbb{R}} f(\lambda) d\mu(\lambda). \tag{3.63a}$$

This is the content of Theorem 1.2, see [146] for the proof.

Now, let f be twice continuously differentiable with compact support. Denoting double differentiation by a double prime,

$$\left| t^2 \hat{f}(t) \right| \leq \|f''\|_1. \tag{3.63b}$$

For these f we have, by (3.63b), that $f \in C_0(\mathbb{R}) \cap L^1(\mathbb{R})$ and $\hat{f} \in L^1(\mathbb{R})$. Thus, by Parseval's formula for measures (see [146, Theorem 2.2, p. 132]),

$$\int_{\mathbb{R}} f(\lambda) d\mu(\lambda) = \frac{1}{2\pi} \int_{\mathbb{R}} \hat{f}(t) \hat{\mu}(-t) dt. \tag{3.63c}$$

By (3.63b) and the fact that $\hat{f}$ is a continuous function, $\hat{f} \in L^2(\mathbb{R})$; together with (3.63c) and Parseval's formula for functions,

$$\left| \int_{\mathbb{R}} f(\lambda) d\mu(\lambda) \right| \leq \text{const. } \|\hat{f}\|_2 = \text{const. } \|f\|_2. \tag{3.64}$$

Since the considered set of functions is dense in $L^2(\mathbb{R})$ in the L^2-norm, (3.64) and the B.L.T. theorem (see, e.g., [207, p. 9]) imply that the map $f \to \int f(\lambda) d\mu(\lambda)$ extends continuously to a bounded map from L^2 to L^2, and is thus of the form

$$\int f(\lambda) d\mu(\lambda) = \int g(\lambda) f(\lambda) d\lambda \tag{3.65}$$

for some $g \in L^2(\mathbb{R})$. Being in $L^2(\mathbb{R})$, g is locally L^1 and defines a regular Borel measure ν by

$$d\nu(\lambda) = g(\lambda) d\lambda. \tag{3.66}$$

Equations (3.65) and (3.66) show that $\mu - \nu$ annihilates the set $C_0(\mathbb{R}) \cap L^2(\mathbb{R})$ which is dense in $C_0(\mathbb{R})$, and thus, by the unicity part in Theorem 1.2,

$$d\mu(\lambda) = g(\lambda) d\lambda. \tag{3.67}$$

We still have to prove that $f \in L^1(\mathbb{R})$. By the assumption that $\hat{\mu} \in L^2$ and Theorem 3.1, $\mu(\{\lambda\}) = 0$ for all $\lambda \in \mathbb{R}$. Thus, for all $a < b$,

$$\mu([a, b)) = \mu((a, b]) = \mu((a, b)) = \mu([a, b]) = \int_a^b g(\lambda) d\lambda. \tag{3.68}$$

The limit as $a \to -\infty$ together with $b \to \infty$ (which should not and does not depend on which of the alternatives on the l.h.s. of (3.68) is chosen) in (3.68) exists and is finite, by the finiteness of μ. Since by (3.68) and non-negativity of μ, $g(\lambda) \geq 0$ a.e., the monotone convergence theorem yields finally that $\int_{-\infty}^{\infty} g(\lambda) d\lambda = \int_{-\infty}^{\infty} |g(\lambda)| d\lambda < \infty$. $\qquad \square$

The case $p = 1$ is the classic result [42, Theorem 6.2.3, p. 155]; the case $p = 2$ corresponds to [42, Exercise 11, p. 159], see also [248]. Both are usually proved by means of (different) applications of the inversion formula (see, e.g., [42, Theorem 6.2.1, p. 153]).

The present proof was inspired by [243, Corollary 3.2], but, by the remark at the beginning of Theorem 3.13, all cases are reduced to $p = 2$, and, therefore, the Hausdorff–Young inequality is not necessary. Moreover, most of the proof is concerned with the existence of a g satisfying (3.67), which is implicitly assumed in Simon's brief sketch in [243].

The sufficient condition in the case $p = 2$ is optimal, see Chap. 4 and the references given there. It is, however, clear that the condition $\hat{\mu} \in L^p$ is far from necessary for μ to be a.c.: an important class of counterexamples in the case $p = 1$ is provided by Proposition C.26 of App. C.

We propose two applications of Theorem 3.13. The third one will be reserved to Sec. 4.2. The first one is due to Kahane and Salem [142].

Theorem 3.15. *For almost every sequence $\{\xi_k\}$ of dissection ratios of a symmetric Cantor set satisfying*

$$1/2 \leq \xi_k \leq 1 \tag{3.69}$$

(see App. C, (C.2)–(C.4)), $\hat{\mu}(u) = \Gamma(u) \in L^2$ and thus μ is a.c. by Theorem 3.13.

Remark 3.2.

(1) Note that Γ is defined by (C.15); by App. C, it equals the F.S. transform of the Lebesgue function (C.12), which is $\hat{\mu}$ in our previous notation (1.28).
(2) The method of proof of Theorem 3.15 is very important: it reappears in the theory of pointwise decay of Sec. 3.3, where we speak of Salem's method.

Proof of Theorem 3.15. We assume that

$$1/2 \leq a_k \leq \xi_k \leq b_k \leq 1 \quad \text{with } k = 0, 1, 2, \cdots \tag{3.70a}$$

and

$$\liminf(a_0 a_1 \cdots a_{p-1})^{1/p} = \alpha > 1/2. \tag{3.70b}$$

Write

$$\xi_k = a_k + \eta_k(b_k - a_k) \tag{3.70c}$$

so that

$$0 \leq \eta_k \leq 1. \tag{3.70d}$$

The hypercube in infinite dimensions may be represented over the segment $0 \leq t \leq 1$, the functions $\eta_0(t) \cdots \eta_p(t) \cdots$ being such that

$$\int_0^1 \Phi(\eta_0(t), \ldots, \eta_p(t))dt = \int_0^1 \cdots \int_0^1 \Phi(\eta_0, \ldots, \eta_p)d\eta_0 \ldots d\eta_p \quad (3.70e)$$

for all measurable Φ. We defer the proof of (3.70e). Assume that the sequences $\{a_k\}$, $\{b_k\}$ are not "too close", i.e.,

$$b_p - a_p \geq \frac{1}{\omega(p)} \quad (3.71a)$$

with

$$\omega(p) \text{ growing with } \log \omega(p) = o(p). \quad (3.71b)$$

The following remark is useful:

$$\int_0^1 \cos^2(l\pi x + m)dx \leq \frac{1 + 1/l}{2} \text{ for } l \geq 1 \text{ and } m \text{ arbitrary.} \quad (3.71c)$$

In order to prove (3.71c), change variables whereby the integral equals

$$\frac{1}{l\pi} \int_m^{m+l\pi} \cos^2 ydy < \frac{1}{l\pi} \int_m^{m+[l]\pi+\pi} \cos^2 ydy = \frac{[l]+1}{2l} < \frac{1+1/l}{2}.$$

In order to prove the last step, use $\cos \alpha - \cos \beta = -2 \sin \frac{\alpha+\beta}{2} \sin \frac{\alpha-\beta}{2}$. By (C.15) and (C.9), with $d = \pi$,

$$\Gamma_t^2(u) = \prod_{k=0}^{\infty} \cos^2(\pi u \xi_0 \cdots \xi_{k-1}(1 - \xi_k))$$

$$\leq \prod_{k=0}^{p} \cos^2(\pi u \xi_0 \cdots \xi_{k-1}(1 - \xi_k)) \equiv f_{u,p}. \quad (3.72)$$

The $f_{u,p}$ being functions of t by (3.70c) and (3.70e),

$$\int_0^1 f_{u,p}dt = \int_0^1 \cdots \int_0^1 f_{u,p}d\eta_0 \cdots d\eta_p$$

$$= \int_0^1 \cdots \int_0^1 f_{u,p-1}d\eta_0 \cdots d\eta_{p-1}$$

$$\times \int_0^1 \cos^2(\pi u \xi_0 \cdots \xi_{p-1}(1 - \xi_p))d\eta_p. \quad (3.73a)$$

The last integral is

$$T_p \equiv \int_0^1 \cos^2(\pi u \xi_0 \cdots \xi_{p-1}[1 - a_p - (b_p - a_p)\eta_p]) d\eta_p \qquad (3.73b)$$

and, due to (3.71c), and $\xi_k \geq a_k$ (see (3.70a)),

$$T_p \leq \frac{1 + (1/p)^2}{2} \qquad (3.73c)$$

under the assumption of having chosen p such that

$$u a_0 \cdots a_{p-1} \frac{1}{\omega(p)} \geq p^2. \qquad (3.73d)$$

Under (3.73d),

$$\int_0^1 f_{u,p} dt \leq \frac{1 + (1/p)^2}{2} \int_0^1 f_{u,p-1} dt. \qquad (3.73e)$$

If (3.73d) is true for an integer p, it is by (3.70a)–(3.70d), (3.71a) and (3.71b) *a fortiori* true upon replacement $p \to p - 1$; hence, iteration of (3.73e) yields

$$\int_0^1 f_{u,p} dt \leq \frac{A}{2^p} \qquad (3.74a)$$

for A fixed constant, hence by (3.72)

$$\int_0^1 \Gamma_t^2(u) dt \leq \frac{A}{2^p}. \qquad (3.74b)$$

Let us now fix α_1 s.t. $1/2 < \alpha_1 < \alpha$, see (3.70b); we have then $a_0 \cdots a_{p-1} > \alpha_1^p$, and (3.73d) will be satisfied if p is chosen s.t.

$$\log u + p \log \alpha_1 - \log \omega(p) \geq 2 \log p. \qquad (3.75a)$$

By (3.71b), (3.75a) will be satisfied if

$$p < \theta \frac{\log u}{|\log \alpha_1|} \qquad (3.75b)$$

for

$$\theta < 1 \text{ fixed.} \qquad (3.75c)$$

We now take p equal to the integer part of $\theta \frac{\log u}{|\log \alpha_1|}$. With this choice

$$2^p > 1/2 \exp\left(\theta \frac{\log 2}{|\log \alpha_1|} \log u\right) \qquad (3.76a)$$

and, since $|\log \alpha_1| = \log \frac{1}{\alpha_1} < \log 2$, one may choose $\theta < 1$ such that

$$\theta \frac{\log 2}{|\log \alpha_1|} = 1 + \epsilon, \quad \text{with } \epsilon > 0. \tag{3.76b}$$

By (3.74b), (3.76a) and (3.76b)

$$\int_0^1 \Gamma_t^2(u)dt < \frac{A}{2u^{1+\epsilon}} \tag{3.77a}$$

from which

$$\int_0^\infty du \int_0^1 \Gamma_t^2(u)dt < \infty \tag{3.77b}$$

whence

$$\Gamma_t(u) \in L^2 \quad \text{for a.e. } t \tag{3.77c}$$

proving Theorem 3.15. □

Theorem 3.15, together with App. C, shows that within this class recurrent absolutely continuous measures are "generic". The idea of looking at interpolating sequences (3.70a)–(3.70d), which ultimately limits the search to "typical" assertions, i.e., holding a.e. in a parameter $0 \le t \le 1$, is very fertile, and will reappear in a variety of situations. Two examples are the Anderson-like transition in Sec. 4.2, and the Salem method of proving pointwise decay in Sec. 3.3 of the present chapter. In Theorem 3.15, the application relies on (3.70e) which precisely reduces the extra set to $[0, 1]$: this ingenious method is due to Steinhaus [139]. In order to prove (3.70e), let $t \in [0, 1]$ have the dyadic expansion (see also App. A) $t = 0, \epsilon_1 \epsilon_2 \epsilon_3 \cdots$ with $\epsilon_i \in \{0, 1\}, i = 1, 2, \cdots$, and define

$$\eta_0(t) = 0.\epsilon_1 \epsilon_3 \epsilon_6 \epsilon_{10} \cdots$$
$$\eta_1(t) = 0.\epsilon_2 \epsilon_5 \epsilon_9 \epsilon_{14} \cdots$$
$$\eta_2(t) = 0.\epsilon_4 \epsilon_8 \epsilon_{13} \epsilon_{19} \cdots$$
$$\eta_3(t) = 0.\epsilon_7 \epsilon_{12} \epsilon_{18} \epsilon_{25} \cdots$$
$$\vdots \tag{3.78}$$

corresponding to splitting the natural sequence $\{1, 2, 3, \ldots\}$ into infinitely many subsequences according to the following

$$
\begin{pmatrix}
1 & 3 & 6 & 10 & 15 & \cdots \\
2 & 5 & 9 & 14 & 20 & \cdots \\
4 & 8 & 13 & 19 & 26 & \cdots \\
7 & 12 & 18 & 25 & 33 & \cdots \\
11 & 17 & 24 & 32 & 41 & \cdots \\
\vdots & \vdots & \vdots & \vdots & \vdots & \ddots
\end{pmatrix}
$$

We see that the natural sequence is precisely reproduced along the diagonals making an angle of $45°$ with the lines of writing. In this way, we have a mapping of the segment $0 \le t \le 1$ onto the hypercube in infinite dimensions $\{0 \le \eta_n \le 1$ with $n = 0, 1, 2, \ldots\}$. Apart from the collection of finite dyadic rationals in t and the η_i's, the mapping is one-to-one. This mapping has two properties: **(P1)** Each $t \in I_t \equiv [0, 1] \to \eta_i(t) \in I_{\eta_i} \equiv [0, 1]$ and is such that to any measurable subset of I_{η_i} corresponds a set of equal measure in I_t, for all $i = 0, 1, \ldots$. **(P2)** The functions $\eta_i(t)$ are independent in the probabilistic sense, i.e., if

$$G_1 \subset I_t \text{ such that } F_1(\eta_{\alpha_1}(t), \eta_{\beta_1}(t), \ldots, \eta_{\mu_1}(t)) \in \Gamma_1 \subset I_{\eta_{\alpha_1}} \times \cdots \times I_{\eta_{\mu_1}}$$

and

$$G_2 \subset I_t \text{ such that } F_2(\eta_{\alpha_2}(t), \eta_{\beta_2}(t), \ldots, \eta_{\mu_2}(t)) \in \Gamma_2 \subset I_{\eta_{\alpha_2}} \times \cdots \times I_{\eta_{\mu_2}}$$

with F_1, F_2 any two measurable functions and Γ_1, Γ_2 any two measurable sets, such that each of the indices $\alpha_1, \beta_1, \ldots, \mu_1$ is distinct from each of the indices $\alpha_2, \beta_2, \ldots, \mu_2$, then

$$|G_1 \cap G_2| = |G_1\|G_2|.$$

For the proof of **(P1)**, let for $\eta_0(t)$, $\eta_0 \in (\frac{p}{2^q}, \frac{p+1}{2^q})$; the first q numbers $\epsilon_1, \epsilon_2, \ldots, \epsilon_q$ are fixed in this way. Consider, for instance, the splitting of $[0, 1]$ into the union of $q = 8$ disjoint intervals $I_{3,p}$ with $p = 0, \ldots, 2^3 - 1$ of the form $I_{3,p} = [\frac{p}{2^3}, \frac{p+1}{2^3})$, i.e., $[0, 1] = [0, 1/8) \cup [1/8, 1/4) \cup [1/4, 3/8) \cup [3/8, 1/2) \cup [1/2, 5/8) \cup [5/8, 3/4) \cup [3/4, 7/8) \cup [7/8, 1)$ which is a subset of I_{η_0}. The interval $I_{3,3} = [3/8, 1/2)$ corresponds to fixing in I_{η_0}, by (3.78), $(\epsilon_1, \epsilon_2, \epsilon_3) = (0, 1, 1)$. The corresponding set in I_t is the set of all "dual t-fractions" $t = 0.\epsilon_1 \cdot \epsilon_2 \cdots \epsilon_6$, where the points between the ϵ are to be filled in an arbitrary way with $\{0, 1\}$. In our example, we have

001001; 001011; 001101; 001111; 011001; 011011; 011101; 011111 for the possibilities, i.e., $8 = 2^3$ intervals, each of length 2^{-6}, with total length 2^{-3}, which equals the length of the original set in I_{η_0}. In general, we have $2^{1+\cdots+(q-1)} = 2^{q(q-1)/2}$ intervals each of length $2^{-q(q+1)/2} = 2^{-(1+\cdots+q)}$, with total length 2^{-q}, which coincides with the length of the set $[\frac{p}{2^q}, \frac{p+1}{2^q})$. From this, **(P1)** follows for all sets consisting of intervals or sets of zero measure for η_0, i.e., for all measurable I_{η_0}, and similarly all I_{η_i}. In order to prove **(P2)** in the simplest case, consider G_1 given by $\eta_{\alpha_1}(t) \in [\frac{p}{2^q}, \frac{p+1}{2^q})$, G_2 by $\eta_{\alpha_2} \in [\frac{r}{2^s}, \frac{r+1}{2^s})$. In this case, by **(P1)**, $|G_1| = 1/2^q$, $|G_2| = 1/2^s$ and $G_1 \cap G_2$ is the set given by both conditions holding simultaneously, i.e., it is the set of dual t-fractions of I_t in which certain $(q+s)$ values are fixed, but we have seen that the measure of this set is $1/2^{q+s} = 1/2^q \cdot 1/2^s = |G_1||G_2|$. The general case follows from these considerations (see also App. A). As a corollary, we have that the "dynamics" defined by $t \in [0,1] \to \eta \equiv (\eta_0(t), \ldots, \eta_p(t)) \in [0,1]^{p+1}$, calling t the "time" coordinate, the point η spends the same time in any two subsets of $\times_{i=0}^p I_{\eta_i}$ of the same measure, i.e., the "motion" is quasi-ergodic (see, again, App. A), which implies the equality of the time and space averages for any integrable function $\Phi(\eta_0, \ldots, \eta_p)$, i.e., (3.70e): indeed, if Φ is the characteristic function of a set $M \subset \{0 \leq \eta_k \leq 1\}$, the $p+1$-fold integral equals $|M|$, while the measure of the t-image of M also equals $|M|$; the extension to integrable Φ is straightforward.

As a second application of Theorem 3.14, we mention the following lemma used by Klein in his seminal proof of the existence of extended states in the Anderson model on the Bethe lattice [153, Lemma 4.3]:

Lemma 3.7. *Let ν be a finite measure on $\mathbb{R}$ and define its Borel transform by*

$$F(z) \equiv \int \frac{d\nu(t)}{t-z}, \quad \text{with } z = E + i\eta, \ E \in \mathbb{R} \text{ and } \eta > 0. \tag{3.79a}$$

Let $\hat{\nu}$ be the F.S. transform of ν. Then,

$$\int_{-\infty}^{\infty} |\hat{\nu}(t)|^2 dt \leq \frac{1}{\pi} \liminf_{\eta \to 0+} \int_{-\infty}^{\infty} |F(E+i\eta)|^2 dE. \tag{3.79b}$$

Problem 3.4. *Prove Lemma 3.7.*

Hint: Let $g_\eta(E) = -(E + i\eta)^{-1}$, so that $F(E + i\eta)) = (g_\eta \star \nu)(E)$, and $g_\eta(E) = -i \int_0^\infty \exp(-\eta t + itE) dt$. Use the Plancherel theorem, $\hat{\nu}(-t) = \overline{\hat{\nu}(t)}$ and Fatou's lemma to prove the result.

Klein [153] used Lemma 3.7 for the restriction ν_a of the measure ν to the interval $(-a, a)$. For the expectation value of the resolvent of the Anderson Hamiltonian $H_\lambda = 1/2\triangle + \lambda V$ (see (1.2b)–(1.3b)) on the Bethe lattice $\mathcal{B}$ (see [153] for the definition of $\mathcal{B}$),

$$\langle x|(H_\lambda - z)^{-1}|x\rangle \equiv G_\lambda(x, x; E + i\eta), \quad \text{with } z = E + i\eta \qquad (3.80a)$$

(cf. (1.33)) with $\Phi = \Psi = |x\rangle$, the Kronecker delta at x, he succeeded in proving that

$$\liminf_{\eta \to 0+} \int_{-E}^{E} |G_\lambda(x, x : E' + i\eta)|^2 dE' < \infty \qquad (3.80b)$$

with probability one for certain values of the parameters E and λ around the center of the band. By (3.79b), (3.80b) and Theorem 3.14, the spectral measure restricted to the interval $(-E, E)$ is purely absolutely continuous, with probability one.

3.2.3. *Sojourn time, Sinha's theorem and time–energy uncertainty relation*

We close this section with two related subjects: a theorem of Sinha [248] relating the finiteness of a (properly defined) "sojourn time" of a particle in a region S to the a.c. subspace, and a related generalized form of the time–energy uncertainty relation. Recall the definition (3.5b) of P_S. Instead of looking at the L^2-norm of $(u_0, u(t))$ (see (3.1a)), we now look at the L^2-norm of (3.1d), with $K = P_S$, i.e., whether

$$J(S; \Psi) \equiv \int_{-\infty}^{\infty} \|P_S V_t \Psi\|^2 dt \qquad (3.81a)$$

is finite, i.e., decay in the L^2-sense of the quantity

$$\sqrt{p(S, \Psi; t)} \equiv \|P_S V_t \Psi\| \qquad (3.81b)$$

with

$$V_t \equiv \exp(-itH), \qquad (3.81c)$$

where $p(S, \Psi; t)$ is the probability that a particle with initial state Ψ (with $\|\Psi\| = 1$) be found in a space region S after time t; recall that Theorem 3.2 is concerned with decay of $p(S, \Psi; t)$ in the average (Cesàro) sense (3.4b). The quantity $J(S; \Psi)$ is called the *sojourn time* of the particle, initially in the state Ψ, in the space region S: it is also related to the notion of time-delay in scattering theory, see [248] and references given there. In [248], Sinha proved the following theorem.

Theorem 3.16. *If there exists a sequence of regions $\{S_n\}_{n=1}^{\infty}$ such that* s-$\lim_{n\to\infty} P_{S_n} = \mathbb{I}$ *and* $J(S_n; \Psi) < \infty$ *for all* $n = 1, 2, \ldots$, *then* $\Psi \in \mathcal{H}_{ac}$ *(see (1.52a) for the definition of $\mathcal{H}_{ac}$).*

Proof. We have that

$$\int |(P_{S_n}\Psi, V_t\Psi)|^2 dt \leq \|\Psi\|^2 \int \|P_{S_n}V_t\Psi\|$$
$$= \|\Psi\|^2 J(S_n; \Psi) \tag{3.82a}$$

by hypothesis. Now, let

$$\Phi_n \equiv P_{S_n}\Psi. \tag{3.82b}$$

Equations (3.82a) and (3.82b) assert that $(\Phi_n, V_t\Psi) \in L^2(\mathbb{R})$, and by Theorem 3.14, the measure $\mu_{\Phi_n, \Psi}$ is a.c. w.r.t. Lebesgue measure, for all $n = 1, 2, \ldots$, and therefore,

$$(\Phi_n, E_B\Psi) = 0, \quad \text{for all } n = 1, 2, \ldots, \tag{3.82c}$$

for any Borel set B s.t. $|B| = 0$, $|S|$ denoting Lebesgue measure of S (see (1.53a) et. seq. for definition of E_B). By (3.82b), (3.82c) and the hypothesis s-$\lim_{n\to\infty} P_{S_n} = \mathbb{I}$, we find $(\Psi, E_B\Psi) = \|E_B\Psi\|^2 = 0$ by (1.53b), hence $E_B\Psi = 0$ for all Borel sets with $|B| = 0$, hence $\Psi \in \mathcal{H}_{ac}$ by (1.52a). $\square$

We have the following corollary to Theorem 3.16:

Corollary 3.1. *If $\Psi \in \mathcal{H}_{sc}$, there exists a region $\tilde{S}$ with*

$$|\tilde{S}| < \infty \tag{3.83a}$$

such that

$$J(\tilde{S}; \Psi) = \infty. \tag{3.83b}$$

Corollary 3.1 leads to a quite precise physical interpretation of $\mathcal{H}_{sc}$ in sparse models (3.13). We consider a lattice version of (3.13), which may be written as ($n = 1, 2, \ldots$):

$$(H_\phi u)(n) = u(n+1) + u(n-1) + V(n)u(n) \tag{3.84a}$$

for $u \in \mathcal{H} = l^2(\mathbb{Z}_+) = l^2(\{0, 1, 2, \ldots\})$ with

$$u(0) \cos \phi - u(1) \sin \phi = 0 \qquad (3.84\text{b})$$

the "ϕ-boundary condition" at $n = 0$, with $\phi \in [0, \pi)$. If $\phi = 0$, we have (lattice) Dirichlet boundary conditions (b.c.) $u(0) = 0$; if $\phi = \pi/2$, we have Neumann b.c. $u(1) = 0$, and $H_\phi = H_0 + \tan \phi \delta_0$, i.e., the operator with ϕ-b.c. is obtained from the Dirichlet Hamiltonian H_0 by a rank-one perturbation: above, δ_i with $i = 0, 1, \ldots$ denotes the Kronecker delta at site i. In analogy to (3.13), we define

$$V(n) = v(n) \text{ if } n = a_j \in \mathbb{A} \text{ and otherwise } 0, \qquad (3.84\text{c})$$

where

$$\mathbb{A} = \{a_j\}_{j \geq 1} \qquad (3.84\text{d})$$

is a set of natural numbers satisfying the "sparseness" condition

$$a_j - a_{j-1} = [\exp(cj^\gamma)], \quad \text{for } j = 2, 3, \ldots \qquad (3.84\text{e})$$

with c and $\gamma > 0$ and $[z]$ denotes as usual the integer part of a real number z. We assume

$$\sum_{n \geq 1} v(n)^2 = \infty \qquad (3.84\text{f})$$

and

$$\sup_j |v(j)| < \infty. \qquad (3.84\text{g})$$

Note that (3.84e) is much weaker than Pearson's condition (3.12e). In Chap. 4, we shall be interested in nondiagonal versions of (3.84c), as well as random versions of the set $\mathbb{A}$ of sites (3.84d). Equation (3.84a) is a *Jacobi matrix*, which is the canonical form of an operator with simple spectrum. The latter assertion is a theorem due to Stone [251]. We shall need (notably in Chap. 4 and App. B) only the simpler statement that a self-adjoint Jacobi matrix has a simple spectrum: the vector δ_0 (see (1.2e)) is a cyclic vector. For a simple proof of this, see [5, Chap. 4, Theorem 4.2.3].

We have seen in our discussion after (1.3c) that in the tight-binding model (1.3d) the particle tunnels from one atom to the next by building up coherences, i.e., the phases (1.3e) which help it to work its way through the crystal. Now, if the potential (3.84c) consists of a series of "bumps" situated at points a_j very far apart (by (3.84e)), this has the effect of causing the particle to "forget" any coherence built up (by accumulation of phases in (1.3e)), and behaves as if successively undergoing *independent* collisions at the points a_j. In the words of Simon [239, p. 6]: "by the Born approximation, for $v(n)$ small in (3.84c), the reflection from this potential is $O(v(n)^2)$; by (3.84f) no particles reach infinity. By tunneling, positivity of V and inability to build up coherence, no bound states are expected and thus particles moving under H wander aimlessly about, and this is precisely what corresponds to singular spectrum."

The last remark refers to the fact that the F.S. transform of the survival probability does not necessarily tend to zero (as seen, e.g., by Proposition C.26), i.e., the s.c. measure is not necessarily a Rajchman measure (see Sec. 3.3) for this concept. It is interesting that Simon's physical picture is, in fact, quite precise under certain additional conditions. Indeed, the spectrum is purely s.c. if (3.84f) and (3.84g) hold and, in addition, $\gamma > 1$ — the superexponential case in (3.84e) — for, in this last case, $\lim_{j \to \infty} a_j / a_{j+1} = 0$ (see [152]). Condition (3.84f) is just what enters in Simon's heuristics, but we now also have Corollary 3.1 as a precise assertion corresponding to the phrase "particles under H wander aimlessly about"! A last interesting topic related to this section, but now concerning the a.c. spectrum, is the *time–energy uncertainty relation*. We come back to the return or survival probability (3.1a) and (3.1b) and consider the quantity, with $\Psi(t) = \exp(-itH)\Psi_0$,

$$\tau_H(\Psi_0) \equiv \int_0^\infty dt |(\Psi_0, \Psi(t))|^2 \tag{3.85a}$$

whenever the integral on the r.h.s. above exists, i.e.,

$$\tau_H(\Psi_0) < \infty. \tag{3.85b}$$

We have

$$\tau_H(\Psi_0) = 1/2 \int_{-\infty}^\infty dt |(\Psi_0, \Psi(t))|^2 = 1/2 \int_{-\infty}^\infty dt |\hat{\mu}_{\Psi_0}(t)|^2, \tag{3.85c}$$

and by (3.85b) and Theorem 3.14, $\Psi_0 \in \mathcal{H}_{ac}$. The later space depends, of course, on H. It is also a well-known result related to Kato-smoothness that, considering the maximal domain of $\tau_H(\cdot)$ (considered as an operator), i.e., $\text{Dom}_{max}(\tau_H(\cdot)) = \{\Phi : \tau_H(\Phi) < \infty\}$, the spectrum of H is purely a.c. iff $\text{Dom}_{max}(\tau_H(\cdot))$ is dense in $\mathcal{H}$ (see [209, Lemma 1, p. 23 and Problem 17, p. 386]). Thus, $\tau_H(\cdot)$ is also important for the spectral theory of the operator H. In the following, we shall assume (3.85b) only for a given Ψ_0, which is, then, necessarily contained in $\mathcal{H}_{\mathrm{ac}}$.

The function $Q(t) \equiv |(\Psi_0, \Psi(t))|^2$ decays from one at $t = 0$ to zero at $t = \infty$ (this will be shown in Sec. 3.3) as a consequence of the sole property (3.85b). The probability that the system has not decayed at time t is $Q(t)$, and the probability that the system decays between t and $t + dt$ is $-Q'(t)dt$. Consequently, the average lifetime [93] of the decaying state is $\tau_H(\Psi_0) = -\int_0^\infty dt t Q'(t) = \int_0^\infty Q(t)dt$, by partial integration. Unfortunately, the above relation is only formal, because it does not follow from (3.85b) that $tQ(t) \to 0$ as $t \to \infty$. It provides, however, a more direct physical interpretation of the sojourn time in case the latter condition is met.

As remarked in [23], it is relevant to obtain lower bounds to the sojourn time τ_H (a problem posed by the late Pierre Duclos) since, in the case of resonances (treated in greater detail from a different point of view in Chap. 5), τ_H assumes very large values: this has been shown rigorously by Lavine in [165]. The heuristic relation between resonances (defined as poles of a suitable meromorphic continuation of the resolvent $r_z(H)$ of H) may be seen from the relation

$$\frac{d\mu_{\Psi_0,\mathrm{a.c.}}(\lambda)}{d\lambda} = \frac{1}{\pi} \lim_{y \to 0+} \Im(\Psi_0, (H - \lambda - iy)^{-1}\Psi_0) \tag{3.85d}$$

(see, e.g., [127, Theorem 3.15]). If the function $(\Psi_0, (H - z)^{-1}\Psi_0)$ may be continued to the lower half-plane with a pole at $E_0 - i\Gamma$, then

$$\hat{\mu}_{\Psi_0}(t) = \int_{-\infty}^\infty d\lambda \exp(-i\lambda t) \frac{d\mu_{\Psi_0,\mathrm{a.c.}}(\lambda)}{d\lambda} \tag{3.85e}$$

becomes, by deforming the contour of integration and using residues,

$$\hat{\mu}_{\Psi_0}(t) \approx \exp(-itE_0 - t\Gamma) \tag{3.85f}$$

which is slowly and exponentially decaying if Γ is small. By (3.85f) and (3.85a), the sojourn time $\tau_H(\Psi_0) \approx 1/\Gamma$ in this case, i.e., one obtains a large lifetime. As remarked in [23], the mathematical justification of this "single-pole approximation" is difficult. It comes naturally about, however, from the point of view of Gamow vectors, which is discussed in Chap. 5, following the work of [76, 91] and particularly [50].

The problem posed by [23] of obtaining lower bounds to $\tau_H(\Psi_0)$ leads to interesting variational problems discussed there. A special lower bound to $\tau_H(\Psi_0)$, which is also related to a peculiar variational problem, appears in the following theorem.

Theorem 3.17 (Time–energy uncertainty relation [93]). *Let* (3.85b) *hold and*

$$\Psi_0 \in D(H), \quad i.e., \ \|H\Psi_0\| < \infty. \tag{3.86}$$

Then

$$I_H(\Psi_0) \equiv \tau_H(\Psi_0)\Delta E \geq \frac{3\pi\sqrt{5}}{25}, \tag{3.87}$$

where

$$(\Delta E)^2 \equiv (\Psi_0, H^2\Psi_0) - (\Psi_0, H\Psi_0)^2 \tag{3.88}$$

is the energy variance (uncertainty) in the state Ψ_0.

Proof. We follow [93] quite literally, except for several points of rigor. By (3.85b), μ_{Ψ_0} is a.c., and we may thus define a.e. in the spectrum $\sigma(H) \subset \mathbb{R}$ of H,

$$f(\lambda) \equiv \frac{d\mu_{\Psi_0}(\lambda)}{d\lambda}. \tag{3.89}$$

By the spectral theorem and (3.89),

$$f(\lambda) \geq 0, \quad \text{for a.e. } \lambda \in \mathbb{R} \tag{3.90a}$$

and by the normalization of Ψ_0,

$$\int_{-\infty}^{\infty} d\lambda f(\lambda) = 1. \tag{3.90b}$$

By (3.85a), (3.85b) and Parseval's formula,

$$\tau_H(\Psi_0) = \pi \int_{-\infty}^{\infty} d\lambda f(\lambda)^2 < \infty. \tag{3.91}$$

By (3.86) and (3.90b), we have

$$(\Psi_0, H^2\Psi_0) = \int_{-\infty}^{\infty} d\lambda \lambda^2 f(\lambda) < \infty \tag{3.92a}$$

and

$$|(\Psi_0, H\Psi_0)| = \left| \int_{-\infty}^{\infty} d\lambda \lambda f(\lambda) \right| \leq \int_{-\infty}^{\infty} d\lambda |\lambda| f(\lambda) < \infty. \tag{3.92b}$$

The last inequality follows from (3.92a) and the Schwartz inequality. Equations (3.91), (3.92a) and (3.92b) imply that $I_H(\Psi_0)$, defined by

(3.87), is a functional of the sole function $f(\cdot)$. Equation (3.92b) allows to define

$$\bar{H} = H - (\Psi_0, H\Psi_0) \tag{3.93a}$$

which satisfies

$$(\Psi_0, \bar{H}\Psi_0) = 0. \tag{3.93b}$$

The replacement $H \to \bar{H}$ does not change $\tau_H(\Psi_0)$ by definition (3.85a). We may further by scaling $f(\lambda) \to \alpha f(\alpha\lambda)$ restrict ourselves to the condition

$$\int_{-\infty}^{\infty} d\lambda \lambda^2 f(\lambda) = 1 \tag{3.94a}$$

which must be added to the condition

$$\int_{-\infty}^{\infty} d\lambda \lambda f(\lambda) = 0 \tag{3.94b}$$

coming from (3.93b). Note that, now $f(\cdot)$ is defined by (3.89), but with μ_{Ψ_0} the a.c. spectral measure associated to $\bar{H}$. The functional $I_H(\Psi_0)$ defined by the l.h.s. of (3.87) now becomes

$$I_H(\Psi_0) = I_f(\Psi_0) = \int_{-\infty}^{\infty} d\lambda f(\lambda)^2 \tag{3.95}$$

i.e., a functional on the set of real-valued $f \in L^2(\mathbb{R})$ satisfying, in addition, the constraints (3.90a), (3.90b) (continuity, as mentioned in [93], should not be assumed, as it is too restrictive). We now turn to the solution of this variational problem.

Let

$$P_t(\lambda) \equiv \begin{cases} \dfrac{3\sqrt{5}}{20}\left(1 - \dfrac{\lambda^2}{5}\right) & \text{if } |\lambda| \leq \sqrt{5}, \\ 0 & \text{otherwise,} \end{cases} \tag{3.96}$$

denote the truncated parabola. By explicit computation, it satisfies the constraints (3.90a), (3.90b). We now prove that any $f_0 \in L^2(\mathbb{R})$ satisfying (3.90a), (3.90b) is such that

$$I_{P_t}(\Psi_0) \leq I_{f_0}(\Psi_0) \tag{3.97a}$$

with equality only if $P_t(\lambda) = f_0(\lambda)$ a.e. in $\mathbb{R}$. In order to show (3.97a), let

$$g_1(\lambda) \equiv f_0(\lambda) - P_t(\lambda). \tag{3.97b}$$

Then

$$I_{f_0}(\Psi_0) = I_{P_t+g_1}(\Psi_0) = I_{P_t}(\Psi_0) + I_{g_1}(\Psi_0) + 2\int_{-\infty}^{\infty} d\lambda P_t(\lambda)g_1(\lambda).$$

$$(3.97c)$$

Note that the last integral in (3.97c) is finite by the Schwarz inequality. From (3.95), $I_{g_1}(\Psi_0) \geq 0$, and, indeed, $I_{g_1}(\Psi_0) > 0$ unless $g_1 = 0$ a.e. in $\mathbb{R}$. Hence, we need only show that the last integral in (3.97c) is non-negative. In order to do this, let

$$\lambda \in \mathbb{R} \to g_2(\lambda) \equiv \frac{3\sqrt{5}}{20}\left(1 - \frac{\lambda^2}{5}\right) \tag{3.98}$$

denote the parabola without truncation. Since both f_0 and P_t obey (3.90a), (3.90b),

$$\int_{-\infty}^{\infty} d\lambda g_1(\lambda)\lambda^n = 0, \quad \text{for } n = 0, 1, 2 \tag{3.99}$$

and, therefore,

$$
\begin{aligned}
0 &= \int_{-\infty}^{\infty} d\lambda g_1(\lambda)g_2(\lambda) \\
&= \int_{-5^{1/2}}^{5^{1/2}} d\lambda g_1(\lambda)g_2(\lambda) + \int_{-\infty}^{-5^{1/2}} d\lambda g_1(\lambda)g_2(\lambda) + \int_{5^{1/2}}^{\infty} d\lambda g_1(\lambda)g_2(\lambda).
\end{aligned}
$$

$$(3.100)$$

The last two integrals on the r.h.s. of (3.100), which are finite by (3.97b) and the assumption that f_0 satisfies (3.94a), are nonpositive because for $|\lambda| \geq \sqrt{5}$, $g_2(\lambda) \leq 0$ by (3.98), and $g_1 = f_0 \geq 0$ a.e. by (3.90a). Therefore, the first integral on the r.h.s. of (3.100) is *non-negative*. However, by definitions (3.96) and (3.98) $g_2(\lambda) = P_t(\lambda)$ if $|\lambda| \leq \sqrt{5}$, and thus this first integral on the r.h.s. of (3.100) equals the last integral in (3.97c). This shows (3.97a), which completes the proof. □

The assertion in [205] that the inequality (3.87) does not make any assumption about the spectrum of H seems to be at least misleading, because, as we have seen above, we have to assume (3.85b), which implies that $\Psi_0 \in \mathcal{H}_{ac}$. Indeed, for p.p. spectrum $\Delta E = 0$, and τ diverges, so that the inequality possibly remains correct, but in the case of s.c. spectrum τ diverges too and nothing seems to be known about ΔE. In any case, the above derivation does not allow any conclusion if $\Psi_0 \in \mathcal{H}_{sc}$. Reference [205] offers, however, a readable and quite complete survey of inequalities of time–energy type, including important contributions by the authors.

The above approach to the time–energy uncertainty relation uses the concept of sojourn time. Here time is treated, as usually in quantum mechanics, as an external parameter; in fact, it is well-known and easy to see that no self-adjoint operator exists which describes the measurement of time [198]. A natural generalization of the concept of operator, i.e., that of a positive operator-valued measure (POVM) (see, e.g., [58, p. 16]), has been used by Brunetti and Fredenhagen [34] to provide a general construction of an observable measuring the time of occurrence of an effect in quantum theory. If α_t denotes the time-evolution of observables, the original proposal starts from the integrals $B(I) \equiv \int_I \alpha_t(A)dt$ where I denotes some bounded interval of $\mathbb{R}$ and $A > 0$ is a bounded operator on a Hilbert space $\mathcal{H}$ which measures the occurrence of an effect at time $t = 0$. The authors construct from the B_I a POVM P, such that $P_{\mathbb{R}} = \mathbb{I}$ exhibiting the countable additivity property and the crucial property of covariance under time translations $\alpha_t(P_I) = P_{I+t}$. This construction has been considerably generalized in [36], to a general framework which includes quantum field theory. Using this concept, a time–energy uncertainty relation of a different nature has been derived [35]: where the state of minimum uncertainty (in the energy) was also explicitly derived, having the form of an Airy function, which is also universal, similar to the truncated parabola in Theorem 3.17. It would be interesting to compare the applicability of these formulas to the study of real physical systems, e.g., resonances (see the remarks in [93]).

One drawback of Theorem 3.17 is that condition (3.86) is often too strong: it expresses that the standard deviation of the "probability distribution" given by the spectral measure is finite, but it may be very large if the probability distribution has a long tail: indeed, for the forthcoming Lorentzian (5.38a) (which corresponds to the standard Breit–Wigner form in the case of a resonance), it is infinite. It is thus of interest to find a time–energy uncertainty in this case: this is Lavine's form (the forthcoming (5.38c)) of the time–energy uncertainty relation, to which we shall come in Chap. 5.

We have seen that decay on the average (Cesàro) sense is related to decay in the L^p-sense, particularly for $p = 2$. One may inquire now whether the latter form of decay implies pointwise decay, i.e., whether the finiteness of

$$J_\alpha^{(n)}(\mu) \equiv \int |\hat{\mu}(t)|^2 |t|^{\alpha-n} d^n t < \infty \tag{3.101a}$$

(see (3.54c)) implies that

$$\lim_{t_i \to \infty} \hat{\mu}(t) = 0, \quad \text{for } i = 1, \ldots, n \tag{3.101b}$$

for $0 < \alpha < n$? Let $0 < \alpha < 1$ and $n = 2$, and $\mu(x) = \delta(x_1) \otimes \chi_{[0,1]}(x_2)$, where δ denotes the Dirac measure and $\chi_{[0,1]}$ the characteristic function of $[0,1]$. Then, by definition (3.54b),

$$\hat{\mu}(t_1, t_2) = \int_0^1 \exp(-it_2 x_2) dx_2 = \exp(-it_2/2) \frac{\sin(t_2/2)}{t_2/2}.$$

We have

$$\int d^2 t |\hat{\mu}(t)|^2 |t|^{\alpha-2} = \int_{-\infty}^{\infty} dt_1 \int_{-\infty}^{\infty} dt_2 (t_1^2 + t_2^2)^{(\alpha-2)/2} \frac{4\sin^2(t_2/2)}{t_2^2}$$

$$\leq c + d \int_1^{\infty} dt_1 \int_1^{\infty} dt_2 t_2^{-2} t_1^{\alpha-2} < \infty$$

for $0 < \alpha < 1$ and some positive finite constants c and d, but: $\lim_{t_1 \to \infty} \hat{\mu}$ $(t_1, t_2) \neq 0$! This special example, due to Mockenhaupt [187], depends crucially on the dimension n being greater or equal to two.

What happens if $n = 1$? For $0 < \alpha < 1$ and $n = 1$, Proposition C.25 of App. C shows that, for a class of symmetric perfect sets (which includes the usual Cantor middle-thirds set), finiteness of J_α^1 does not imply $\lim_{t \to \infty} \hat{\mu}(t) = 0$ as well.

We conclude that pointwise decay of $\hat{\mu}(t)$ is a *stronger* property than decay in the L^2-sense (3.101a). We now turn to the interesting but difficult topic of pointwise decay.

3.3. Pointwise Decay

3.3.1. *Does decay in the L^p-sense and/or absolute continuity imply pointwise decay?*

We have seen, at the end of last section, that finiteness of $J_\alpha^{n=1}(\mu)$ does not, *in general*, imply pointwise decay of $\hat{\mu}$, i.e.,

$$\lim_{|t| \to \infty} \hat{\mu}(t) = 0. \tag{3.102}$$

This is true for $0 < \alpha < 1$. If $\alpha = 1$,

$$J_1(\mu) = J_{\alpha=1}(\mu) = \int_{-\infty}^{\infty} dt |\hat{\mu}(t)|^2,$$

and we know from Theorem 3.14 that if $J_1(\mu) < \infty$, μ is a.c. w.r.t. Lebesgue measure $\mathcal{L}$. The latter property, which is stronger than L^2-decay, does imply (3.102) by the famous Riemann–Lebesgue lemma, with which the theory of decay actually began (see [175] for some of the history):

Lemma 3.8 (Riemann–Lebesgue lemma). *Let $f \in L^1(\mathbb{R})$. Then*

$$\lim_{|t| \to \infty} \hat{f}(t) = 0, \qquad (3.103)$$

where $\hat{f}$ is the Fourier transform of f (defined as in (3.37), with $d\mu(x) = dx$).

We shall not prove Lemma 3.8 here because we shall be proving in Lemma 3.10 a generalization of this result. As a corollary of Lemma 3.8, if ν is a.c. w.r.t. $\mathcal{L}$, i.e., $\nu = f\mathcal{L}$ with $f \in L^1(\mathbb{R})$, by the Radon–Nikodym theorem, and (3.102) holds for ν. Thus, absolute continuity implies pointwise decay, and therefore, so does L^2-decay in the form (3.62). One may ask if the latter result also holds for functions of type (3.1d), for instance $\|K \exp(-itH)\Psi_0\|^2$ with $K = P_S$, which occurs in the integrand of (3.81a) for the sojourn time $J(S; \Psi_0)$. We have the following lemma.

Lemma 3.9. *Let f be a complex-valued function on $\mathbb{R}$ such that, for some $1 \leq p < \infty$, $f \in L^p(\mathbb{R}, \mathcal{L})$. Then $\exists A \subset \mathbb{R}$ s.t. $\mathcal{L}(A) = 0$ and $\lim_{|t| \to \infty : t \notin A} f(t) = 0$.*

Proof. $\int_0^\infty dt |f(t)|^p = \sum_{n=1}^\infty \int_0^1 dy |f(y + n)|^p$. By the monotone convergence theorem and the assumption, $\sum_{n=1}^\infty |f(y+n)|^p < \infty$ for a.e. $y \in [0, 1]$, which implies that $\lim_{n \to \infty} f(y + n) = 0$ for a.e. $y \in [0, 1]$. This proves the assertion for $t \to \infty$, for $t \to -\infty$ the proof is, of course, identical. $\quad\square$

One might ask, in view of Lemma 3.8, how to obtain the stronger assertion $\lim_{|t| \to \infty} f(t) = 0$ in Lemma 3.9. Is continuity necessary to obtain this stronger statement? The following condition, used in Tauberian theory, is natural in this context:

Definition 3.5 ([219, Definition 9.6]). A function $\Phi \in L^\infty(\mathbb{R})$ is *slowly oscillating at $+\infty$* if, to every $\epsilon > 0$ corresponds an $A < \infty$ and a $\delta > 0$ such that $|\Phi(x) - \Phi(y)| < \epsilon$ if $x > A$ and $|x - y| < \delta$.

We now have the following proposition.

Proposition 3.1. *Let* $f \in L^p(\mathbb{R}_+, \mathcal{L})$ *where* $1 \le p < \infty$. *If* f *is slowly oscillating at* $+\infty$, *then*

$$\lim_{t \to \infty} f(t) = 0. \tag{3.104a}$$

Of course, analogous results hold at $-\infty$.

Problem 3.5. *Prove Proposition* 3.1.

Since a slowly oscillating function at infinity is not necessarily continuous, Proposition 3.1 demonstrates that continuity is not a necessary condition for (3.104a). It follows, however, from Definition 3.5 that a uniformly continuous bounded function is slowly oscillating, and we thus obtain immediately the following corollary.

Corollary 3.2. *Let* $f \in L^p(\mathbb{R}, \mathcal{L})$, $1 \le p < \infty$, *be a bounded, uniformly continuous function of* $t \in \mathbb{R}$. *Then* (3.104a) *holds in the form*

$$\lim_{|t| \to \infty} f(t) = 0. \tag{3.104b}$$

Problem 3.6. *For a strongly continuous group* $V_t = \exp(-itH)$ *prove that the function*

$$t \in \mathbb{R} \longmapsto \|P_S V_t \Psi\|^2 \tag{3.105}$$

is, for any $\Psi \in \mathcal{H}$, *a uniformly continuous function of* $t \in \mathbb{R}$.

Since $\|P_S V_t \Psi\|^2 \le \|\Psi\|^2$, applying Corollary 3.2 to the function (3.105), we obtain that, under the condition (3.85b) of finiteness of the sojourn time, the probability $p(S, \Psi; t)$, with (3.81c), that the particle be found in S, decreases to zero as time evolves; this is true for any bounded region S.

One may ask whether Corollary 3.1 has a converse, i.e., whether there *does not exist any* $\Psi \in \mathcal{H}_{ac}$ such that $J(S; \Psi) = \infty$ for some bounded S. In the positive case, the latter condition would imply that $\Psi \in \mathcal{H}_s$. As observed in [248], the answer is, unfortunately, negative, due to Pearson's [199] example of a Hamiltonian such that the wave-operators $\Omega_\pm$ exist but the condition of weak asymptotic completeness (2.17) does not hold, i.e., $\text{Ran}(\Omega_+) \ne \text{Ran}(\Omega_-)$: in that example, $\lim_{t \to \infty} \|P_S V_t \Psi\|^2 \ne 0$ if $\Psi \in \mathcal{H}_{ac} \backslash \text{Ran}(\Omega_+)$, with a similar assertion holding at $-\infty$. Thus, for any such Ψ, $J(S; \Psi) = \infty$ for all bounded S by Corollary 3.2. In the sense of not fulfilling weak asymptotic completeness, Pearson's example may be considered unusual or "pathological", and one might expect that "in ordinary situations" Theorem 3.16 and Corollary 3.1 provide a dynamical distinction between states in $\mathcal{H}_{ac}$ and $\mathcal{H}_{sc}$. However, examining Pearson's potential, the reader will conclude that it looks physically quite reasonable, so that this counterexample is actually very important to show that (and why) a clear-cut geometric distinction between the two spectra is a difficult open problem. In spite of that, we have seen that Corollary 3.1 still sufficed to yield a precise physical interpretation of states in the s.c. spectrum in the case of the sparse models (3.84a)–(3.84g).

3.3.2. Rajchman measures, and the connection between ergodic theory, number theory and analysis

If we write, in correspondence to (3.2a),

$$\hat{\mu}(t) = \hat{\mu}_{ac}(t) + \hat{\mu}_{sc}(t) + \hat{\mu}_{pp}(t) \tag{3.106}$$

then, for some sequence $\{\xi_j\} \subset \mathbb{R}$,

$$\hat{\mu}_{pp}(t) = \sum_j p_j \exp(it\xi_j), \quad \text{with} \sum_j p_j = 1 \text{ and } p_j \geq 0 \tag{3.107}$$

so that $\hat{\mu}_{pp}$ is an *almost-periodic function* of $t \in \mathbb{R}$ (see [146, Definitions 5.1 and 5.2, p. 155]):

Definition 3.6. Let f be a complex-valued function on $\mathbb{R}$ and let $\epsilon > 0$. An ϵ-almost period of f is a number τ such that $\sup_x |f(x - \tau) - f(x)| < \epsilon$.

Definition 3.7. A function f is (uniformly) almost-periodic on $\mathbb{R}$ if it is continuous and if, for every $\epsilon > 0$, there exists a number $\Lambda = \Lambda(\epsilon, f)$ such that every interval of length Λ in $\mathbb{R}$ contains an ϵ-almost period of f.

By [146, Theorem 5.7, p. 158], any uniform limit of trigonometric polynomials such as (3.107) is almost-periodic, and by (3.107) and Definitions 3.6 and 3.7,

$$\limsup_{|t| \to \infty} |\hat{\mu}_{pp}(t)| = 1. \tag{3.108}$$

One might be led to ask if the validity of (3.108) for a probability measure μ implies that μ is p.p.. This is false! Indeed, Jessen and Wintner

[131, Theorem 11] show that $f(t) \equiv \prod_{n=1}^{\infty} \cos(t/n!)$ is singular continuous. For integer k,

$$1 - f(2\pi k!) = O\left(\sum_{n=k+1}^{\infty} (k!)^2/(n!)^2\right) = o(1),$$

as $k \to \infty$, so that $\limsup_{|t|\to\infty} |f(t)| = 1$. Thus, by (3.108), the only *general* result is that μ is *continuous* if $\limsup_{|t|\to\infty} |\hat{\mu}(t)| = 0$, i.e., if (3.102) holds. The class of (continuous) measures satisfying (3.102) has a name:

Definition 3.8. A finite Borel measure satisfying (3.102) is called a *Rajchman measure*. We denote by R the set of Rajchman measures.

By Proposition C.25 of App. C, the Lebesgue function constructed on a symmetric Cantor set over $[-\pi, \pi]$ with dissection ratio $\xi = 1/p$ with p any odd integer such that $p \geq 3$, has a F.S. transform which does not satisfy (3.102), and thus such measures are s.c. measures not in R. We have [175]:

$$M_{\mathrm{ac}}(\mathbb{R}) \subsetneq R \subsetneq M_{\mathrm{c}}(\mathbb{R}), \tag{3.109}$$

where M_{ac} (respectively, M_{c}) denote the set of a.c. (respectively, continuous) measures. The first inclusion in (3.109) follows from Lemma 3.8, and the second inclusion from (3.108). The second inequality sign in (3.109) is a consequence of the previously cited TAMS paper of Jessen and Wintner. The first inequality sign in (3.109) follows from the following classical result due to Salem and Zygmund (see [144, Chap. VI, Theorems VI and VII] and references given there):

Theorem 3.18. *Let C_ξ denote the symmetric Cantor set of ratio ξ (see App. C), and μ_ξ (equal to $dL = dL_\xi$ in the notation of App. C) be the corresponding Cantor measure on C_ξ, with $0 < \xi < 1/2$. The following are equivalent:*

(1) $\hat{\mu}_\xi(t) = o(1)$ *as* $|t| \to \infty$, *i.e.,* $\hat{\mu}_\xi \in R$;
(2) ξ^{-1} *is not a P.V. number.*

(The notion of P.V. number will be given shortly.)

The above theorem is quite typical of the deep interplay between analysis, number theory and ergodic theory, further developed in App. A. We shall try to explain some of the connections in Theorem 3.18, without presenting a complete proof. Some definitions are needed, the first one concerns number theory:

Definition 3.9. A real number $\alpha > 1$ is called a Pisot–Vijayaraghavan number (P.V. number) if α is an algebraic integer all of whose conjugates — apart from α itself — lie in the open unit circle $\{z \in \mathbb{C} : |z| < 1\}$. In other words, $\alpha > 1$ is a P.V. number if there exists a polynomial

$$P(x) = x^m + a_{m-1}x^{m-1} + \cdots + a_0 \quad \text{with } a_i \in \mathbb{Z} \quad \text{for } i = 0, \ldots, m-1$$

$$\text{(3.110a)}$$

irreducible over $\mathbb{Q}$ (i.e., it is not a product of two nonconstant polynomials with rational coefficients) such that

$$P(x) = (x - \alpha_1) \cdots (x - \alpha_m) \quad \text{with } \alpha_1 = \alpha \text{ and } \alpha_j < 1 \text{ for } 2 \leq j \leq m.$$

$$\text{(3.110b)}$$

Trivially, every integer > 1 is a P.V. number, but in the next exercise, a nontrivial example appears.

Problem 3.7. *Show that* $\alpha = \frac{1+\sqrt{5}}{2}$ *is a P.V. number.*

For some additional rudiments of number theory which will be needed in the book, we refer the reader to App. A; for a readable account of P.V. numbers, Salem numbers and Meyer sets in $\mathbb{R}^n$ (the latter also called Penrose tilings), to which we shall briefly come back in Sec. 3.4 in connection to quasicrystals, see [186]. We now come back to some definitions which are a necessary preliminary to the third concept mentioned above, that of ergodic theory, which is also expounded to some length in App. A.

For all x, let $\mathrm{fr}(x)$ denote its fractional part $x - [x]$, where $[x]$ is the greatest integer $\leq x$.

Definition 3.10. Given a sequence $\omega = (x_k)_{k \geq 1}$ of real numbers and a subset A of $[0, 1)$, let $N(A, n; \omega)$ denote the number of terms x_k, $1 \leq k \leq n$, in ω whose fractional part belongs to A, i.e., $\mathrm{fr}(x_k) \in A$. ω is said to be *uniformly distributed* modulo 1 (abbr. u.d. mod 1) if

$$\lim_{n \to \infty} \frac{1}{n} N([a, b), n; \omega) = b - a \qquad \text{(3.111a)}$$

for each couple a, b s.t. $0 \leq a < b \leq 1$.

Denoting by $\chi_A(x)$ the characteristic function of A, i.e., $\chi_A(x) = 1$ if $x \in A$ ($= 0$ otherwise), then (3.111a) may be written as

$$\lim_{n \to \infty} \frac{1}{n} \sum_{k=1}^{n} \chi_{[a,b)}(\mathrm{fr}(x_k)) = \int_0^1 \chi_{[a,b)}(x)dx. \tag{3.111b}$$

This remark, together with the construction of the Riemann integral of a function, leads to the following proposition.

Proposition 3.2. *The necessary and sufficient condition for $\omega = (x_k)_{k\geq 1}$ to be u.d. mod 1 is that*

$$\lim_{n \to \infty} \frac{1}{n} \sum_{k=1}^{n} f(x_k) = \int_0^1 f(x)dx \tag{3.111c}$$

be satisfied for all f continuous (and thus Riemann integrable) and periodic of period 1.

A fundamental theorem in the subject is given below.

Theorem 3.19 (Weyl's criterion of u.d.). *A sequence $\omega = (x_k)_{k\geq 1}$ is u.d. mod 1 iff*

$$\lim_{n \to \infty} \frac{1}{n} \sum_{k=1}^{n} \exp(2\pi i h x_k) = 0 \tag{3.111d}$$

for all integers $h \neq 0$.

Proof. Necessity: apply Proposition 3.2 for the function $f(x) = \exp(2\pi i h x)$. Sufficiency: depends on the Weierstrass approximation theorem (see [146, Corollary to Theorem 2.12]): given $\epsilon > 0$, there exists a polynomial $P(z)$ such that $\sup_{x \in [0,1)} |P(\exp(2\pi i x)) - f(x)| < \epsilon/3$. Now, use the fact that by (3.111d) $\int_0^1 P(\exp(2\pi i x))dx - \frac{1}{n}\sum_{k=1}^{n} P(\exp(2\pi i x_k))$ is arbitrarily small if n is sufficiently large. Together with an $\epsilon/3$ argument, this proves (3.111c). □

An immediate corollary of Theorem 3.19 is that the sequence $\omega = (x_k)_{k\geq 1}$ with $x_k = kx$ is u.d. mod 1 for all $x \in (0,1)$ irrational (why?). A generalization of this result for a.e. $x \in [0,1)$ is possible by the following theorem, which is the first of a series of metric Weyl theorems (for others, see [156] and Sec. 4.2):

Theorem 3.20. *The sequence $\omega = (a_k x)_{k\geq 1}$ is u.d. mod 1 for a.e $x \in (0,1)$ if $(a_k)_{k\geq 1}$ is a sequence of distinct integers.*

Proof. Define the "Weyl sum" (see (3.111d)):

$$S(x, n) \equiv \frac{1}{n} \sum_{k=1}^{n} \exp(2\pi i h a_k x). \qquad (3.111e)$$

Then,

$$|S(x, n)|^2 = S(x, n)\overline{S(x, n)} = \frac{1}{n^2} \sum_{k,l=1}^{n} \exp[2\pi i h(a_k - a_l)x]$$

from which, by hypothesis $\int_0^1 |S(x, n)|^2 dx = 1/n$. By Fatou's lemma, this leads to

$$\sum_{m=1}^{\infty} \int_0^1 |S(x, m^2)|^2 dx = \sum_{m=1}^{\infty} \frac{1}{m^2} = \pi^2/6 < \infty$$

which implies that $\sum_{m=1}^{\infty} |S(x, m^2)|^2 < \infty$ for a.e. $x \in (0, 1)$. An immediate consequence of the latter is that $\lim_{m \to \infty} |S(x, m^2)| = 0$ for a.e. $x \in (0, 1)$. Now, given n, there exists m such that $m^2 < n \le (m + 1)^2$, with $m = m(n) \to \infty$ as $n \to \infty$, and $(m + 1)^2 - (m^2 + 1) = 2m$. It follows that

$$|S(x, n)| \le |S(x, m^2)| + \frac{1}{n} \left| \sum_{k=m^2+1}^{n} \exp(2\pi i h a_k x) \right|$$

$$\le |S(x, m^2)| + \frac{1}{n} \sum_{k=m^2+1}^{(m+1)^2} |\exp(2\pi i h a_k x)|$$

$$\le |S(x, m^2)| + \frac{2m}{n} \le |S(x, m^2)| + \frac{2}{\sqrt{n} - 1}$$

(with $n \ge 2$), from which $\lim_{n \to \infty} |S(x, n)| = 0$ for a.e. $x \in (0, 1)$. $\square$

Taking $a_n = b^n$, with b an integer, it follows from Theorem 3.20 that $(b^k x)_{k \ge 1}$ is u.d. mod 1 for a.e. $x \in (0, 1)$. By Koksma's metric theorem (see [156, Theorem 4.3]) the sequence $\omega = (x^k)_{k \ge 1}$ is u.d. mod 1 for a.e. $x > 1$: the numbers $x > 1$ for which $\omega = (x^k)_{k \ge 1}$ is *not* u.d. mod 1 include the class of P.V. numbers. For the integers $x = b \ge 2$, which are P.V. numbers, this is obvious because b^k mod 1 equals zero, for all $k = 1, 2, \ldots$. What about the Golden Rule of Problem 3.7?

Problem 3.8. *Show that $\omega = (\alpha^k)_{k \ge 1}$ is not u.d. mod 1, if $\alpha = \frac{1+\sqrt{5}}{2}$.*

Hint: The conjugate of α is $-1/\alpha$; define $G_k \equiv \alpha^k + (-1/\alpha)^k$ and prove that it satisfies the recurrence relation

$$G_{k+2} = G_{k+1} + G_k \tag{3.112}$$

with $G_0 = 2$, $G_1 = 1$. Thus G_k is an integer for all $k \geq 1$. Since $1/\alpha < 1$, $\lim_{k\to\infty}(-1/\alpha)^k = 0$. This means that α^k approaches an integer as $k \to \infty$, i.e., α^k mod 1 tends to zero or one as $k \to \infty$.

In order to show (3.112), it may also be useful to remark that α is the irrational number worse approximated by rationals (see App. A) $\alpha = 1/(1 + 1/(1 + \cdots))$ and hence satisfies the equation $r = 1/(1 + 1/r)$ or $r^2 - r - 1 = 0$.

As an addendum to Problem 3.8, it may be seen that $\alpha^k \to 1$ as $k \to \infty$ with k even, and $\alpha^k \to 0$ as $k \to \infty$ with k odd: indeed, for k even,

k	α^k
10	$0.992\cdots$
20	$0.9999\cdots$
30	$0.999999\cdots$

and for k odd,

k	α^k
11	$0.005\cdots$
21	$0.00004\cdots$
31	$0.000000\cdots$

We now come back to (3.109) and Theorem 3.18. Clearly, (1) and (2) of the latter establish a connection, to which we come back soon, between harmonic analysis and number theory. On the other hand, Problem 3.8 suggests a "sort of" connection of (2) of Theorem 3.18 with ergodic theory. We now try to see directly a connection of (1) of Theorem 3.18, i.e., the property that $\mu_\xi \in R$, with ergodic theory. For simplicity, with Lyons [175], we restrict ourselves to Borel measures μ on the circle $\mathbb{T} = \mathbb{R}/\mathbb{Z} = [0, 1)$, with Fourier transform

$$\hat{\mu}(n) = \int_{\mathbb{T}} \exp(-2\pi i n t) d\mu(t)$$

and define the class R of Rajchman measures, in correspondence with Definition 3.8:

$$R \equiv M_0(\mathbb{T}) \equiv \left\{ \mu \in M(\mathbb{T}) : \lim_{|n|\to\infty} \hat{\mu}(n) = 0 \right\}.$$

By Lemma 3.8, all a.c. measures are in R. In analogy to (3.109),

$$L^1(\mathbb{T}) \subsetneq R \subsetneq M_c(\mathbb{T}).$$

Now, $L^1(\mathbb{T})$ and $M_c(\mathbb{T})$ are defined in terms of their common null sets: $L^1(\mathbb{T})$ consists of those measures which annihilate (give measure zero to) all sets of Lebesgue measure zero, and $M_c(\mathbb{T})$ consists of those measures which annihilate all countable sets. In [174], the question was posed: is there a class of sets intermediate between those of Lebesgue measure zero and countable ones such that a measure is a Rajchman measure iff it annihilates all sets in this class? As we shall see, the answer to this question will provide a precise link between the property $\mu \in R$ (part (1) of Theorem 3.18) with ergodic theory. We follow [174, 175].

For a class $\mathcal{E}$ of subsets of $\mathbb{T}$, let

$$\mathcal{E}^\perp \equiv \{\mu \in M(\mathbb{T}) : \mu(E) = 0, \forall E \in \mathcal{E}\}.$$

The above-posed question may be written: is there a class $\mathcal{E}$ such that

$$R = \mathcal{E}^\perp?$$

We refer to [175] for references and (interesting) historical remarks.

We begin with an elementary but interesting generalization of Lemma 3.8; although the proof is written for the circle, it generalizes immediately to measures on the real line:

Lemma 3.10. *The class R is a band, i.e., it is closed under absolute continuity*:

$$\nu \ll \mu \in R \Rightarrow \nu \in R.$$

Proof. By the Weierstrass approximation theorem (see [146, Corollary to Theorem 2.12]), the trigonometric polynomials P are uniformly dense in the set of continuous functions on $\mathbb{T}$, which in turn are norm-dense in $L^1(\mu)$. Thus, the trigonometric polynomials are norm-dense in $L^1(\mu)$. By the Radon–Nikodym theorem, there exists $f \in L^1(\mu)$ s.t. $\nu = f\mu$. Since it is immediate that $P\mu \in R$ if $\mu \in R$, we have

$$\limsup_{|n|\to\infty} |\widehat{f\mu}(n)| = \limsup_{|n|\to\infty} |\widehat{(f\mu - P\mu)}(n)|$$
$$\leq \|f - P\|_{L^1(\mu)}.$$

Since the r.h.s. above may be made arbitrarily small, it follows that $\nu \in R$. □

It is clear that the property of being a band is necessary for the existence of a class $\mathcal{E}$ s.t. $R = \mathcal{E}^\perp$, and that Lemma 3.8 is a special case of Lemma 3.10, where $\mu = \mathcal{L}$.

Several possible choices of the set $\mathcal{E}$ have been proposed by various people [175]. In order to discuss some of the most important ones, which is both interesting and enlightening in order to understand some of the significance of the final solution given by Lyons [174], we need some preliminary definitions:

Definition 3.11. A number $x \in (0,1)$ is called *normal* in base 2 if in its base 2 expansion, every block of digits occurs as often as any other block of the same length; alternatively, if $x = 0.\epsilon_1\epsilon_2\epsilon_3 \cdots = \sum_{i=1}^\infty \frac{\epsilon_i}{2^i}$ with $k \geq 1$ and $\alpha_1, \ldots, \alpha_k \in \{0,1\}$ then

$$\lim_{n \to \infty} \frac{1}{n} |\{j \in [1,n] : \epsilon_j = \alpha_j, \epsilon_{j+1} = \alpha_{j+1}, \ldots, \epsilon_{j+k-1} = \alpha_{j+k-1}\}| = 2^{-k}.$$

The connection between normal numbers and ergodic theory is clear by comparing Definition 3.11 with Definition 3.10: together with Theorem 3.20 (with $a_k = 2^k$), *Borel's theorem* follows: a.e. $x \in (0,1)$ is normal of base 2. An instructive alternate proof of Borel's theorem using Birkhoff's ergodic theorem and some properties of classical dynamical systems is presented in App. A.

Definition 3.11 suggests another one:

Definition 3.12. A Borel set $E \subset \mathbb{T}$ is called a $W^\star$ set if there is an increasing set of integers $\{n_k\}$ such that, for all $x \in E$, the sequence $\{n_k x\}$ is *not* u.d.

By Theorem 3.20, every $W^\star$ set has Lebesgue measure zero. The question was posed (by Kahane and Salem [143], and Kahane [140]) whether $R = (W^\star)^\perp$? It turned out that the answer is negative, i.e., the class $W^\star$ is too big to satisfy the above (see [175] and references given there), but the following definition does the job. We need, again, some preliminaries: a sequence $\{x_k\} \subset \mathbb{T}$ is said to have an *asymptotic distribution* $\nu \in M(\mathbb{T})$ if

$$\lim_{K \to \infty} \frac{1}{K} \sum_{k=1}^K \delta(x_k) = \nu$$

in the weak-∗ topology $\sigma(M(\mathbb{T}, C(\mathbb{T}))$, where $\delta(x)$ denotes the Dirac point mass at x; alternatively, for every real-valued continuous function f on $\mathbb{T}$,

$$\lim_{K\to\infty} \frac{1}{K} \sum_{k=1}^{K} f(x_k) = \int_0^1 f(x)d\nu(x)$$

(see [156, Theorem 7.2, p. 54]). In case $\nu = \mathcal{L}$, the sequence is u.d. as in Definition 3.9. The analogue of Weyl's theorem on uniform distribution, Theorem 3.19, may be proved, i.e., a sequence $\{x_k\} \subset \mathbb{T}$ has an asymptotic distribution function (a.d.f.) ν iff

$$\lim_{K\to\infty} \frac{1}{K} \sum_{k=1}^{K} \exp(2\pi i h x_k) = \int_0^1 \exp(2\pi i h x)d\nu(x) \qquad (3.113)$$

for all integers $h \neq 0$ (see [156, Theorem 7.3]). We may now state:

Definition 3.13. A Borel set $E \subset \mathbb{T}$ is a *W-set* iff there is an increasing sequence of integers $\{n_k\}$ such that, for every $x \in E$, the sequence $\{n_k x\}$ has an asymptotic distribution, but is *not* uniformly distributed.

In [174], Lyons proved:

Theorem 3.21 ([174, Theorem 3]). *A measure $\mu \in R$ iff $\mu(E) = 0$ for all W-sets E.*

Proof of necessity. This is the easier part of Theorem 3.21. Let $\mu \in R$ and E be a W-set with corresponding sequence $\{n_k\}$. Corresponding to (3.113), we may define:

$$c_h(x) = \begin{cases} \lim_{K\to\infty} \dfrac{1}{K} \displaystyle\sum_{k=1}^{K} \exp(2\pi i h n_k x) & \text{if } x \in E, \\[2mm] 0 & \text{otherwise.} \end{cases}$$

Let F be any Borel subset of E, and $\nu \equiv \mu_F$ the restriction of μ to F. By Lemma 3.10, $\nu \in R$ and, therefore, for $h \neq 0$,

$$\int_F c_h(x)d\mu(x) = \int_F c_h d\nu$$

$$= \lim_{K\to\infty} \int_F \frac{1}{K} \sum_{k=1}^{K} \exp(2\pi i h n_k x)\, d\nu(x)$$

$$= \lim_{K \to \infty} \int_{\mathbb{T}} \frac{1}{K} \sum_{k=1}^{K} \exp(2\pi i h n_k x) \, d\nu(x)$$

$$= \lim_{K \to \infty} \frac{1}{K} \sum_{k=1}^{K} \hat{\nu}(-h n_k) = 0.$$

Since F is arbitrary, it follows that $c_h(x) = 0$ for $|\mu|$-almost all $x \in E$. But, by definition, if $x \in E$, then $c_h(x) \neq 0$ for some $h \neq 0$. Hence $|\mu|(E) = 0$. $\square$

The converse is much more difficult and we must refer to the very readable article [174]. The previous proof of necessity shows, however, the flavor of a general relationship between the property $\mu \in R$ and the (geometric) characterization of $\mu \in R$ by annihilation of a set related to an ergodic property, which includes the property of being not u.d.. We have seen that μ annihilates sets defined by sequences which are not u.d.: the P.V. numbers comprise a set of the latter type, by Problem 3.8 (the general proof follows the same lines). Now, *special* P.V. numbers, viz. ξ^{-1} equal to an integer ≥ 2, correspond to *special* measures μ_ξ supported by symmetric Cantor sets with dissection ratio ξ which are not in R by Proposition C.25, and thus we have, together with Theorem 3.18, finally exhibited (albeit in rather special cases) the interplay between harmonic analysis (associated to the property $\mu \in R$), number theory and ergodic theory. For other aspects of this deep interplay, not necessarily related to pointwise decay, we refer to the beautiful monograph by Kac [138].

3.3.3. *Fourier dimension, Salem sets and Salem's method*

The connection between the property $\mu \in R$ and other areas of mathematics yielded elegant general results, such as Theorems 3.18 and 3.21. The actual proof that a given measure is Rajchman presents us, however, with a formidable problem. In particular, we would like to know at which rate $\hat{\mu}(t) \to 0$ as $|t| \to \infty$ when $\mu \in R$. There are very few results of this kind in mathematical physics, and Sec. 4.3 is reserved to one of them. The remainder of this section shall be, however, primarily concerned with general results and concepts in mathematics related to the rate of pointwise decay of singular measures. For a.c. measures, Lemma 3.8 does not establish any rate of decay, but for certain a.c. measures of interest in mathematical physics, related to the return probability, this rate may be determined by the stationary phase method, as reviewed in Chap. 2.

We recall the Definition 3.1 of α-dimensional Hausdorff measure h^α on $\mathbb{R}$, and its properties (Lemma 3.2). It will be useful to consider the generalization of h^α to $\mathbb{R}^n$, $0 < \alpha < n$, recall definitions (3.54b)–(3.54d), and Theorem 3.13. The n-dimensional analogue of Lemma 3.6 is, by (3.58), easily seen to hold: if $J_\alpha^{(n)}(\mu) < \infty$, then μ is α-continuous. We shall need this property in the following.

A very important concept related to pointwise decay of singular measures is that of *Fourier dimension* [141, 187]. Recall in this connection Frostman's lemma (see (3.59) and (3.60)).

Definition 3.14. The Fourier dimension, denoted by $\dim_F E$, of a compact set $E \subset \mathbb{R}^n$ is

$$\dim_F E = \sup\{\beta : 0 \leq \beta < \infty \text{ s.t. for some probability measure (p.m.) } d\mu$$
$$\text{supported on } E, |\hat{d\mu}(k)| \leq C|k|^{-\beta/2}\}. \tag{3.114a}$$

Lemma 3.11.

$$\dim_H E \geq \dim_F E. \tag{3.114b}$$

Proof. We first show that if $J_\alpha^{(n)}(\nu) < \infty$ for ν a p.m. with support equal to E, then $\dim_H E \geq \alpha$. As mentioned in the paragraph just before Definition 3.12, ν is α-continuous, which implies that $h^\alpha(E) = 0 \Rightarrow \nu(E) = 0$. We have

$$\dim_H E = \alpha \Leftrightarrow \begin{cases} \forall \beta' < \alpha & \text{we have } h^{\beta'}(E) = \infty, \\ \forall \beta' > \alpha & \text{we have } h^{\beta'}(E) = 0. \end{cases}$$

Assume that $\dim_H E = \alpha' < \alpha$; then if $\alpha' < \beta' < \alpha$, $h^{\beta'}(E) = 0$ from which $h^\alpha(E) = 0$ and $\nu(E) = 0$. But, by hypothesis $\nu(\mathbb{R}^n \backslash E) = 0$, which yields a contradiction; thus $\dim_H E \geq \alpha$. Now, if $\beta = \dim_F E$ as defined by (3.114a), and μ is a p.m. with support on E, $J_\alpha^{(n)}(\mu) \equiv \int dk |k|^{\alpha-n} |\hat{d\mu}(k)|^2 \leq C \int dk |k|^{-\beta+\alpha-n} < \infty$ if $\beta > \alpha$. Thus μ is α-continuous, which implies $\dim_H E \geq \alpha$ *for any* $\alpha < \beta$, whence (3.114b). $\qquad \square$

If E is a compact smooth α-dimensional submanifold of $\mathbb{R}^n$, we expect isotropic decay $|x|^{-\alpha/2}$ only under special conditions on the curvature and dimension of E. The unit sphere in $\mathbb{R}^n$ has Hausdorff dimension $n-1$:

Problem 3.9. *Show that for the unit sphere in $\mathbb{R}^n$, the Hausdorff dimension equals the Fourier dimension.*

Hint: The Bessel function of order ν has the integral representation

$$J_\nu(z) = C z^\nu \int_0^1 (1 - t^2)^{\nu - 1/2} \cos(zt) dt$$

for a suitable constant C (see [1, (9.120)]).

As remarked by Stein [250, p. 348], decay estimates of type (3.114a) are of much more general nature than the above exercise may lead the reader to expect: (they) "are not limited to the fortuitous circumstance connecting rotational symmetry with Bessel functions". Indeed, they may be obtained for Fourier transforms of measures carried by smooth hypersurfaces in $\mathbb{R}^n$ whose Gaussian curvature is nonzero everywhere (see [250, Theorem 1, p. 348], and further developments in that chapter). The following definition is naturally related to Definition 3.14:

Definition 3.15. A compact subset $E \subset \mathbb{R}^n$ is called a Salem set iff

$$\dim_F E = \dim_H E. \tag{3.114c}$$

The above definition extends to $\mathbb{R}^n$ the original property first shown by Salem [226] for certain subsets of the real line. By Problem 3.9, the unit sphere in $\mathbb{R}^n$ with $n \geq 2$ is a Salem set. On the other hand, the Cantor middle-thirds set has Fourier dimension zero (see Proposition C.3), but Hausdorff dimensional has $\frac{\log 2}{\log 3}$. As remarked by Mockenhaupt [187], the fact that Hausdorff dimension and Fourier dimension do not agree in general is not surprising, because Hausdorff dimension measures a metric property of a set, while Theorem 3.18 and Definition 3.14 make it clear that Fourier dimension measures an arithmetic property of a set.

In a beautiful paper, Salem [226] showed that, for any given $\alpha \in (0, 1)$ random Cantor sets can be constructed which are random Salem sets of dimension α.

Let $I \subset [0, 1]$ be a segment of length $|I| = l$ and let $d \in \mathbb{N}$. Let $\alpha_1, \ldots, \alpha_d$ be numbers such that

$$0 < \alpha_1 < \alpha_2 < \cdots < \alpha_d < 1.$$

Let each of the points α_j with $j \in \{1, \ldots, d\}$ be the origin of an interval $I_\eta(\alpha_j) = (\alpha_j l, (\alpha_j + \eta) l)$, with η satisfying (with $\alpha_{d+1} \equiv 1$)

$$0 < \eta < \alpha_{j+1} - \alpha_j, \quad j = 1, \ldots, d.$$

The d disjoint intervals $I_\eta(\alpha_1), \ldots, I_\eta(\alpha_d)$ of length ηl obtained in this way will be called "white" intervals and the $d + 1$ complementary intervals with respect to I will be called "black" intervals. Black intervals will be dissected and the dissection of the interval I will be said to be of the type $(d, \alpha_1, \ldots, \alpha_d, \eta)$.

Starting from the interval $[0, 1]$ and fixing the numbers $\alpha_1^{(1)}, \ldots, \alpha_{d_1}^{(1)}$ for some d_1, we operate a dissection of the type $(d_1, \alpha_1^{(1)}, \ldots, \alpha_d^{(1)}, \eta_1)$ and remove the black intervals. On each white interval $I_{\eta_1}(\alpha_j^{(1)})$ left, we operate a dissection $(d_2, \alpha_1^{(2)}, \ldots, \alpha_d^{(2)}, \eta_2)$ and we remove the black intervals, and so on. After p operations we have $d_1 \cdots d_p$ white intervals, each of length $\eta_1 \cdots \eta_p$. When $p \to \infty$, we obtain a perfect set E, nowhere dense, which is of Lebesgue measure zero if $\prod_{k=1}^p d_k \eta_k \to 0$.

The sequence $(\eta_k)_{j \geq 1}$ is arbitrary while it satisfies the constraint $(\alpha_{d_k+1}^{(k)} = 1)$

$$0 \leq \eta_k < \alpha_{j+1}^{(k)} - \alpha_j^{(k)}, \quad j = 1, \ldots, d_k. \tag{3.115}$$

On the pth stage of the dissection,

$$x_{\mathbf{j}}^{(p)} = \alpha_{j_1}^{(1)} + \eta_1 \alpha_{j_2}^{(2)} + \cdots + \eta_1 \cdots \eta_{p-1} \alpha_{j_p}^{(p)} \tag{3.116}$$

with $\mathbf{j} = (j_1, \ldots, j_p) \in \Lambda_p \equiv \{1, \ldots, d_1\} \times \cdots \times \{1, \ldots, d_k\}$, gives the left extremity coordinate in $[0, 1]$ of the white intervals $I_{\eta_p}(\alpha_j^{(p)})$ of length $\eta_1 \cdots \eta_p$. Each point x of the set E can thus be written as

$$x = \alpha_{j_1}^{(1)} + \eta_1 \alpha_{j_2}^{(2)} + \cdots + \eta_1 \cdots \eta_{k-1} \alpha_{j_k}^{(k)} + \cdots,$$

where $j_k \in \{1, \ldots, d_k\}$, $k \geq 1$.

Let $F_p(x)$ be a continuous nondecreasing function such that $F_p(x) = 0$ for $x \leq 0$, F_p increases linearly by $(\prod_{k=1}^p d_k)^{-1}$ on each $\prod_{k=1}^p d_k$ white intervals obtained in the pth step, F_p is constant in every black interval and $F_p(x) = 1$ for $x \geq 1$. The limit function $F(x)$ is a singular continuous nondecreasing function, having the perfect set E as spectrum — the uniform probability distribution function on E.

The Fourier–Stieltjes transform of F

$$\gamma(t; F) = \int_{-\infty}^{\infty} e^{ixt} dF(x)$$

is the limit, as p tends to infinity, of

$$\gamma_p(t) = \sum_{\mathbf{j} \in \Lambda_p} \left(\prod_{k=1}^p d_k \right)^{-1} e^{ix_{\mathbf{j}}^{(p)} t}. \tag{3.117}$$

Thus, replacing (3.116) into (3.117) and writing

$$Q_k(u) = \frac{1}{d_k} \sum_{j \in \{1, \ldots, d_k\}} e^{i\alpha_j^{(k)} u}$$

we have (with $\eta_0 \equiv 1$)

$$\gamma(t; F) = \prod_{k=1}^{\infty} Q_k(\eta_0 \cdots \eta_{k-1} t) \ .$$

Lemma 3.12. *Let* $P(u) = \lambda_1 e^{i\alpha_1 u} + \cdots + \lambda_d e^{i\alpha_d u}$, *where the* α_j *are linearly independent*:

$$n_1 \alpha_1 + \cdots + n_d \alpha_d = 0 \quad \text{with } n_j \in \mathbb{Z} \Longleftrightarrow \alpha_1 = \cdots = \alpha_d = 0$$

and $|\lambda_1| + \cdots + |\lambda_d| \leq 1$. *Let* $s > 0$. *There exists a positive number* $T_0 = T_0(s, d, \{\alpha_j\}, \{\lambda_j\})$ *such that for* $T > T_0$ *and for all values of a*,

$$\frac{1}{T} \int_a^{a+T} |P(u)|^{s+2} \, du < 2 \left(\frac{s}{2} + 1\right)^{s/2} \left(\lambda_1^2 + \cdots + \lambda_d^2\right)^{s/2} . \qquad (3.118)$$

Proof. Let $2q$ be the even integer such that $s \leq 2q < s + 2$. Then,

$$|P(u)|^{2q} = \sum_{n_1 + \cdots + n_d = q} \lambda_1^{2n_1} \cdots \lambda_d^{2n_d} \left(\frac{q!}{n_1! \cdots n_d!}\right)^2 + R,$$

where R is a sum of terms of the form $Ae^{i\mu u}$ with nonvanishing μ. Note that

$$\frac{1}{T} \int_a^{a+T} e^{i\mu u} du = e^{i\mu a} \frac{e^{i\mu T} - 1}{i\mu T}$$

tends to zero as $T \to \infty$. We thus have

$$\lim_{T \to \infty} \frac{1}{T} \int_a^{a+T} |P(u)|^{2q} du = \sum_{n_1 + \cdots + n_d = q} \lambda_1^{2n_1} \cdots \lambda_d^{2n_d} \left(\frac{q!}{n_1! \cdots n_d!}\right)^2$$

$$\leq q!(\lambda_1^2 + \cdots + \lambda_d^2)^{s/2}$$

$$\leq q^q(\lambda_1^2 + \cdots + \lambda_d^2)^{s/2}$$

uniformly in a.

For $T \geq T_0$, where T_0 is independent of a,

$$\left(\frac{1}{T} \int_a^{a+T} |P(u)|^{2q} du \right)^{1/2q} < 2^{1/2q} q^{1/2} (\lambda_1^2 + \cdots + \lambda_d^2)^{1/2}$$

$$< 2^{1/2q} \left(\frac{r}{2} + 1 \right)^{1/2} (\lambda_1^2 + \cdots + \lambda_d^2)^{1/2}$$

and, together with

$$\left(\frac{1}{T} \int_a^{a+T} |P(u)|^{s+2} du \right)^{1/s} \leq \left(\frac{1}{T} \int_a^{a+T} |P(u)|^{2q} du \right)^{1/2q},$$

gives (3.118), concluding the proof. $\qquad\qquad\qquad\qquad\qquad\qquad$ □

Disordered sets. For simplicity, we shall first consider Salem's sets where the $\alpha_j^{(k)}$, $j = 1, \ldots, d_k$, are the same for all k. For fixed $\alpha \in (0,1)$ and $d \in \mathbb{N}$, let η be determined by the condition

$$\alpha = \frac{\log d}{\log 1/\eta} \tag{3.119}$$

so that $0 < \eta < 1/d$. Let the d linearly independent numbers $\alpha_1, \ldots, \alpha_d$ satisfy

$$0 < \alpha_1 < \frac{1}{d} - \eta$$

and, for $j = 1, \ldots, d - 1$,

$$\eta < \alpha_{j+1} - \alpha_j < \frac{1}{d}$$

from which, it follows that

$$1 - \alpha_d = 1 - \alpha_1 - \sum_{j=1}^{d-1} (\alpha_{j+1} - \alpha_j) > 1 - \frac{1}{d} + \eta - \frac{d-1}{d} = \eta.$$

That is to say, all inequalities (3.115) are satisfied for the η_k less or equal η. Let η_k be chosen so that

$$\xi_k \leq \eta_k \leq \eta, \quad j \geq 1, \tag{3.120a}$$

where

$$\xi_k = \eta \left(1 - \frac{1}{(k+1)^2} \right) \tag{3.120b}$$

and let $E(\eta_1, \eta_2, \ldots)$ denote the perfect set obtained by successive dissections of type $(d, \alpha_1, \ldots, \alpha_d, \eta_k)$. To every sequence $(\eta_k)_{k \geq 1}$ satisfying (3.120a) and (3.120b) corresponds a set E. It is clear from (3.119), (3.120a) and (3.120b) that all such sets have Hausdorff dimension α. To every set E, we associate the corresponding distribution function F having E as spectrum.

By writing

$$\eta_k = \xi_k + (\eta - \xi_k)\zeta_k, \tag{3.121}$$

a probability measure π in the space $\mathcal{S}$ of Salem's sets $E(\eta_1, \eta_2, \ldots)$ is introduced by letting $\zeta_1, \zeta_2, \ldots, \zeta_k \ldots$ be independent random variables

$$\zeta_j : [0, 1] \longrightarrow \mathbb{R}$$

defined in the probability space $([0, 1], \mathcal{B}, \mu)$ — $\mathcal{B}$ stands for Borel sets and μ is the Lebesgue measure — uniformly distributed in $[0, 1]$: $\mathbf{P}(a < \zeta_j \leq b) = \int_a^b d\tau = b - a$, for every $0 < a < b < 1$.

The set $E(\eta_1, \eta_2, \ldots) = E_\tau$ depends now on the variable $\tau \in [0, 1]$ and we shall show that for almost all τ the Fourier–Stieltjes transform $\gamma_\tau(t) \equiv \gamma(t; F_\tau)$ with respect to F_τ, the uniform distribution function on E_τ, decays as $t^{-\alpha/2+\varepsilon}$ for any $\varepsilon > 0$. A useful well-known property, inherited from independence, is that

$$\int_0^1 \Phi(\zeta_1(\tau), \ldots, \zeta_p(\tau))d\tau = \int_0^1 \cdots \int_0^1 \Phi(\zeta_1, \ldots, \zeta_p)d\zeta_1 \cdots d\zeta_p \tag{3.122}$$

for any measurable function Φ, whenever either side exists.

Fixing parameters. The numbers $0 < \alpha < 1$ and $\varepsilon > 0$ are given and set

$$s' = 2 + s = \frac{2 + \alpha}{\varepsilon}. \tag{3.123}$$

Let d be the smallest integer such that

$$\sqrt{d} \geq 2 \left(\frac{s}{2} + 1\right)^{s/2} = 2(s'/2)^{s'-2}. \tag{3.124}$$

Fix T_0 such that

$$\frac{1}{T} \int_a^{a+T} |Q(u)|^{s'} du < 2 \left(\frac{s'}{2}\right)^{s'-2} d^{(-s'-2)/2}$$

holds for all a and $T \geq T_0$ with

$$Q(u) = \frac{1}{d}(e^{i\alpha_1 u} + \cdots + e^{i\alpha_d u})$$

as clearly possible by Lemma 3.12. Then, in view of (3.124), we have

$$\frac{1}{T} \int_a^{a+T} |Q(u)|^{s'} du < \frac{1}{d^{(s'-3)/2}} \qquad (3.125)$$

for all a and $T \geq T_0$.

Decay of Fourier–Stieltjes transform of dF. We now have the following theorem.

Theorem 3.22. *Given any number $\alpha, 0 < \alpha < 1$, and a positive ε, arbitrarily small but fixed, there exists a perfect set E, with Hausdorff dimension α, and a nondecreasing function F, singular, with spectrum E, such that the Fourier–Stieltjes transform of dF is of order $t^{-\alpha/2+\varepsilon}$.*

Proof. Since $|Q(u)| \leq 1$,

$$|\gamma_\tau(t)|^{s'} \leq \prod_{k=1}^{p} |Q(\eta_1 \cdots \eta_k t)|^{s'} \equiv f(t; p)$$

with p being any positive number. By (3.122), we have

$$\int_0^1 |\gamma_\tau(t)|^{s'} d\tau$$

$$\leq \int_0^1 \cdots \int_0^1 f(t; p) d\zeta_1 \cdots d\zeta_p$$

$$= \int_0^1 \cdots \int_0^1 f(t; p-1) d\zeta_1 \cdots d\zeta_{p-1} \int_0^1 |Q(\eta_1 \cdots \eta_p t)|^{s'} d\zeta_p. \quad (3.126)$$

The integral with respect to ζ_p, in view of (3.121), is equal to

$$\int_0^1 |Q(\eta_1 \cdots \eta_{p-1}(\xi_p + (\eta - \xi_p)\zeta_p)t)|^s d\zeta_p = \int_0^1 |Q(T\zeta_p + a)|^s d\zeta_p$$

$$= \frac{1}{T} \int_a^{a+T} |Q(u)|^{s'} du$$

with $a = \eta_1 \cdots \eta_{p-1}\xi_p t$ and

$$T = \eta_1 \cdots \eta_{p-1}(\eta - \xi_p)t > \eta^p \frac{1}{(p+1)^2} t \geq T_0$$

provided p is chosen dependent of t in such way that

$$\log t - p \log \frac{1}{\eta} - 2\log(p+1) \geq \log T_0.$$

It is sufficient to take

$$p = p(t) = \left[\theta \frac{\log t}{\log 1/\eta} \right] + 1, \tag{3.127}$$

where the brackets stand for the integer part, $\theta < 1$ being fixed, but arbitrarily close to 1, if t is large enough.

Having chosen $p = p(t)$, the inequality

$$\eta^q \frac{1}{(q+1)^2} t \geq T_0$$

holds for every $q \leq p$. Hence,

$$\int_0^1 \cdots \int_0^1 f(t;p)d\zeta_1 \cdots d\zeta_p \leq \frac{1}{d^{(s-1)/2}} \int_0^1 \cdots \int_0^1 f(t;p-1)d\zeta_1 \cdots d\zeta_{p-1}$$

and successive integration of (3.126), yields

$$\int_0^1 |\gamma_\tau(t)|^{s'} d\tau \leq \frac{1}{d^{p(s'-3)/2}}$$

$$\leq \frac{1}{t^{\theta\alpha(s'-3)/2}} \tag{3.128}$$

by (3.127) and (3.119), for $t > t_0(\theta)$.

Fixing

$$\theta \frac{s'-3}{2} = \frac{s'}{2} - 2, \tag{3.129}$$

we have

$$\int_0^1 |\gamma_\tau(t)|^{s'} d\tau \leq \frac{1}{t^{\alpha(s'/2-2)}}. \tag{3.130}$$

Writing

$$\alpha\left(\frac{s'}{2} - 2\right) = 2 + \gamma \tag{3.131}$$

the integral with respect to t

$$\int_1^\infty t^\gamma \left(\int_0^1 |\gamma_\tau(t)|^{s'} d\tau \right) dt < \infty \tag{3.132}$$

converges and, consequently,

$$t^\gamma \, |\gamma_\tau(t)|^{s'} \to 0 \qquad (3.133)$$

holds for almost every $\tau \in [0,1]$, i.e., with probability 1 with respect to the measure π on the space of Salem's sets by the forthcoming lemma (Kahane's lemma (Lemma 3.13)). For such sets, the Fourier–Stieltjes transform of the corresponding distribution function F_τ satisfies

$$|\gamma_\tau(t)| < \frac{1}{t^{\gamma/s'}} = \frac{1}{t^{\alpha/2-\varepsilon}}$$

in view of (3.123) and (3.131):

$$\frac{\gamma}{s'} = \frac{\alpha}{2} - \frac{2\alpha+2}{s'} = \frac{\alpha}{2} - \varepsilon,$$

concluding the proof of the theorem. $\qquad\square$

In order to conclude (3.133) from (3.132), a special lemma is necessary. We write (3.130) replacing $t \in \mathbb{R}$ by $n \in \mathbb{Z}$, with $|n| \geq T_0$. By dividing the resulting inequality by its right-hand side, multiplying both sides by $|n|^{-2}$, and summing over all $n \in \mathbb{Z}$ with $|n| \geq T_0$, we get

$$\int_0^1 d\tau \sum_{n \in \mathbb{Z}:|n| \geq T_0} |\gamma_\tau(n)|^{s'} |n|^{\alpha(s'/2-2)-2} < \infty \qquad (3.134)$$

which immediately implies

$$\lim_{|n|\to\infty} |\gamma_\tau(n)|^{s'} |n|^{\alpha(s'/2-2)-2} = 0 \quad \text{a.e. } \tau \in [0,1]. \qquad (3.135)$$

Equation (3.135) implies that

$$|\gamma_\tau(n)| = O\left(\frac{\phi(n)}{\Psi(n)}\right),$$

where

$$\phi(n) = |n|^{-\alpha(s'/2-2)/s'},$$
$$\Psi(n) = |n|^{-2/s'}.$$

Both ϕ and Ψ are (for $|n| \geq 1$) positive decreasing functions of $|n|$. We now have the following lemma.

Lemma 3.13 (Kahane's lemma [141, Lemma 1, p. 252]). *Let μ be a measure carried by a compact set interior to $[0,1]$, and $\phi(t)$, $\Psi(t)$ be two* *positive decreasing functions of $t > 0$ such that $\phi(t/2) = O(\phi(t))$, $\Psi(2t) = O(\Psi(t))$, as $t \to \infty$. Write $\hat{\mu}(u) = \int \exp(2\pi i a x) d\mu(x)$. If $\hat{\mu}(n) = O(\frac{\phi(|n|)}{\Psi(|n|)})$ as $|n| \to \infty$, then $\hat{\mu}(u) = O(\frac{\phi(|u|)}{\Psi(|u|)})$ as $|u| \to \infty$.*

Proof. Let γ be a function of class C^∞ carried by a compact set interior to $[0, 1]$, and equal to one on the support of μ. For each a and x in $[0, 1]$, write

$$\gamma_a(x) = \exp(2\pi i a x)\gamma(x) = \sum_{n \in \mathbb{Z}} \hat{\gamma}_a(n) \exp(2\pi i n x).$$

Since each derivative of arbitrarily high order of γ_a is uniformly bounded with respect to a, we have, for each $q > 0$,

$$|\hat{\gamma}_a(n)| \leq C|n|^{-q},$$

where C depends only on γ and q. Now,

$$\hat{\mu}(m + a) = \int \exp(2\pi i a x) \exp(2\pi i m x) d\mu(x)$$

$$= \int \gamma_a(x) \exp(2\pi i m x) d\mu(x) = \sum_{n \in \mathbb{Z}} \hat{\gamma}_a(n)\hat{\mu}(n + m).$$

We divide the above sum into two parts, $\sum_{|n| \leq |m|/2}$ and $\sum_{|n| > |m|/2}$. Suppose that $|\hat{\mu}(n)| < \frac{\phi(|n|)}{\Psi(|n|)}$ and $|\hat{\mu}(n)| < \kappa$ for each n. The first part is majorized by $\frac{\phi(|m|/2)}{\Psi(2|m|)} \sum_{n \in \mathbb{Z}} |\hat{\gamma}_a(n)|$, the second by $\kappa \sum_{|n| > |m|/2} |\hat{\gamma}_a(n)|$, and hence by hypothesis (C_1, C_2 independent of a and m)

$$|m\hat{u}(m + a)| \leq C_1 \frac{\phi(|m|)}{\Psi(|m|)} + C_2|m|^{-q}$$

for q arbitrary, therefore

$$|\hat{\mu}(m + a)| \leq C_3 \frac{\phi(|m|)}{\Psi(|m|)}$$

and, finally,

$$\hat{\mu}(u) = O\left(\frac{\phi(|u|)}{\Psi(|u|)}\right)$$

concluding the proof. $\qquad\qquad\qquad\qquad\qquad\qquad\qquad\qquad\qquad\qquad\qquad$ $\square$

As we see above, Salem's method uses step-by-step an increasing number of contractions with randomized Lipschitz factors and fixed translation vectors (3.120b), (3.121). It seems difficult to find a direct generalization of his construction to $\mathbb{R}^n$ with $n > 1$. Bluhm [28], by means of an interesting novel technique, whereby the Lipschitz factors were fixed and the translation vectors were randomized instead, was able to construct statistically self-similar Salem sets in $\mathbb{R}^n$. He was able to prove as well that, in Theorem 3.18, condition (2) may be replaced by a random perturbation of the Cantor set C_ξ, which, moreover, makes it into a statistically self-similar Salem set of dimension $\frac{\log 2}{\log(1/\xi)}$ (see [28, Theorem 5]). In contrast to Salem's work, the effect of reducing the strength of the random perturbation in [28] leads to realizations which are arbitrarily "close" to the deterministic set. In a similar token, Kahane [141, p. 257] has considered images of measures (see, e.g., [183, p. 15]) by random n-dimensional Gaussian Fourier series F and shown that if E is a compact set of Hausdorff dimension $\alpha < n/2$, $F(E)$ is a.s. a Salem set of dimension 2α (see [141, Theorems 1, p. 251, and 3, p. 258]).

One may ask what would be the possible applications in mathematical physics of the ideas in this section of pointwise decay. We believe that Kahane's methods may be applied to models in mathematical physics, in order to show that there exist random Hamiltonians arbitrarily "close" to certain deterministic Hamiltonians (such as the sparse model of Sec. 4.3) with the same spectral characteristics of the latter (e.g. s.c. spectrum with the same Hausdorff dimension), but such that the F.S. transform of their spectral measure has a.s. the best possible decay, i.e., of the type of Theorem 3.22.

3.4. Quantum Dynamical Stability

Stability considerations play a central role both in classical mechanics (classical dynamical systems) [257] and in quantum mechanics. In the latter, one may distinguish various types of stability, according to one's focus. When dealing with a finite number of degrees of freedom (N-body systems), there are the concepts of stability of the first and second kinds, which are relevant to the stability of matter in quantum mechanics, including the interaction with the electromagnetic field (quantized or not) and gravitating systems (including white dwarfs and neutron stars) [170]. If a system with an infinite number of degrees of freedom is being considered, such as quantum statistical mechanics and quantum field theory, one will be dealing with the thermodynamic limit: see [170, Chap. 14] for a beautiful introduction with view of application to Coulomb systems. One important notion of stability in this context is that of local thermodynamic stability [235].

The above-mentioned stability concepts do not involve the dynamics. For systems with infinite number of degrees of freedom, several important notions of dynamical stability exist [235]. Only two of them will be briefly reviewed in Sec. 6.4 in connection to the quantum mixing property. They are both related to the approach to the equilibrium state: local dynamical stability and return to equilibrium.

The present section will be devoted to the concepts of dynamical stability for quantum systems with finite number of degrees of freedom. While they will be stated and formulated for one-particle systems, their conceptual extension to the N-body problem presents no problem.

Most of the concepts of dynamical stability in quantum mechanics are related to the growth of expectation values

$$\langle A(t) \rangle \tag{3.136}$$

of certain observables $A = A(0)$ as time t evolves. *Boundedness* of $\langle A(t) \rangle$ is related to quantum stability, and this problem has been studied extensively in connection to "quantum chaos", because classical chaos manifests itself through a diffusive growth of energy, e.g., in the kicked rotor, which is of the form (3.23a), with

$$H_0 = -\beta \frac{\partial^2}{\partial \theta^2} \tag{3.137a}$$

and

$$V(t) = \kappa \cos \theta \sum_{n \in \mathbb{Z}} \delta(t - nT) \tag{3.137b}$$

on $\mathcal{H} = L^2(0, 2\pi)$, formally speaking; rigorously speaking, the time-evolution operator is a product of unitaries, i.e., the monodromy matrix $U(T, 0)$ (see (3.24d) et ff.) is given by

$$U(T, 0) = \exp(iTH_0) \exp(iV), \tag{3.137c}$$

where V is the multiplication operator $(Vf)(\theta) = \kappa \cos \theta f(\theta)$, $f \in \mathcal{H}$. This model is a prototype for "quantum chaos", for reasons well expounded in [39]: experiments by Bayfield and Koch in 1974 on the hydrogen atom in an external microwave field were, indeed, among the most interesting to motivate the field of "quantum chaos". We now briefly describe their main features.

Single atoms prepared in very elongated states with high principal quantum number ($n_0 \approx 63-69$) were injected into a microwave cavity and the ionization rate was measured. The microwave frequency was 9.9 GHz, corresponding to a photon energy well below the ionization energy of level 66 and even lower than the transition energy from state 66 to 67. Very surprisingly, the ionization rate suffered a jump when the electrical field intensity exceeded a threshold value of about 20 V/cm (for $n_0 = 66$), much lower than the static Stark value, but astoundingly close to the transition to chaotic behavior in a corresponding *classical* model of the hydrogen atom in an external microwave field (this classical transition may be determined by the Chirikov criterion of overlapping resonances, see [39, App. C]). This is true as long the microwave frequency does not exceed a critical value, see [39, Fig. 5]. The main theoretical tool to investigate this problem is based on Kepler's map (see [39, 5.2]), which establishes a nice connection with the kicked rotor model (3.137b) and (3.137c).

What is particularly remarkable about this problem is that the perturbation theory is useless: due to the high value of n_0, a very large number of terms in the perturbation expansion (of order 100) would be required. We have, therefore, a highly nonperturbative, strongly coupled quantum system which is, in a certain region of parameters, extremely well described by classical mechanics!

The multiphoton ionization occurring in this system has several universal features, which will reappear in Chap. 5, which deals (in part) with ionization of electrons bound to an atom by an extremely short-range potential (e.g., H^-).

With the notation $I^2 = H_0$, and denoting by $I_N^2 = U(NT, 0)I^2U$ $(NT, 0)^\dagger$ the evolution of I^2 after N "kicks", the diffusion coefficient for the kicked rotor may be defined in analogy to the classical case by:

$$D \equiv \lim_{N \to \infty} \frac{\langle I_N^2 \rangle}{NT} \qquad (3.138a)$$

when the limit above exists. The classical model is the well-known standard map [39], for which, for $\beta = 1/2$, there are good numerical indications that $D > 0$ inside the region of "hard chaos" of the model, viz. $\kappa > \kappa_{\mathrm{cr}} \equiv 0.9716\ldots$, (see [39, p. 305]). For the quantum model, $\alpha = 2$ in (3.31), but V is not smooth, and thus a version of Howland's Theorem 3.6 is still open. Indeed, no rigorous results are known for this model, either on the spectrum or decay of correlations.

In the present example, $A = I^2$. In general, for a lattice model (1.1), (1.2b), (1.3b) (which includes the Anderson model if the potential is random, or the case of an almost-periodic potential), it is common as seen in Theorem 3.9, to adopt $A = |X|^2$ ($m = 2$ in (3.47b)), and one distinguishes the *localization regime* in which $\langle |X|^2 \rangle < C$ for all t or the *ballistic regime* where

$$\langle |X|^2 \rangle = O(t^2), \quad \text{as } t \to \infty \tag{3.138b}$$

which are supposed to correspond, respectively, to p.p. and a.c. spectrum; see Chap. 2 for (3.138b). Intermediate regimes are characterized by the diffusion constant

$$D_\Psi = \lim_{t \to \infty} \frac{\langle |X|^2 \rangle}{t} \tag{3.138c}$$

when the above limit exists — (3.138a) is the discrete time analogue, for a different observable, and the dependence on the initial state Ψ (see Theorem 3.9) is explicitly indicated. As we have seen in Theorem 3.11 and the considerations preceding it, bounds on D_Ψ are related to decay (at least on the average) of the return probability for Ψ.

If $\langle |X|^2 \rangle < C$ with C a positive constant for all t, then Ψ has no continuous component. This is [162, Corollary 2.3.1.] which may be proved by the same methods as Theorem 3.11, and implies that $D_\Psi = 0$. It is remarkable that the converse, i.e., that if $\Psi \in \mathcal{H}_{\text{pp}}$, then $D_\Psi = 0$, is *not* true. Indeed, the only general result about the converse is Simon's paper [242] on the absence of ballistic motion (3.138b), i.e., $\lim_{t \to \infty} \langle |X|^2(t) \rangle = 0$ if $\Psi \in \mathcal{H}_{\text{pp}}$, which is far from the expected $D_\Psi = 0$. The reason for this is the *instability* of the two exotic spectra, thick p.p. and s.c.; indeed, even a rank one perturbation with arbitrarily small norm is able to induce a transition from one type to the other! [119, 245]. Since one does not expect the dynamics to be strongly affected by such perturbations, it is plausible that the absence of ballistic motion — the latter characterizing a.c. spectra — is both a feature of s.c. spectra *and* the (thick) p.p. spectra obtained from s.c. spectra by such perturbations: therefore Simon's result might be optimal! This has indeed been shown in the remarkable paper [67], where a potential was constructed such that the Hamiltonian H on $l^2(\mathbb{Z})$ has a complete set of exponentially decaying eigenfunctions but, for any $\delta > 0$, $\|X \exp(-itH)\delta_0\|^2 / t^{2-\delta}$ is unbounded as $t \to \pm\infty$. Note that $\|x \exp(-itH)\delta_0\|^2 = (\exp(-itH)\delta_0, X^2 \exp(-itH)\delta_0)$. In the words of Jitomirskaya [132], this example showed that mere "exponential localization" of eigenfunctions need not have any consequence for the dynamics. Thus, [67] was pioneer in demonstrating the importance of *dynamical localization*, which, since then, was proved for various random models in the form

$$\sup_{t} \| \, |X|^{q} E_{I}(H_{\omega}) \exp(-it H_{\omega} \Psi) \| < \infty \qquad (3.139)$$

with probability one. $I \subset \mathbb{R}$ is an energy interval in the localization region and E_{I} are spectral projections for I (see (1.50a)). For H_{ω} the Anderson Hamiltonian (3.14), or the deterministic almost-Mathieu model (3.15), several results are known, and we refer to [132, p. 621] for a discussion and (numerous) references.

For the models of atoms and oscillators in external (quasi)-periodic fields (3.21), (3.22), stability of the p.p. spectrum has been proved either by perturbative (Theorem 3.4) or nonperturbative methods (Theorem 3.6). As in classical KAM theory (see [40, 257]), there are conditions on the coupling constants which exclude resonances, as discussed in connection with Theorems 3.4 and 3.5 — otherwise the perturbed system has a.c. spectrum (see [119, 272]). As remarked by Howland [119], *resonance* is a phenomenon of cooperation between two or more elements, while *randomness* is based on independence — noncooperation — between elements. It should not prove surprising, therefore, that resonance in a system can be removed by randomizing parameters in the system, as in the theory of Schrödinger operators with random potentials — of which the Anderson model (3.14) is a major example — where the randomness may prevent the cooperation necessary for traveling waves (tunneling, see remarks after (1.3b), resulting in the phenomenon of localization). The papers of Jona-Lasinio, Martinelli and Scoppola [160] and Simon [241] develop this concept further: tunneling is very sensitive to minimal changes in a double-well potential, even those localized very far away from the minima (a phenomenon called "flea on the elephant" by Simon). Thus, from this point of view, the existence of p.p. spectra in random media may be seen as a result of instability of tunneling! Of course, this instability presupposes a comparison with a reference state, in the present case the free lattice Hamiltonian (1.2d).

Later, several types of *dynamical* instability (or metastability) of tunneling, dealing with two-level atoms in external periodic or quasi-periodic fields (3.21), were found: they may be considered as explicit models of *decoherence*, [99, 276], and will be treated in Chap. 5 (see (5.56a) et seq.).

Concerning now the diffusion constant D_Ψ given by (3.138c), one may define the so-called diffusion exponents (see, e.g., [22]):

$$\beta_m^\pm(\Psi) = \lim_{T\to\pm\infty} \frac{\log\langle|X|^m\rangle_T}{\log(T^m)}, \qquad (3.140a)$$

where $\lim_\pm \equiv \lim\sup(\lim\inf)$, and the notation is otherwise the same as in (3.50f). Both $\beta_m^\pm(\Psi)$ are nonincreasing functions of m and obey

$$0 \le \beta_m^-(\Psi) \le \beta_m^+(\Psi) \le 1. \qquad (3.140b)$$

By Theorem 3.11,

$$\beta_m^-(\Psi) \ge \frac{\alpha}{n} = \frac{\dim_H(\mu_\Psi)}{n}. \qquad (3.140c)$$

For μ_Ψ a.c., $\alpha = 1$; hence, by (3.140b), in dimension $n = 1$, a.c. spectrum implies ballistic transport (cf. (3.138b)), while in dimension $n = 3$, a.c. spectrum and *subdiffusive motion* ($\beta_m < 1/2$) may coexist. An example of this was constructed in [22], which we now briefly review.

A quantum motion is called *anomalous* if it is neither ballistic, nor regular diffusive, nor localized, that is, the diffusion exponents may take arbitrary values in the interval $(0, 1)$, different from $\beta_2 = 1/2$, i.e., it is either *subdiffusive*, if

$$0 < \beta_2 < 1/2 \qquad (3.141a)$$

or *superdiffusive*:

$$1/2 < \beta_2 < 1. \qquad (3.141b)$$

A *quasicrystal* is a crystal with "forbidden symmetry", i.e., exhibiting the five-fold symmetry thought to be excluded in solid state physics by a "crystallographic restriction theorem" (for a nice discursive exposition, see the article by Marjorie Senechal in Vol. 53, p. 886 of the Notices of the AMS: "What is a quasicrystal?"). They were discovered in 1984, when Schechtman synthesized aluminum–manganese crystals with icosahedral symmetry [230]: he received the Nobel prize for chemistry in 2011 for this seminal work. It turns out that the observed structures — "almost lattices" — had been predicted in 1969 by Meyer (see his interview in the IAMP News Bulletin October 2011, p. 14), and later rediscovered independently by Penrose (in 1976), therefore, now called, Meyer sets or Penrose pavings. In Definition 3.9, we defined P.V. numbers by requiring all the conjugates to lie in the *open* unit circle. If this requirement is replaced by the closed unit circle, the numbers with the resulting property are called *Salem numbers*, in recognition of the seminal work of Salem. Following the short but quite illuminating summary in Meyer's interview (for details consult his book [186]) an almost lattice $\Lambda \subset \mathbb{R}^n$ is defined by three conditions: (1) there exists a positive number r such that every ball with radius r, whatever be its center, contains at most one point in Λ; (2) there exists a positive number $R > r$ such that every ball with radius R, whatever be its center, contains at least one point in Λ; (3) there exists a finite set F such that $\Lambda - \Lambda \subset \Lambda + F$. Above, $\Lambda - \Lambda$ is the set consisting of all the differences $\lambda - \lambda'$, with λ and λ' both in Λ. If Λ is an ordinary lattice and if $\theta \in \mathbb{R}$, then the dilated lattice $\theta\Lambda$ is contained in Λ iff θ is an integer. What is quite remarkable is that for almost lattices, one has something similar: if Λ is an almost lattice and if $\theta\Lambda$ is contained in Λ, then θ is either a P.V. number or a Salem number. Conversely, if θ is a P.V. number or a Salem number, there exists an almost lattice Λ such that $\theta\Lambda \subset \Lambda$.

Some models which have been employed to describe the properties of quasicrystals are lattice models with Hamiltonians of type $H = H_0 + V$, where H_0 is the lattice Laplacian (in one dimension) and V is a potential which takes only the values V and $-V$ arranged in a Fibonacci sequence (3.112) (recall the relation with the P.V. numbers in Problem 3.8). For this model, a slow decay on the average of the return probability with a generalized dimension (smaller or equal to the Hausdorff dimension) was predicted in [148], and rigorously proved in [116] using wavelets. Thus, quasicrystals provide a physical illustration of the phenomenon of slow decay of Sec. 3.1.

The above-mentioned model is, however, one-dimensional, and three-dimensional models are sought which describe the interesting dynamical properties of quasicrystals. Perhaps, the most striking of these is anomalous motion (see [22] and references given there). The anomalous Drude formula for the direct conductivity $\sigma_\parallel$ (see [233]):

$$\sigma_\parallel(\tau) \approx \tau^{2\beta_2 - 1} \quad \text{as } \tau \to \infty, \tag{3.142}$$

where τ is the relaxation time due to impurity and electron–phonon scattering, is consistent with experiments in quasicrystals only if (3.141a) is assumed, i.e., subdiffusive motion. Bellissard and Schulz-Baldes [22] constructed models in dimension $n \geq 3$ defined in terms of Jacobi matrices (3.84a) — but with nondiagonal entries $t(n+1)$ with $t(n) > 0$ for $n \geq 1$, and diagonal entries $V(n)$, in a way similar to the model we study in Sec. 4.2, but with $t(\cdot)$, $V(\cdot)$ chosen in a special way, so that the corresponding one-dimensional Jacobi matrices have *self-similar* spectral measures (see App. C for the definition). The latter are the first examples of models with a.c. spectrum and subdiffusive dynamics, as expected for real (three-dimensional) quasicrystals.

Very interesting lower bounds on $\beta_m^-(\Psi)$ (respectively, $\beta_m^+(\Psi)$) have been obtained by Damanik and Tcheremchantsev [56], see also Theorems 7.4 and 7.5 (for an upper bound) in Damanik's comprehensive review [55], to which we refer for discussion and references.

We close this section (and the chapter) with some remarks relating dynamical stability to the finer splitting, due to Avron and Simon [16], of the a.c. spectrum into transient and recurrent a.c. spectrum (see Definition C.22 and Proposition C.27 in App. C). By (C.20),

$$\mathcal{H}_{\text{tac}} = \overline{\{\phi : \hat{\mu}_\phi \in L^1(\mathbb{R})\}}$$

which should be compared with [209, Lemma 1, p. 23, and Problem 17, p. 386]:

$$\mathcal{H}_{\text{ac}} = \overline{\{\phi : \hat{\mu}_\phi \in L^2(\mathbb{R})\}}.$$

As remarked by Last [164, p. 701], since $|\hat{\mu}_\phi(t)|^2$ is the survival probability of ϕ, $\mathcal{H}_{\text{tac}}$ is a closed subspace of $\mathcal{H}_{\text{ac}}$ which is made of vectors which have the *fastest escape rate* from their initial position under the Schrödinger time evolution. Besides the free cases of Chap. 2, an interesting example of this may be given in the case of model (3.22), with

$$f(t) = \lambda_1 \cos \omega_1 t + \lambda_2 \cos \omega_2 t. \qquad (3.143)$$

We define the *autocorrelation function* by

$$C_\Psi(t) = \lim_{T \to \infty} \frac{1}{2T} \int_{-T}^{T} ds \, (\Psi(s), \Psi(s+t)) \qquad (3.144)$$

when the limit exists; above, $\Psi(t) = U(t, 0)\Psi$ where $U(t, s)$ is the (unique) solution of (3.25a), (3.25b). When C_Ψ exists, it is of positive type, i.e.,

$$\sum_{i,j=1}^{n} C_\Psi(t_i - t_j) a_i \bar{a}_j \geq 0 \quad \text{for all } a_i \in \mathbb{C} \text{ with } i = 1, \ldots, n$$

and any finite integer n. Hence, by Bochner's theorem (see [193, Theorem 10.4]), there exists a Stieltjes measure μ_Ψ such that

$$C_\Psi(t) = \int \exp(-it\lambda) d\mu_\Psi(\lambda). \qquad (3.145)$$

That (3.145) is of positive type is clear. Let $\{\Psi_i\}_{i=1}^\infty$ be a countable dense set in $\mathcal{H}$ (separability of $\mathcal{H}$ is always assumed). We define the autocorrelation spectrum as the closure of the union of the supports of the measures μ_{Ψ_i}. It is a.c. (t.a.c., r.a.c.) if each μ_{Ψ_i} is a.c. (t.a.c., r.a.c.); it is p.p. if each of the measures μ_{Ψ_i} is p.p.. We have (see [272, Theorem 1; 18]):

Proposition 3.3. *In the resonant case, in* (3.22) *and* (3.143), $\omega_0 = \omega_1$, ω_2 *incommensurate with* ω_1, *the autocorrelation function* $C_\Psi(t)$ *given by* (3.144) *exists and satisfies the inequality*

$$|C_\Psi(t)| \le a \exp\left(-\frac{\lambda_1^2 t^2}{4} + bt\right) \qquad (3.146)$$

with $a > 0$, b *independent of* t, *for* Ψ *in a dense set of coherent states (for coherent states see [182, pp. 14–16]). In the nonresonant case, under diophantine conditions analogous to* (3.30q), *i.e., for* $c > 0$,

$$|\omega \cdot m| \ge c|m|^\sigma \quad \text{with } |m| \ne 0 \quad \text{for some } \sigma \in \mathbb{R}, \qquad (3.147)$$

where $\omega \cdot m \equiv \omega_0 m_0 + \omega_1 m_1 + \omega_2 m_2$ *and* $|m| = |m_0| + |m_1| + |m_2|$, *with* m_i, $i = 0, 1, 2$, *arbitrary in* $\mathbb{Z}$, C_Ψ *is a special almost-periodic function (Definition 3.6), which is not identically zero for* Ψ *in a dense set of coherent states.*

For the proof, we refer to [18], and for a discussion of the main points, we refer to [272].

As a consequence of definition (C.25) of App. C, in the resonant case the autocorrelation spectrum is t.a.c. (by (3.146)) (covering the whole line (see Chap. 5)); in the nonresonant case under the assumption (3.147), by [146, p. 155 et seq.], it is p.p.. The above illustrates well that in the resonant case, there is instability expressed as the fastest escape rate from the initial state Ψ: in special models of atoms in external fields this leads to ionization, see Chap. 5. In the nonresonant case, here achieved by a diophantine condition instead of randomization, for a large but porous set of parameters (in analogy to a set of initial conditions in classical mechanics) the spectrum remains p.p. and there is stability. Note that if $\sigma > 3$ in (3.147), the

Lebesgue measure of the complement of the set of $\omega = (\omega_0, \omega_1, \omega_2)$ which satisfies (3.147) is zero, see App. A.

Examples of recurrent a.c. spectrum have already been given by Avron and Simon [16] in connection with almost-periodic Schrödinger operators, see also Last's review [164] for a discussion and references.

Chapter 4

Time Decay for a Class of Models with Sparse Potentials

The study of pointwise decay for s.c. measures (Sec. 3.3) provided a bridge between ergodic theory, number theory and analysis. The present chapter exploits further ergodic theory's link with spectral analysis and is divided into three sections.

In Sec. 4.1, we prove a theorem (Theorem 4.1) connecting the uniform distributions of Prüfer angles and spectral transitions by which, together with the Strichartz–Last theorem (Theorem 3.8) on time decay in average, we prove in Sec. 4.2 the existence of Anderson-like transition to a class of sparse models in dimensions higher than two (Theorem 4.6).

In Sec. 4.3, we study pointwise decay of the Fourier–Stjeltjes transform of Pearson's (fractal) measures which includes, in particular, the spectral measure μ of sparse Schrödinger operators.

We consider in this chapter a lattice version of sparse Schrödinger operators (3.12a)–(3.12e), similar to the Hamiltonian H_Φ on $\mathcal{H} = l^2(\mathbb{Z}_+)$ defined by (3.84a)–(3.84c), given by a Jacobi operator J^ϕ (see (4.1a) et seq.). In Sec. 3.2, we used Corollary 3.1 to give a precise physical interpretation of the subspace $\mathcal{H}_{\text{sc}}$ in sparse models under a weaker (3.84e) than Pearson's condition (3.12e). Here, we investigate the least possible sparse condition that leads the Fourier–Stieltjes transform $\hat{\mu}_\Psi$ of the spectral measure μ_Ψ of J^ϕ, for Ψ in a dense set of $\mathcal{H}$, to decay like $\hat{\mu}_\Psi \sim t^{-1/2}$. This is just at the borderline for the class of Jacobi operators in consideration, i.e., if $\hat{\mu}_\Psi \sim t^{-1/2-\varepsilon}$ for some $\varepsilon > 0$, then μ_Ψ would be absolutely continuous w.r.t. Lebesgue measure $\mathcal{L}$ by Theorem 3.14 but, as seen in App. B, they have purely singular continuous spectrum.

4.1. Spectral Transition for Sparse Models in $d = 1$

We shall prove both the existence of p.p. spectrum and the sharpness of the transition with (s.c.) spectrum by a novel method. The existence proof involves the ergodicity of the Prüfer angles, corresponding to a solution u of the Schrödinger equation $(J^\phi - \lambda\mathbb{I})u = 0$. As remarked by Remling [212] in his review of our nonlinearity paper [180], which introduced our method, our new idea was to *fix* the energy λ and assume the Prüfer angles $(\theta_j)_{j\geq 1}$ at a_j are uniformly distributed (u.d.) as a function of j instead of the traditional approach which exploits the u.d. of the Prüfer angles in the energy variable λ, at fixed a_j. In [180], we were only able to fix the energy in the s.c. spectrum. In [65], we were able to bring our ideas to full fruition looking at an ensemble of Jacobi matrices $J^{\phi,\omega}$, $\omega \in \Omega$, of the form

$$J = \begin{pmatrix} 0 & p_1 & 0 & 0 & \cdots \\ p_1 & 0 & p_2 & 0 & \cdots \\ 0 & p_2 & 0 & p_3 & \cdots \\ 0 & 0 & p_3 & 0 & \cdots \\ \vdots & \vdots & \vdots & \vdots & \ddots \end{pmatrix}, \tag{4.1a}$$

in which $(p_n)_{n\geq 1}$ is given by

$$p_n = \begin{cases} p & \text{if } n = a_j^\omega \in \mathcal{A}^\omega, \\ 1 & \text{otherwise,} \end{cases} \tag{4.1b}$$

$p \in (0,1)$, for a random set $\mathcal{A}^\omega = \{a_j^\omega\}_{j\geq 1}$,

$$a_j^\omega = a_j + \omega_j \tag{4.1c}$$

with a_j satisfying the "sparseness condition":

$$\beta_j \equiv a_j - a_{j-1} = \beta^j, \tag{4.1d}$$

$j \in \mathbb{N}$, with $a_1 + 1 = \beta \geq 2$ and $\{\omega_j\}_{j\geq 1}$, independent random variables defined on a probability space $(\Omega, \mathcal{B}, \nu)$, uniformly distributed on $\Lambda_j = \{-j, \ldots, j\}$.

The Schrödinger equation associated with $J^{\phi,\omega}$:

$$\left((J^{\phi,\omega} - \lambda\mathbb{I})u\right)_n = p_n u_{n+1} + p_{n-1} u_{n-1} - \lambda u_n = 0 \tag{4.2}$$

for $n \geq 1$ with $p_0 \equiv 1$ and $\lambda \in \mathbb{R}$, acts on the $l_2(\mathbb{Z}_+)$ space of square-summable sequences $u = (u_n)_{n\geq 0}$ satisfying ϕ-phase boundary condition at 0:

$$u_0 \cos\phi - u_1 \sin\phi = 0, \tag{4.3}$$

for some $\phi \in [0, \pi)$. Our method applies as well for ensemble of Jacobi matrices $J^{\phi,\omega} = J_0^\phi + V^\omega$, where

$$(J_0^\phi u)_n = u_{n+1} + u_{n-1} \tag{4.4a}$$

is the (free) discrete Laplacian, satisfying ϕ-phase b.c. (4.3) at 0, and

$$(V^\omega u)_n = v_n^\omega u_n \tag{4.4b}$$

the perturbation potential

$$v_n^\omega = \begin{cases} v & \text{if } n = a_j^\omega \in \mathcal{A}^\omega, \\ 0 & \text{otherwise,} \end{cases} \tag{4.4c}$$

for $v \in (0, \infty)$ and $\mathcal{A} = \{a_j^\omega\}_{j \geq 1}$ as before. Such models are nowadays called Poisson models (see [132] and references therein).

We refer the reader to App. B, in particular to Definition (B.24a) of the Prüfer angles $(\theta_j^\omega)_{j \geq 1}$ as well as Definitions B.3, B.4 and Propositions B.6, B.7, where we deduce some useful expressions. For convenience, the Weyl–Titchmarsh m-function, the Gilbert–Pearson theory and Last–Simon's (transfer matrix) criteria are reviewed in App. B (see (B.1a)–(B.14), Theorem B.1, Proposition B.10 and its Corollary B.3 et seq.).

4.1.1. *Existence of "mobility edges"*

The essential spectrum of $J^{\omega,\phi}$ equals $[-2, 2]$: it will be represented as $\lambda = 2\cos\varphi$ with $\varphi \in [0, \pi)$. Zlatoš [279] proved that the model (4.4a)–(4.4c) exhibits a sharp transition from s.c. to p.p. spectrum. This was shown independently in [65].

Theorem 4.1. *Let $J^{\omega,\phi}$ be as above. Let*

$$I \equiv \left\{ \lambda \in [-2, 2] \backslash 2\cos\pi\mathbb{Q} : \frac{1}{v^2}(\beta - 1)(4 - \lambda^2) > 1 \right\} \tag{4.5}$$

with $v \in (0, \infty)$ ($v = (1 - p^2)/p$ for model (4.1a)–(4.1d)) and $\beta \in \mathbb{N}$, $\beta \geq 2$.

Then, for almost all ω w.r.t. the uniform product measure on $\Lambda = \times_{j=1}^\infty \{-j, \ldots, j\}$:

(a) *there exists a set A_1 of Lebesgue measure zero such that the spectrum restricted to the set $I \backslash A_1$ is purely singular continuous, with (exact)* *local Hausdorff dimension*

$$\alpha(\lambda) = 1 - \frac{\log r(\lambda)}{\log \beta}, \tag{4.6}$$

where $r(\lambda) = 1 + v^2/(4 - \lambda^2)$ (i.e., $\forall \varepsilon > 0$, $\exists \delta > 0$ so that $h^{\alpha(\lambda)+\varepsilon}((\lambda - \delta, \lambda + \delta) \cap \sigma(J^\phi)) = 0$ and $h^{\alpha(\lambda)-\varepsilon}((\lambda - \delta, \lambda + \delta) \cap \sigma(J^\phi)) = \infty$);

(b) *the spectrum of $J^{\omega,\phi}$ is dense pure point when restricted to $I^c = ([-2,2]\backslash 2\cos \pi\mathbb{Q})\backslash I$ for almost every $\phi \in [0, \pi)$, where ϕ characterizes the boundary condition. Thus it is* purely *p.p.*

The proof of (a) of the theorem may be found in [279] but we shall prove it here for the reader's convenience. The proof of (b) is a new result in what concerns the purity of the p.p. spectrum. Because of its relation to the ergodic theory methods which appear in Sec. 4.1.2, we postpone it to Sec. 4.1.3.

For superexponential sparseness, e.g. $a_j - a_{j-1} = [e^{cn^\gamma}]$ ($[z]$ the integer part of z), with $c > 0$ and $\gamma > 1$, it may be proved that $\sigma(J^{\phi,\omega})$ is purely singular continuous (s.c.) for almost every $\omega \in \times_{j=1}^\infty \Lambda_j$ (see [65, Theorem 5.2]; see also Sec. 3.2.3 for a precise physical interpretation of the states in $\mathcal{H}_{sc}$). On the other hand, for subexponential sparseness, e.g., $a_j - a_{j-1} = [e^{cn^\gamma}]$ with $\gamma < 1$, $\sigma_{ess}(J^{\phi,\omega}) = \sigma_{pp}(J^{\phi,\omega}) = [-2,2]$ for a.e. boundary phase $\phi \in [0, \pi]$ and for a.e. $\omega \in \times_{j=1}^\infty \Lambda_j$ (see [65, Theorem 5.1]). These results join smoothly to the one (corresponding to $\gamma = 0$) for the standard Anderson model (see definition (4.22)) in $d = 1$, according to which all states are localized [95, 157]. The latter is believed to be physically related to the subtle instability of tunneling [160, 241] which is strongest in $d = 1$.

The seminal proof of Klein [153] of the transition (from a.c. to p.p. spectrum) in the Bethe lattice uses the loopless character of the graph and is thus, as remarked by Jitomirskaya [132], the Bethe lattice, while infinite-dimensional, is in a sense quasi-one-dimensional. We refer to [3, 4] for recent important work on the Anderson transition on graphs. Figure 4.1 depicts the one-dimensional spectral transition described in Theorem 4.1, whose "mobility edges" $\lambda^\pm = 2\cos\varphi^\pm$, given by the implicit solutions of

$$1 - \frac{\lambda^2}{4} = \sin^2\varphi = \frac{v^2}{v_c^2},$$

exist provided $v < v_c = 2\sqrt{\beta - 1}$.

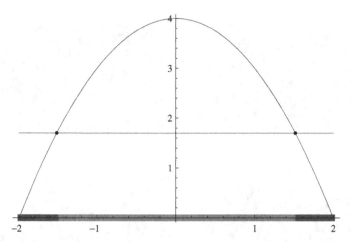

Fig. 4.1. Singular continuous (light gray) and pure point (dark gray) spectra separated by the "mobility edges" $\lambda^{\pm} = \pm 2\sqrt{1 - v^2/v_c^2}$; $v/v_c = 1.3038\ldots$.

We remark that the transition depicted in Theorem 4.1 is *robust*: by [67], the transition is stable because the Hausdorff dimension of the s.c. spectrum is nonzero. The surprising part is, however, the existence of a pure point spectrum in a regime of strong sparsity! This should be viewed as a direct manifestation of tunneling instability in this special situation.

Unfortunately the s.c. spectrum does not possess either the dynamic [67, 248], nor the perturbation-theoretic (stability) properties [118, 245] which are commonly associated with the physical picture of delocalized states.

4.1.2. *Uniform distribution of Prüfer angles*

We begin presenting compelling numerical evidence that the Prüfer angles are u.d. mod π for almost every $\varphi \in [0, \pi)$. We recall that $(\theta_k)_{k \geq 1}$ (disregarding the random variables ω_j; see (B.24a)) is a map

$$\varphi_j = \beta \varphi_{j-1},$$
$$\theta_j = g(\varphi, \theta_{j-1}) - \beta \varphi_{j-1}, \quad j \in \mathbb{N}$$
$$g(\varphi, \theta) = \tan^{-1} \left(\frac{1}{p^2} (\tan \theta + \cot \varphi) - \cot \varphi \right), \tag{4.7}$$

with $\theta_0 \in [0, \pi]$, $\varphi_0 = \varphi \in [0, \pi)$ and β_j given by (4.1d). We compute the density function for $\beta = 2$, φ typically a dyadic irrational and p^2 (respectively, v) varying over the values 0.1 (2.84), 0.3 (1.27), 0.5 (0.7)

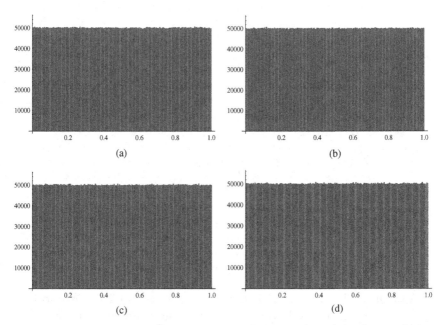

Fig. 4.2.   Histograms with 10^7 iteration points of the map (B.24a) for $\beta = 2$, φ/π is a dyadic irrational: (a) $p^2 = 0.1$, (b) $p^2 = 0.3$, (c) $p^2 = 0.5$ and (d) $p^2 = 0.9$, respectively.

and 0.9 (0.1). The corresponding histograms depicted in Fig. 4.2, each containing 10^7 iteration points of this map, indicate that the sequence $(\theta_k)_{k \geq 1}$ has u.d. mod π, i.e.,

$$\frac{1}{n} N([0, \theta/\pi), n) \quad \text{for } n \text{ large,}$$

where $N(A, n)$ counts the number of terms in $(\theta_k/\pi)_{k=1}^{n}$ for which θ_k/π mod 1 belongs to A (see Definition 3.10), approaches uniformly the identity function $x = \theta/\pi$.

It seems to us extremely hard to prove u.d. of the Prüfer angles without the incertainties ω_j's in the "bump" positions (4.1c). This task has been avoid in [180] by introducing another sequence, $(\xi_k)_{k \geq 1}$, which is uniformly distributed and remains close to $(\theta_k)_{k \geq 1}$, in the sense that will be made precise shortly. We consider first random Prüfer angles $(\theta_j^\omega)_{j \geq 1}$ and prove that they are uniformly distributed for any fixed energy value λ, by exploiting a modification of the metric extension of Weyl's criterion (Theorem 3.19).

Let $(S_N)_{N \geq 1}$ be the sequence of Weyl sums

$$S_N = \frac{1}{N} \sum_{k=1}^{N} e^{2ih\theta_n}. \tag{4.8}$$

According to Weyl criterion, the sequence $(\theta_n)_{n \geq 1}$ is u.d. mod π if and only if $\lim_{N \to \infty} S_N = 0$ for every integer $h \neq 0$. As $\theta_n = \theta_n(\varphi)$ is defined for each φ lying in some interval $[a, b] \subset [0, \pi)$, then $S_N = S_N(\varphi)$ is also defined for each $\varphi \in [a, b]$. Since we deal with sequence of random Prüfer angles $(\theta_k^\omega(\varphi))_{k \geq 1}$, we make use of a slight modification of the (optimal) metric extension of Weyl's criterion for uniform distribution by Davenport, Erdös and LeVeque [57] (see [156, Theorem 4.2] for a proof).

Theorem 4.2. *Let* $(\theta_n(x))_{n \geq 1}$ *be a sequence of real-valued random variables defined on a probability space* $(\Omega, \mathcal{B}, \mu)$. *For integers* $h \neq 0$, $N \geq 1$ *and* $A \subset \Omega$, *we set*

$$S_h(N, x) = \frac{1}{N} \sum_{n=1}^{N} e^{2ih\theta_n(x)} \tag{4.9a}$$

and

$$I_h(N, A) = \int_A |S_h(N, x)|^2 d\mu(x). \tag{4.9b}$$

If the series $\sum_{N=1}^{\infty} I_h(N, A)/N$ *converges for each* $h \neq 0$, *then the sequence* $(\theta_n(x))$ *is u.d. mod* π *for almost all* $x \in A$ *with respect to* μ.

To apply Theorem 4.2 to the sequence $(\theta_k^\omega)_{k \geq 1}$ of Prüfer angles, the statistically independent random variables $\{\omega_j, j \geq 1\}$ need to be defined on probability space $(\Omega, \mathcal{B}, \mu)$:

$$\Omega \ni x \longmapsto \omega_j(x) \in \Lambda_j := \{-j, \ldots, j\} \quad \text{for } j = 1, 2, \ldots. \tag{4.10}$$

Let $\Lambda = \times_{j=1}^{\infty} \Lambda_j$ be the configuration space of the ω_j's. The probability measure ν on the measurable space $(\Lambda, \mathcal{F})$, where $\mathcal{F}$ is the natural product σ-algebra, is given by the product measure

$$\nu(B) = \prod_{j=1}^{\infty} \nu_j(B_j)$$

on the closure of all cylinder sets $B = \times_{j=1}^{\infty} B_j \subset \Lambda$ with ν_j the uniform measure in Λ_j: $\nu_j(k) = 1/(2j+1)$, $\forall k \in \Lambda_j$, and it is, by (4.10), induced by the measure μ:

$$\nu(B) = \mu(\omega^{-1}(B))$$

where $\omega^{-1}(B) = \bigcap_{j=1}^{\infty} \omega_j^{-1}(B_j)$. Denoting by $\theta_k(x)$ the Prüfer angle θ_k^ω as a random variable on $(\Omega, \mathcal{B}, \mu)$, we have the following theorem.

Theorem 4.3. *The sequence of Prüfer angles $(\theta_k(x))_{k \geq 1}$ is u.d. mod π for all $\varphi/\pi \in [0,1] \backslash \mathbb{Q}$ and all $x \in \Omega$ apart from a set with μ measure 0.*

Proof. By Theorem 4.2, we must show that the series $\sum_{N=1}^{\infty} I_h(N, \Omega)/N$ converges, $I_h(N, \Omega)$ defined by (4.9b). It is, nevertheless, sufficient to show that the series converges absolutely. By (4.9a),

$$I_h(N, \Omega) = \int_\Omega |S_h(N, x)|^2 d\mu(x)$$

$$\leq \frac{1}{N} + \frac{2}{N^2} \sum_{1 \leq m < n \leq N} \left| \int_\Omega e^{2ih(\theta_m(x) - \theta_n(x))} d\mu(x) \right|. \quad (4.11a)$$

Define $\tilde{\theta}_n^\omega$ by the equation

$$\theta_n^\omega = g(\theta_{n-1}^\omega, \varphi) - (\beta_n + \omega_n - \omega_{n-1})\varphi =: \tilde{\theta}_n^\omega - \omega_n\varphi,$$

satisfied for $(\theta_n^\omega)_{n \geq 1}$ given by (B.24a) with $\mathcal{A} = \mathcal{A}^\omega$ defined by (4.1c). Since θ_m^ω with $m < n$ and $\tilde{\theta}_n^\omega$ are statistically independent of ω_n, we have

$$\left| \int_\Omega e^{2ih(\theta_m(x) - \theta_n(x))} d\mu(x) \right| \leq \left| \int_{\Lambda_n} e^{2ih\omega_n\varphi} d\nu_n(\omega_n) \right|. \quad (4.11b)$$

The r.h.s. of (4.11b) is the characteristic function of ν_n:

$$\Phi_n(t) := \int_{\Lambda_n} e^{i\omega_n t} d\nu_n(\omega_n) = \frac{1}{2n+1} \sum_{k=-n}^{n} e^{ikt} = \frac{1}{2n+1} \frac{\sin(n+1/2)t}{\sin t/2}.$$

Evaluating $\Phi_n(t)$ at $t = 2h\varphi$, we have

$$\left| \int_{\Lambda_n} e^{2ih\omega_n\varphi} d\nu_n(\omega_n) \right| = |\Phi_n(2h\varphi)|$$

$$\leq \frac{1}{2n+1} |\sin(h\varphi)|^{-1} < \infty \quad \forall h \in \mathbb{Z} \backslash \{0\}, \quad (4.11c)$$

for $\varphi \in [0, \pi]$ different from a rational multiple of π, i.e., $\varphi/\pi \notin \mathbb{Q}$.

Putting (4.11a), (4.11b) and (4.11c) together, yields

$$I_h(N,\Omega) \le \frac{1}{N} + \frac{2}{N^2} |\sin h\varphi|^{-1} \sum_{1 \le m < n \le N} \frac{1}{2n+1} < \frac{1}{N}(1 + |\sin h\varphi|^{-1}),$$

which implies that $\sum_{N=1}^{\infty} I_h(N,\Omega)/N$ is finite for each $h \ne 0$ and $\varphi/\pi \notin \mathbb{Q}$, concluding the proof. $\qquad\square$

Remark 4.1. The uniform assumption on ν_j is not necessary. Moreover, to assure that the Prüfer angles $(\theta_k^\omega)_{k \ge 1}$ are u.d. mod π, it is sufficient to define the random variables ω_j supported in an interval $\Lambda_j \equiv \{-n_j, \dots, n_j\}$, for a subsequence $(n_j)_{j \ge 1}$ with $n_j \to \infty$ as $j \to \infty$.

We come back to the Prüfer angles as a deterministic mapping (4.7). We replace $(\theta_n)_{n \ge 1}$ by a sequence $\{\xi_n\}_{n \ge 1}$ of continuous piecewise linear functions $\xi_n = \xi_n(\varphi)$ of $\varphi \in [0, \pi]$ that coincide with $\theta_n = \theta_n(\varphi)$ on every value $\varphi_{n,j}$ at which $\theta_n(\varphi_{n,j})$ is a multiple of π. Zlatoš [279] introduced this sequence to estimate the Hausdorff dimension of invariant subsets of $[0, \pi]$ for which the Birkhoff averages do not exist (see [279, App. A]). We here show that the general metric theorem can be applied with $\Omega = [0, \pi]$, $\mathcal{B}$ the Borel sets of $[0, \pi]$ and μ the Lebesgue measure. Before that, we need the following (see [180, Proposition 5.2, Eqs. (5.9) and (5.10)]).

Proposition 4.1. *The function $\theta_k = \theta_k(\varphi)$ is strictly monotone increasing function of $\varphi \in [0, \pi]$ satisfying*

$$\theta_k'(\varphi)/\beta^k \in [d_1, d_2] \tag{4.12}$$

with $d_1 = d_1(p, \beta, \varphi) = c_1 p^2/\beta > 0$ and $d_2 = d_2(p, \beta, \varphi) = c_2/(p^2 \beta) < \infty$.

For each $k \in \mathbb{N}$ and interval $I = (a, b) \subset [0, \pi]$, let

$$\varphi_{k,0} \le a < \varphi_{k,1} < \varphi_{k,2} < \cdots < \varphi_{k,n_k} \le b < \varphi_{k,n_k+1},$$

$n_k = n_k(I)$, be such that

$$\theta_k(\varphi_{k,i}) = \left(\left[\frac{\theta_k(a)}{\pi} \right] + i \right) \pi \tag{4.13a}$$

holds for $i = 0, 1, \dots, n_k + 1$ with $[x]$ the integer part of $x \in \mathbb{R}$. We set $I_{k,0} = (a, \varphi_{k,1}]$, $I_{k,i} = (\varphi_{k,i}, \varphi_{k,i+1}]$ for $i = 1, \dots, n_k - 1$, $I_{k,n_k} = (\varphi_{k,i}, b]$ and notice that $\{I_{k,i}\}_{i=0}^{n_k}$ is a partition of I satisfying

$$\frac{\pi}{d_2 \beta^k} \le |I_{k,i}| \le \frac{\pi}{d_1 \beta^k} \tag{4.13b}$$

by (4.12), for every $i \neq 0$, n_k. Let $\{\xi_k\}_{k \geq 1}$ be a sequence defined by the linear interpolation of θ_k at the points $\{\varphi_{k,i}\}$:

$$\xi_k(\varphi) = \theta_k(\varphi_{k,i}) + \frac{\varphi - \varphi_{k,i}}{|I_{k,i}|}\pi, \quad \varphi \in I_{k,i}. \tag{4.13c}$$

Theorem 4.4. *The sequence* $\{\xi_k(\varphi)\}_{k \geq 1}$ *is u.d. mod* π *for almost every* $\varphi \in [0, \pi]$, *any* $p \in (0, 1)$ *and* $\beta \geq \beta_0(p, \varphi)$ *large enough.*

Proof. It is enough to verify the assumption of Theorem 4.2. For $h \in \mathbb{Z} \setminus \{0\}$ and $I = (a, b] \subset [0, \pi]$, we have

$$I_h(N, (a, b]) = \frac{1}{n^2} \sum_{k,m=1}^{n} \int_a^b \exp\{2ih(\xi_m(\varphi) - \xi_k(\varphi))\}\, dx$$

$$\leq \frac{|b - a|}{n} + \frac{2}{n^2} \sum_{1 \leq k < m \leq n} \left| \int_a^b \exp\{2ih(\xi_m(\varphi) - \xi_k(\varphi))\}\, d\varphi \right|. \tag{4.14a}$$

Note that, by (4.13b), $|I_{m,j}| < |I_{k,l}|$ holds uniformly for $k < m$ if β is large enough and there is at least one $l = l(j)$ such that $I_{k,l} \cap I_{m,j} \neq \emptyset$. Using

$$\int_{I_{m,j}} \exp\{2ih\xi_m(\varphi)\}\, d\varphi = \int_{I_{m,j}} \exp\left\{2ih\frac{\varphi - \varphi_{m,j}}{|I_{m,j}|}\pi\right\}\, d\varphi$$

$$= |I_{m,j}| \int_0^1 e^{2ihx\pi}\, dx = 0$$

together with $|e^{iy} - 1| \leq |y|$, $y \in \mathbb{R}$, and (4.13b), we have

$$\left| \int_{I_{m,j}} \exp\{2ih(\xi_m(\varphi) - \xi_k(\varphi))\}\, d\varphi \right| \leq \frac{2|h|\pi}{|I_{k,l}|} \int_{I_{m,j}} \varphi\, d\varphi = |h|\pi \frac{|I_{m,j}|^2}{|I_{k,l}|}$$

and, consequently,

$$\left| \int_a^b \exp\{2ih(\xi_m(\varphi) - \xi_k(\varphi))\}\, d\varphi \right|$$

$$\leq \sum_{j=0}^{n_k} \left| \int_{I_{m,j}} \exp\{2ih(\xi_m(\varphi) - \xi_k(\varphi))\}\, d\varphi \right|$$

$$\leq |h|\pi \sup_i \frac{|I_{m,i}|}{|I_{k,l(i)}|} \sum_{j=0}^{n_k} |I_{m,j}|$$

$$\leq \frac{|h|\pi d_2}{d_1} \beta^{m-k} |b-a|. \tag{4.14b}$$

Substituting (4.14b) into (4.14a), yields

$$I_h(N,(a,b]) \leq \left(1 + \frac{2|h|\pi d_2}{d_1(\beta-1)}\right) \frac{|b-a|}{n},$$

which proves the assumption of Theorem 4.2 and concludes the proof of Theorem 4.4. □

Let $g_n(\varphi) \equiv g(\theta_n(\varphi), \varphi)$, where g is defined in (4.7). The functions $\xi_k(\varphi)$ and $\theta_k(\varphi)$ are closed in the following sense (see [180, Proposition 5.5]).

Proposition 4.2.

$$|\theta_k(\varphi) - \xi_k(\varphi)| \leq \frac{|I_{k,j}|}{4} \left(\max_{\varphi \in I_{k,j}} g'_{k-1}(\varphi) - \min_{\varphi \in I_{k,j}} g'_{k-1}(\varphi)\right) \tag{4.15}$$

holds for $\varphi_{k,j} < \varphi \leq \varphi_{k,j+1}$ *and* $j = 0, \ldots, n_k$.

Equation (4.15) is a refined (local) version of an upper bound given in [279 App. A]. Proposition 4.2 implies that the difference

$$E_N = \frac{1}{N} \sum_{k=1}^{N} (f(\theta_k) - f(\xi_k)) \tag{4.16}$$

of two Birkhoff averages, related with the sequences $(\theta_k)_{k\geq 1}$ and $(\xi_k)_{k\geq 1}$, can be made small as one wishes by taking β large enough for any periodic smooth function f, with period π. Equation (4.16) together with Proposition 4.2 was used in [180] to establish part (a) of Theorem 4.1 when the incertainties ω_j's in (4.1c) are not present and β tends to infinity.

4.1.3. *Proof of Theorem 4.1*

We refer now to App. B, Eq. (B.29), where the Prüfer radius R_N is written as a Birkhoff sum

$$R_N^2 = \exp\left(\sum_{k=1}^{N} f(\varphi, \theta_k)\right) R_0^2 \tag{4.17}$$

for f defined by (B.30) and (B.24d).

We need an equivalent to Weyl's criterion (Theorem 3.19) for uniformly distributed sequences mod π.

Theorem 4.5 (Corollary 1.1 [156, Chap. 2]). *A sequence* $(\theta_k)_{k\geq 1}$ *of real numbers is u.d. mod π if and only if the "discrepancy" of its distribution about the uniform one*

$$D_n^* := \sup_{0<\theta\leq\pi} \left| \frac{1}{n}N([0,\theta/\pi),n) - \theta/\pi \right| \tag{4.18}$$

tends to zero as $n \to \infty$. Here, for any subset E of $[0,1)$, $N(E,n) = N(E,n;(\theta_k/\pi)_{k\geq 1})$ is the number o terms in the sequence $(\theta_k/\pi)_{k\geq 1}$, up to n, for which $fr(\theta_k/\pi) := \theta_k/\pi - [\theta_k/\pi]$ belongs to E (see Definition 3.10).

In Theorem 4.3 we have shown that the angles of Prüfer $(\theta_k^\omega)_{k\geq 0}$, given by (B.24a), are u.d. mod π a.e. φ. Koksma's inequality (see, e.g., [156, Theorem 5.1]) then implies:

$$\left| \frac{1}{N}\sum_{k=1}^{N} f(\theta_k) - \frac{1}{\pi}\int_0^\pi f(\theta)\,d\theta \right| \leq D_N^* \frac{1}{\pi}\int_0^\pi |f'(\theta)|\,d\theta. \tag{4.19}$$

By (4.17), (B.31) and (B.32)

$$R_0 \exp\left(-N\left| \frac{1}{N}\sum_{k=1}^{N} f(\theta_k) - \bar{f} \right| \right) r^N$$

$$\leq R_N^2 \leq R_0 \exp\left(N\left| \frac{1}{N}\sum_{k=1}^{N} f(\theta_k) - \bar{f} \right| \right) r^N, \tag{4.20a}$$

which, together with (4.19) and (4.18), yields

$$C_N^{-1} r^N \leq R_N^2 \leq C_N r^N \tag{4.20b}$$

with

$$C_N \equiv (1 + |\cos\varphi|) \exp\left(ND_N^* \frac{1}{\pi}\int_0^\pi |f'(\theta)|\,d\theta \right). \tag{4.20c}$$

By u.d. of the sequence $(\theta_k)_{k\geq 1}$ (Theorem 4.3) and Theorem 4.5, D_N^* tends to zero for $\lambda \in [-2,2]\backslash 2\cos\pi\mathbb{Q}$. This, together with the estimate:

$$\sup_{0\leq\theta\leq\pi} |f'(\theta)| \leq \sup_{0\leq\theta\leq\pi} \left| \frac{2b\sin 2\theta - 2c\cos 2\theta}{a + b\cos 2\theta + c\sin 2\theta} \right| \leq C < \infty$$

uniformly in $0 \leq \varphi \leq \pi$, by (B.25), implies

$$\lim_{N \to \infty} C_N^{1/N} = 1 \tag{4.21a}$$

for almost every $\varphi \in [0, \pi]$.

By (B.42a), (B.42c) and (4.20b) there is an A with zero Lebesgue measure such that, for any $\lambda \in [-2, 2] \backslash A$ and any $n \in \mathbb{Z}_+$ such that $a_N \leq n < a_{N+1}$, we have

$$\tilde{C}_N^{-1} r^N \leq \|T(n; \lambda)\|^2 \leq \tilde{C}_N r^N \tag{4.21b}$$

with r given by (B.32) and $\tilde{C}_N$ satisfying (4.21a).

By Theorem B.1 together with (4.21b) and the fact that $r > 1$, the essential support Σ_{ac} of the a.c. part μ_{ac} of spectral measure $\mu = d\rho$ is an empty set (see Theorem B.1). By (4.21b), (4.1d), (4.21a) and the ratio test, the sum

$$\sum_{n=1}^{\infty} \|T(n; \lambda)\|^{-2} \geq \sum_{N=1}^{\infty} \tilde{C}_N^{-1} \frac{\beta_N}{r^N}$$

diverges provided

$$\frac{\beta}{r} > 1,$$

which, together with (B.44) and since we have excluded both the a.c. and the p.p spectrum, proves the first part of item (a).

Another necessary ingredient is the following result (see [180] and references therein, for a proof).

Proposition 4.3. *Suppose*

$$[a, b] \subset \sigma(J^\phi) \tag{4.21c}$$

holds for all $\phi \in [0, \pi]$ and there exists $A \subset [a, b]$ of Lebesgue measure zero independent of ϕ such that, if $\lambda \in [a, b] \backslash A$,

$$(J^\phi - \lambda I)u = 0 \tag{4.21d}$$

has a solution $u \in l_2(\mathbb{Z}_+)$, for some ϕ. Then $\sigma(J^\phi)$ has only dense p.p. spectrum in $[a, b]$ for a.e. (w.r.t. Lebesgue measure) ϕ-boundary condition (4.3).

By (4.21b), (B.46) and (B.48) are simultaneously satisfied provided $\beta/r < 1$ holds, i.e., provided $\lambda \in ([-2, 2] \backslash 2 \cos \pi \mathbb{Q}) \backslash I$, by the ratio test. This, together with Proposition 4.3, concludes the proof of item (b).

To prove the second part of item (a), we evoke Corollaries B.1 and B.2. By (4.21b) and assumptions of Theorem 4.1, there is an A with zero Lebesgue measure such that, for any $\lambda \in [-2, 2] \backslash A$ and any $n \in \mathbb{Z}_+$ such that $a_k \leq n < a_{k+1}$, we have

$$\|T(n; \lambda)\|^2 \leq \tilde{C}_k r^k \leq \tilde{C}_k' a_k^\gamma \leq \tilde{C}_k'' n^\gamma,$$

with $\gamma = \gamma(\lambda) \equiv \ln r(\lambda) / \ln \beta$ satisfying $0 < \gamma < 1$ by $\beta/r > 1$ and $\lim_{n \to \infty} (\tilde{C}_n'')^{1/n} = 1$. This implies that

$$\sum_{n=1}^{l} \|T(n; \lambda)\|^2 \leq c l^{1+\gamma} \tag{4.21e}$$

holds for some $c > 0$ and every $\lambda \in [-2, 2] \backslash A$.

Corollaries B.1 and B.2 guarantee the existence of a strong subordinate solutions $u_k^{\phi^*}$ satisfying

$$\|\mathbf{u}_k^{\phi^*}\|^2 \leq \tilde{C}_k''' a_k^{-\gamma}.$$

Since every solution of $J^{\phi, \omega}$ has constant modulus on the interval $[a_k + 1, a_{k+1}]$, we have

$$\|u^{\phi^*}\|_l^2 \leq c' l^{1-\gamma}, \tag{4.21f}$$

for some $c' > 0$.

Thus, by (4.21e) and (4.21f)

$$\limsup_{l \to \infty} \frac{1}{l^{2-\alpha}} \sum_{n=0}^{l} \|T(n; \lambda)\|^2 < \infty \tag{4.21g}$$

and

$$\liminf_{l \to \infty} \frac{\|u^{\phi^*}\|_l^2}{l^{\alpha'}} = 0 \tag{4.21h}$$

hold provided $2 - \alpha \geq 1 + \gamma$ and $\alpha' > 1 - \gamma$.

As ρ has no absolutely continuous part, by Proposition B.5(iv), the spectral measure ρ is supported on the set of λ's for which u^ϕ satisfies the boundary condition ϕ. By the theory of rank-one perturbation (see [133,

Theorem 1.3]), we have $u^\phi = u^{\phi^*}$ for a.e. $\lambda \in I$ with respect to $\mu = d\rho$ and for a.e. boundary condition ϕ.

By Corollary B.1, if (4.21g) is satisfied, the restriction $\rho(((\lambda - \delta, \lambda + \delta)\backslash A) \cap \cdot)$ is α-continuous. Clearly, $\alpha = 1 - \gamma(\lambda) = 1 - \ln r(\lambda)/\ln \beta$ satisfies the requirement:

$$\limsup_{l \to \infty} \frac{1}{l^{2-\alpha}} \sum_{n=1}^{l} \|T(n; \lambda)\|^2 \leq \limsup_{l \to \infty} \frac{1}{l^{1+\gamma}} \sum_{n=1}^{l} \|T(n; \lambda)\|^2 < \infty,$$

which implies that (4.21g) holds for $\lambda \in I \backslash A$, concluding the proof. $\qquad \square$

4.2. Decay in the Average

We now come to the third application of Theorem 3.14.

As emphasized in Sec. 3.4, *resonance* is a phenomenon of cooperation between two or more elements, while *randomness* is based on independence — noncooperation — between elements. It should not prove surprising, therefore, that resonance in a system can be removed by randomizing parameters in the system, as in the theory of Schrödinger operators with random potentials — of which the Anderson model is a major example — where the randomness may prevent the cooperation necessary for traveling waves (tunneling), resulting in the phenomenon of localization. Thus, from this point of view, the existence of p.p. spectra in random media may be seen as a result of instability of tunneling occurring for large disorder; for weak disorder one may expect delocalization to persist, and a sharp transition between the two regimes, characterized by the so-called *mobility edge* was predicted by Anderson.

Tunneling instability, which may also be of dynamical nature, is also at the basis of the existence of the chiral superselection rules induced by the environment which account for the shape of molecules such as Ammonia NH_3, see the review by Arthur Wightman, "Superselection sectors: old and new" [269].

We dedicated this section to show that the Kronecker sum of $d \geq 2$ copies of random one-dimensional sparse models, studied in Sec. 4.1, display a spectral transition of the type predicted by Anderson, from absolutely continuous around the center of the band to pure point around the boundaries. The proof of our main result (Theorem 4.6) shows that ideas of Kahane and Salem [142, 144] (see Sec. 3.3.3) combine with Theorem 3.8

in a neat way, yielding a result of quite general nature, i.e., showing the existence of a.c. spectrum for any Kronecker sum of operators $A \otimes I + \theta I \otimes A$ for a.e. $\theta \in [0,1]$ whenever A has s.c. spectrum in some nonempty interval with local Hausdorff dimension greater than $1/2$.

4.2.1. *Anderson-like transition for "separable" sparse models in* $d \geq 2$

We study a class of models whose relationship to the original Anderson [8] model will now be briefly explained. The Anderson Hamiltonian

$$H^\omega = \Delta + \lambda V^\omega \tag{4.22}$$

on $\mathcal{H} = l^2(\mathbb{Z}^d)$, the space of complex-valued square-integrable sequences, $u = (u_n)_{n \in \mathbb{Z}^d}$, in dimension $d \geq 1$, has a kinetic part given by a (centered) discrete Laplacian

$$(\Delta u)_n = \sum_{n' : |n-n'|=1} u_{n'}, \tag{4.23}$$

and a random potential

$$(V^\omega u)_n = V_n^\omega u_n, \tag{4.24}$$

where $\{V_n^\omega\}_{n \in \mathbb{Z}^d}$ is a family of independent identically distributed random variables (i.i.d.r.v.) on the probability space $(\Omega, \mathcal{B}, \mu)$, with a common distribution $F(x) = \mu(\{\omega : V_n^\omega \leq x\})$; $\lambda > 0$ is the disorder parameter also called coupling constant. The spectrum of H^ω is, by the ergodic theorem, almost surely a nonrandom set $\sigma(H^\omega) = [-2d, 2d] + \lambda \operatorname{supp} dF$.

Anderson [8] conjectured that there exists a critical coupling constant $0 < \lambda_c < \infty$ such that for $\lambda \geq \lambda_c$ the spectral measure of (4.22) is pure point (p.p.) for μ-almost every ω, while, for $\lambda < \lambda_c$ the spectral measure of H^ω contains two components, separated by so-called *mobility edge* $E^\pm$: if $E \in [E^-, E^+]$, the spectrum of H^ω is pure absolutely continuous (a.c.); in the complementary set $\sigma(H^\omega) \backslash [E^-, E^+]$, H^ω has pure point spectra. We refer to [132] for a comprehensive review on the status of the problem and references, and only wish to remark that for $d = 1$ the spectrum is p.p. for all λ for almost every ω (see [95, 157]), while, for $d \geq 2$ the existence of a.c. spectrum is open, except for the version of (4.22) on the Bethe lattice, where it was first proved by Klein in a seminal paper [153] (see also [132, Sec. 2.31]).

Given the above-mentioned difficulties, one might be led to study the limit $\lambda \to 0$ of (4.22), for which the spectrum is pure a.c. We shall instead follow a different approach to the Anderson conjecture suggested by Molchanov: the limit of zero concentration, i.e., taking V^ω in (4.22) such that

$$V_n^\omega = \sum_i \varphi_i^\omega (n - a_i), \qquad (4.25)$$

with elementary potential ("bump") $\varphi^\omega : \mathbb{Z}^d \longrightarrow \mathbb{R}$ satisfying a uniform integrability condition

$$|\varphi^\omega(z)| \leq \frac{C_0}{1 + |z|^{d+\varepsilon}} \qquad (4.26)$$

for some $\varepsilon > 0$ and $0 < C_0 < \infty$ and

$$\lim_{R \to \infty} \frac{\#\{i : |a_i| \leq R\}}{R^d} = 0. \qquad (4.27)$$

Due to condition (4.27) of zero concentration, potentials such as (4.25) are called sparse and have been intensively studied in recent years since the seminal work by Pearson in $d = 1$ [200], notably by Kiselev, Last and Simon [152] for $d = 1$ and by Molchanov in the multidimensional case [188] (see also [189, 190] for complete proofs and additional results). As a consequence of (4.26), for $d \geq 2$ the interaction between bumps is weak [188] while for $d = 1$ the phase of the wave after propagation between distant bumps become "stochastic" [200]. This is the right moment to introduce our one-dimensional model.

Instead of (4.22) we shall adopt the models $J^{\omega,\phi}$ given by (4.1a) et seq. and the corresponding diagonal version (4.4a)–(4.4c), introduced by Zlatoš [279]. Note that (4.4c) satisfies trivially (4.26), since $\varphi_i^\omega(n) = \delta_{\omega_i,n}$ is just a Kronecker delta at ω_i. The nondiagonal version (4.1a) has some advantages in addition to the initial motivation coming from [119]: that the spectrum $\sigma(J^\omega)$ of J^ω interpolates between purely absolutely continuous for $p = 1$ and dense pure point for $p = 0$ (in the latter case, J^ω is a direct sum of finite matrices; the dense character is due to (4.1d)). Since, by Theorem 4.1, the absolutely continuous spectrum for $p = 1$ does not stand for any $p < 1$ we try therefore to attain higher dimensions.

Multidimensional version. We follow our work in [179].

Consider the Kronecker sum of $J^{\omega,\phi}$'s as an operator on $\mathcal{H} \otimes \mathcal{H}$:

$$J_\theta^{(2)} := J^{\omega^1,\phi} \otimes I + \theta I \otimes J^{\omega^2,\phi}, \qquad (4.28)$$

where $\omega^1 = (\omega_j^1)_{j \geq 1}$ and $\omega^2 = (\omega_j^2)_{j \geq 1}$ are two independent sequences of independent random variables defined in $(\Omega, \mathcal{B}, \nu)$, as before (we omit ω^1 and ω^2 in the l.h.s above for brevity). Above, the parameter $\theta \in [0,1]$ is included to avoid resonances. We ask for properties of $J_\theta^{(2)}$ (e.g. the spectral type) which hold for *typical* configurations, i.e., a.e. $(\omega^1, \omega^2, \theta)$ with respect to $\nu \times \nu \times \mathcal{L}$ where $\mathcal{L}$ is the Lebesgue measure in $[0,1]$. $J_\theta^{(2)}$ is a special two-dimensional analog of $J^{\omega,\phi}$; if the latter was replaced by $-\Delta + V$ on $L^2(\mathbb{R}, dx)$ where $\Delta = d^2/dx^2$ is the second derivative operator, and V a multiplicative operator $V\psi(x) = V(x)\psi(x)$ (potential), the sum above would correspond to $(-d^2/dx_1^2 + V_1) + (-d^2/dx_2^2 + V_2)$ on $L^2(\mathbb{R}^2, dx_1 dx_2)$, i.e., the "separable case" in two dimensions. Accordingly, we shall also refer to $J_\theta^{(n)}$, $n = 2, 3, \ldots$, as the separable case in n dimensions.

4.2.2. *Uniform α-Hölder continuity of spectral measures*

In order to formulate and prove our main result (Theorem 4.6) we evoke Definition 3.4 of uniformly α-Hölder continuous (UαH) Borel measure μ.

Let $\{E(\lambda)\}$ denote the spectral family associated to $J^{\phi,\omega}$ (we omit the indices for simplicity) and $\{E_{\text{sc}}(\lambda)\}$, $\{E_{\text{pp}}(\lambda)\}$ its singular continuous and pure point parts. As usual (see e.g. [207]), we define $\mathcal{H}_{\text{sc}}$ and $\mathcal{H}_{\text{pp}}$ so that, if $\psi \in \mathcal{H}_{\text{sc}}$ the spectral measure,

$$\mu_\psi^{\text{sc}}(\lambda) \equiv (\psi, E(\lambda)\psi), \qquad (4.29a)$$

is purely singular continuous and, if $\psi \in \mathcal{H}_{\text{pp}}$,

$$\mu_\psi^{\text{pp}}(\lambda) \equiv (\psi, E(\lambda)\psi) \qquad (4.29b)$$

is purely pure point. Here $\mathcal{H}_{\text{sc}}$ and $\mathcal{H}_{\text{pp}}$ are closed (in norm), mutually orthogonal subspaces: $\mathcal{H} = \mathcal{H}_{\text{sc}} \oplus \mathcal{H}_{\text{pp}}$, and invariant under J_ϕ^ω.

We now choose an arbitrary $\varepsilon > 0$ and pick $(\lambda_i)_{i=1}^{N_\varepsilon}$ with $\lambda_i \in I$ and $(\delta_\varepsilon^i)_{i=1}^{N_\varepsilon}$, with

$$0 < \delta_\varepsilon^i < 1, \qquad (4.30a)$$

for some $N_\varepsilon < \infty$, in such way that

$$\lambda_1 - \delta_\varepsilon^1 = -\sqrt{4 - \frac{v^2}{\beta - 1}}, \qquad (4.30b)$$

$$\lambda_i + \delta_\varepsilon^i = \lambda_{i+1} - \delta_\varepsilon^{i+1}, \quad i = 1, \ldots, N_\varepsilon - 1, \qquad (4.30c)$$

$$\lambda_{N_\varepsilon} + \delta_\varepsilon^{N_\varepsilon} = \sqrt{4 - \frac{v^2}{\beta - 1}}. \tag{4.30d}$$

We set

$$A_\varepsilon^i = [\lambda_i - \delta_\varepsilon^i, \lambda_i + \delta_\varepsilon^i), \tag{4.30e}$$

for $1 \le i < N_\varepsilon$, with $A_\varepsilon^{N_\varepsilon} = [\lambda_{N_\varepsilon} - \delta_\varepsilon^{N_\varepsilon}, \lambda_{N_\varepsilon} + \delta_\varepsilon^{N_\varepsilon}]$, and

$$\tilde{A}_\varepsilon^i = (\lambda_i - \delta_\varepsilon^i, \lambda_i + \delta_\varepsilon^i), \tag{4.30f}$$

for $1 \le i \le N_\varepsilon$, and write I as a mutually disjoint union:

$$I = \bigcup_{i=1}^{N_\varepsilon} A_\varepsilon^i. \tag{4.31}$$

Observe that (4.30b) and (4.30d) represent the boundary points $\lambda_\pm$ of I, given by (4.5). The choice of $(\lambda_i)_{i=1}^{N_\varepsilon}$ is arbitrary but the quantities δ_ε^i, $i = 1, \ldots, N_\varepsilon$, are chosen in correspondence to ε according to the definition of exact Hausdorff dimension $\alpha(\cdot)$ given by (4.6), and satisfy

$$\bar{\delta}_\varepsilon \equiv \max_i \delta_\varepsilon^i \to 0, \tag{4.32}$$

by continuity, as ε tends to 0. As a consequence, the spectral measure of J_ϕ^ω restricted to $\tilde{A}_\varepsilon^i$

$$\mu_\psi^{\mathrm{sc}} \restriction \tilde{A}_\varepsilon^i \tag{4.33}$$

is $(\alpha(\lambda_i) - \varepsilon)$-continuous and $(\alpha(\lambda_i) + \varepsilon)$-singular, for $i = 1, \ldots, N_\varepsilon$.

Proposition 4.4. *Under the hypotheses of Theorem 4.1 and (4.30a)– (4.33), there exists a dense set D in $\mathcal{H}_{\mathrm{sc}}$ such that, $\forall \psi \in D$, $\mu_\psi^{\mathrm{sc}} \restriction \tilde{A}_\varepsilon^i$ is, for each $i \in \{1, \ldots, N_\varepsilon\}$, uniformly $(\alpha(\lambda_i) - \varepsilon)$-Hölder continuous.*

Proof. We write

$$\mathcal{H} = \bigoplus_{i=1}^{N_\varepsilon} \mathcal{H}_i,$$

where $\mathcal{H}_i$ is the subspace of $\mathcal{H}_{\mathrm{sc}}$ generated by

$$\{E_I \psi : \psi \in \mathcal{H}_{\mathrm{sc}}, \text{ for every } I = (\lambda, \lambda'] \subset \tilde{A}_\varepsilon^i\}$$

where $E_I = \int_I dE(\lambda)$ is the spectral projection on I. By (4.6), (4.30a)– (4.33) and [162, Theorem 5.2] for each $\mathcal{H}_i$ we may choose D_i dense in $\mathcal{H}_i$

such that $\forall \psi \in D_i$, μ_ψ is uniformly $(\alpha(\lambda_i) - \varepsilon)$-Hölder continuous. Since the subspace $\mathcal{M}$ generated by $\{E(\lambda_i + \delta_\varepsilon^i)\psi : \psi \in \mathcal{H}\}$ for $i = 1, \ldots, N_\varepsilon - 1$ is such that $\mathcal{M} \subset \mathcal{H}_{\mathrm{sc}}^\perp$, we have by (4.30d), (4.30f) and (4.32) that $\bigoplus_{i=1}^{N_\varepsilon} D_i$ is dense in $\mathcal{H}_{\mathrm{sc}}$ and satisfies the assertion by (4.33). $\qquad\square$

Corollary 4.1. *Let $I_0 \subseteq I$ and $\psi \in D$. Then $\mu_\psi^{\mathrm{sc}} \restriction I_0$ is $U\alpha H$, where*

$$\alpha = \min_{i:\tilde{A}_\varepsilon^i \cap I_0 \neq \emptyset} \alpha(\lambda_i) - \varepsilon. \tag{4.34}$$

Proof. This follows immediately from Proposition 4.4, Definition 3.4 and additivity of μ_ψ^{sc}. $\qquad\square$

In the rest of the subsection we assume ε and $(\delta_\varepsilon^i)_{i=1}^{N_\varepsilon}$ is a given fixed set of numbers, with $\varepsilon > 0$ arbitrarily small (but with $N_\varepsilon < \infty$).

4.2.3. *Formulation, proof and comments of the main result*

We now go back to Theorem 4.1. Let

$$\lambda^\pm = \pm 2\sqrt{1 - \frac{v^2}{v_c^2}} \tag{4.35a}$$

under the condition

$$0 < v < v_c = 2\sqrt{\beta - 1} \tag{4.35b}$$

so that

$$0 < \lambda^+ < 2. \tag{4.35c}$$

Theorem 4.6. *Let $J_\theta^{(2)}$ be defined by (4.28) and let*

$$v^2 < a(\sqrt{\beta} - 1) < v_c^2 \tag{4.36a}$$

with $a < 4$. Then, for almost every $(\omega^1, \omega^2, \theta)$ with respect to $\nu \times \nu \times \mathcal{L}$,

(a) *there exist $\tilde{\lambda}^\pm$ with $\tilde{\lambda}^+ = -\tilde{\lambda}^-$ and*

$$0 < \tilde{\lambda}^+ < \lambda^+ \tag{4.36b}$$

such that

$$(\tilde{\lambda}^-(1 + \theta), \tilde{\lambda}^+(1 + \theta)) \subset \sigma_{\mathrm{ac}}(J_\theta^{(2)}); \tag{4.36c}$$

(b)

$$[-2(1+\theta), \lambda^-(1+\theta)) \cup (\lambda^+(1+\theta), 2(1+\theta)] \subset \sigma_{\mathrm{pp}}(J_\theta^{(2)}); \quad (4.36\mathrm{d})$$

(c)

$$\sigma_{\mathrm{sc}}(J_\theta^{(2)}) \cap (\lambda^-(1+\theta), \lambda^+(1+\theta)) \quad (4.36\mathrm{e})$$

may, or may not, be an empty set.

Proof. We first choose I_0 in Corollary 4.1 such that

$$I_0 = [-\tilde{\lambda}^+, \tilde{\lambda}^+]$$

and

$$\alpha = \min_{i:\tilde{A}_\varepsilon^i \cap I_0 \neq \emptyset} \alpha(\lambda_i) - \varepsilon > \frac{1}{2}. \quad (4.37\mathrm{a})$$

We quote [179, Proposition A.1] where inequalities (4.36b) and (4.37a) have been established for any choice of parameters p, β satisfying (4.36a) and ε depending on p, β and a.

We define

$$f_{\mathrm{sc}}^i(s) := \int_{I_0} e^{-is\lambda} d\mu_{\varphi_i}^{\mathrm{sc}}(\lambda), \quad i = 1, 2 \quad (4.37\mathrm{b})$$

$\varphi_i \in D_i$ and let

$$I_i(T) := \int_0^T \left| f_{\mathrm{sc}}^i(s) \right|^2 ds. \quad (4.37\mathrm{c})$$

By Theorem 3.8 and (4.37a)

$$I_i(T) \leq C_i T^{1-\alpha} \leq C T^{1-\alpha} \quad (4.37\mathrm{d})$$

for $0 < C_i < \infty$, $i = 1, 2$, T-independent constants and $C = \max(C_1, C_2)$. By (4.37b) and (4.37d) and a change of variable, we have

$$\int_0^1 |f_{\mathrm{sc}}^2(\theta t)|^2 d\theta = \frac{1}{t} I_2(t) \leq C t^{-\alpha}$$

which implies

$$\int_1^T dt |f_{\mathrm{sc}}^1(t)|^2 \int_0^1 |f_{\mathrm{sc}}^2(\theta t)|^2 d\theta \leq C \int_1^T |f_{\mathrm{sc}}^1(t)|^2 t^{-\alpha} dt. \quad (4.37\mathrm{e})$$

We now perform an integration by parts on the r.h.s. of (4.37e)

$$\int_1^T |f_{\text{sc}}^1(t)|^2 t^{-\alpha} dt = I_1(t) t^{-\alpha}\big|_1^T + \alpha \int_1^T dt I_1(t) t^{-\alpha-1}. \qquad (4.37\text{f})$$

By (4.37e), (4.37f) and Fubini's theorem $(T \geq 1)$

$$\int_0^1 d\theta \int_1^T |f_{\text{sc}}^1(t)|^2 |f_{\text{sc}}^2(\theta t)|^2 dt \leq CT^{1-2\alpha} + \alpha C \int_1^T dt t^{-2\alpha}$$

$$\leq C\frac{1}{2\alpha-1}(\alpha - (1-\alpha)T^{1-2\alpha}). \qquad (4.37\text{g})$$

By (4.37a) and (4.37g), the limit

$$\int_0^1 d\theta \int_0^\infty |f_{\text{sc}}^1(t)|^2 |f_{\text{sc}}^2(\theta t)|^2 dt = \lim_{T\to\infty} \int_0^1 d\theta \int_0^T |f_{\text{sc}}^1(t)|^2 |f_{\text{sc}}^2(\theta t)|^2 dt$$

that exists is finite and

$$\int_0^\infty |f_{\text{sc}}^1(t)|^2 |f_{\text{sc}}^2(\theta t)|^2 dt < \infty \qquad (4.37\text{h})$$

for a.e. $\theta \in [0,1]$. By [209, Theorem VIII.33] (or Ichinose's Theorem [123]), for A_k bounded, its corollary, and (4.28), the spectrum of $J_\theta^{(2)}$ is the arithmetic sum of the spectrum of $J_\phi^{\omega^1}$ and $\theta J_\phi^{\omega^2}$. Together with Theorem 4.1, Theorem 3.14 and (4.37h), this proves (4.36c).

In order to prove (4.36d), we need to consider

$$f_{\text{pp}}^i(s) := \int_{I_0} e^{-is\lambda} d\mu_{\rho_i}^{\text{pp}}(\lambda), \quad i = 1, 2, \qquad (4.38)$$

where $\rho_i \in \mathcal{H}_{\text{pp}}$. By [146, Theorem 5.6], $\mathbb{R} \ni t \longmapsto f_{\text{pp}}^i(t)$ is an almost-periodic function on $\mathbb{R}$, i.e., $f_{\text{pp}}^i \in \text{AP}(\mathbb{R})$ (see [146, Definitions 5.1 and 5.2]) and, therefore,

$$k(t,\theta) = f_{\text{pp}}^1(t) f_{\text{pp}}^2(\theta t)$$

belongs to $\text{AP}(\mathbb{R})$ by [146, Theorem 5] and, again by [146, Theorem 5.6] μ defined by

$$\mu = \mu_{\rho_1}^{\text{pp}} * \tilde{\mu}_{\rho_2}^{\text{pp}},$$

where $\tilde{\mu}_{\rho_2}^{\text{pp}}(\lambda) = \mu_{\rho_2}^{\text{pp}}(\lambda/\theta)$, $\theta \neq 0$, is pure point. Together with Ichinose's theorem and Theorem 4.1, this proves (4.36d).

By the definition of measure convolution, it follows that

$$\mu_{\rho_1}^{\rm pp} * \tilde{\mu}_{\varphi_2}^{\rm sc}(\{\lambda\}) = \mu_{\varphi_1}^{\rm sc} * \tilde{\mu}_{\rho_2}^{\rm pp}(\{\lambda\}) = 0$$

for any singleton $\{\lambda\}$. Hence, by Ichinose's theorem and Theorem 4.1, the spectrum of $J_\theta^{(2)}$ restricted to $[(1+\theta)\lambda^-, (1+\theta)\lambda^+]$ is necessarily continuous — but may be singular continuous — showing part (c) and concluding the proof of Theorem 4.6.   □

Remark 4.2. Some of the ideas used in the proof of Theorem 4.6 have also employed by Kahane and Salem [142, 144] in more specific contexts (see Sec. 3.3.3). In our case the parameter θ (the analog of ζ) appears in (4.28), and the F.S. transform of the corresponding measure is L^2 for a.e. $\theta \in [0,1]$, which implies that it tends to zero at infinity by the Riemann–Lebesgue lemma.

Remark 4.3. The a.c. part of the spectrum of $J_\theta^{(2)}$ is not, of course, promoted by the randomness on the "bump" positions. It makes, however, the Hausdorff dimension of the spectral measures $\mu_{\varphi_1}^{\rm sc}$ and $\tilde{\mu}_{\varphi_2}^{\rm sc}$ and, consequently, the intervals I_0 and I appearing in Theorems 4.6 and 4.1, be determined exactly. Items (a) and (b) of Theorem 4.6 thus hold for a bidimensional model (4.28) with the J^{ϕ,ω_i} replaced by deterministic sparse models studied in [180] since their local Hausdorff dimension may be determined as accurately as one wishes, provided the sparse parameter β is large enough. The p.p. part of the spectrum cannot, however, be established except for the random model (see comment after [65, Theorem 2.3] and [180, Remark 5.9.1]).

Remark 4.4. It is important to employ our version of Zlatoš's theorem (see [65, Theorem 2.4]), which shows the purity of the p.p. spectrum. For, in case that the p.p. spectrum contains admixture of s.c. spectrum, the latter may, by convolution, generate an a.c. part in $J_\theta^{(2)}$. Since a (possibly dense) p.p. superposition to the a.c. spectrum of $J_\theta^{(2)}$ cannot be excluded (originated, e.g., from the convolution of two — again possibly dense — p.p. spectra which may be superposed to the s.c. spectrum of [65, Theorem 2.4]), we would, in this special case, have no transition at all in the spectral type from one region to another.

Remark 4.5. In the special case of exactly self-similar spectral measures μ and μ_θ ($\mu_\theta(\lambda) = \mu(\lambda/\theta)$), a theorem of Hu and Taylor [120] implies that their convolution is a.e. $\theta \in [0,1]$ absolutely continuous. This fact has been used by Bellissard and Schulz-Baldes [22] to construct the first models in $d \geq 2$ dimensions with a.c. spectrum and subdiffusive quantum transport (thought to describe properties of quasicrystals) — see their theorem in [22] and a previous remark that it cannot be true for all θ due to resonance phenomena; see also [204]. It is to be remarked that exact self-similarity is a rare property. In particular, Combes and Mantica [46] proved that this property does not hold for sparse models, such as ours (see [46, Theorem 2]).

Remark 4.6. It is clear that the proof of Theorem 4.6 generalizes to dimensions $d > 2$, for even a wider range of parameter values, since the corresponding condition on the r.h.s. of (4.37a), given by $\alpha > 1/d$, becomes successively weaker for increasing d.

Remark 4.7. We have not proved pointwise decay of the F.S. transform $\hat{\mu}$ of the spectral measure μ of J_ϕ^ω; i.e., a bound of the form

$$|\widehat{|f|^2\mu}(t)| \leq C_f t^{-\alpha/2} \tag{4.39a}$$

for $C_0^\infty([-2,2])$ functions f. Indeed, such a bound (4.39a) has never been proved except for classes of sparse models with superexponential sparsity, for which the spectrum is purely s.c. and the Hausdorff dimension equal to one; in this case, (4.39a) assumed the form: $\forall \varepsilon > 0$, $\exists 0 < C_\varepsilon < \infty$ such that

$$|\widehat{|f|^2\mu}(t)| \leq C_{f,\varepsilon} t^{-1/2+\varepsilon} \tag{4.40}$$

(see Sec. 4.3.1 and [64, 155, 243]). It is a challenging open problem to prove (4.40) for the present model, with $1/2$ replaced by $\alpha/2$ on the r.h.s. with α being the local Hausdorff dimension.

4.3. Pointwise Decay

4.3.1. *Pearson's fractal measures: Borderline time-decay for the least sparse model*

Introduction. We come back to the lattice version of sparse Schrödinger operator considered in Sec. 4.1 (Eqs. (4.1a)–(4.1d) and (4.4a)–(4.4c), but now with sequence of "bump" distances $(\beta_j)_{j\geq 1}$ increasing faster than exponential:

$$\frac{\beta_{j-1}}{\beta_j} \leq \delta < \frac{1}{2}, \ \forall j > 1 \quad \text{and} \quad \lim_{j\to\infty} \frac{\beta_{j-1}}{\beta_j} = 0. \tag{4.41}$$

Under sparseness condition (4.41), the model properties discussed in this section will not be affected by the presence of randomness ω and it will be omitted henceforth from our notation. For instance, a sparseness of this type, together with $\sum_{n=1}^{\infty} v_n = \infty$, by (4.4c), prevents $l_2(\mathbb{Z}_+)$ solutions of $(J^\phi - \lambda\mathbb{I})u = 0$ to exist (see Eqs. (B.44) and (B.45), for a criterion). As a consequence, the essential spectrum $\sigma_{\text{ess}}(J^\phi) = [-2, 2]$ is purely singular continuous (see [152, Theorem 1.7(2)]) and has Hausdorff dimension equal to 1 (see [279, Theorem 1.4(2)]). For the reader's convenience, these and some other results are proven in App. B, with the randomness ω being an important simplifying ingredient (see, e.g., Proposition B.9).

The method of proving singular continuous spectra for sparse Schrödinger operator was introduced by Pearson in his celebrated paper [200]. By Pearson's method, subsequently improved and extended notably in [152, 190, 211], a singular continuous measure is attained from limiting sequence of absolutely continuous measures, not necessarily related with differential (or difference) operators — the exposition in [200] started with a Cantor measure.

Pearson's sparseness (see (3.12e)), however, is far from being optimal and we search, in the present subsection, for the least possible sparseness so that the Fourier–Stieltjes transform of the spectral measure μ_Ψ of J^ϕ has pointwise decay

$$(\Psi, \exp(itJ^\phi)\Psi) = \hat{\mu}_\Psi(t) \sim |t|^{-1/2} \tag{4.42}$$

for Ψ in a dense set of $\mathcal{H}$. Time-decay (4.42) is just at the borderline for the class of Jacobi operators J^ϕ in consideration, i.e., if $\hat{\mu}_\Psi(t) \sim |t|^{-1/2-\varepsilon}$ for some $\varepsilon > 0$, then, by Theorem 3.14, μ would be absolutely continuous. In this sense, by Definitions 3.14 and 3.15, the support Σ_Ψ of the spectral measures μ_Ψ for J^ϕ are Salem's sets.

Earlier pointwise decay investigation has been carried by Simon [243], for continuous Schrödinger operators with generic and sufficiently sparse potentials, and Krutikov–Remling [155], for a model similar to one considered here, have found a resonant set $\mathcal{R} \subset \mathbb{R}$ in which (4.42) holds if $t \in \mathcal{R}$, otherwise $\hat{\mu}_\Psi(t)$ decays faster than any arbitrary large but finite power m. As far as sparseness is concerned, (4.42) decays as $(1 + |t|)^{-1/2+\varepsilon}$ for $t \in \mathcal{R}$ by Krutikov–Remling's method if $a_{j-1} < Ca_j^{1/2}$, i.e., $a_j \sim \exp(\exp(cj))$, $c > \log 2$; in [243], the same decay holds provided $a_j \sim \exp(C_\varepsilon j^{3/2})$ for some $C_\varepsilon > 0$.

We shall follow the exposition presented in [64], whose sparseness condition is less restrictive than those stated in [155, 243]. We also refer to App. B, in particular to Definitions B.3, B.4 and Propositions B.6, B.7, where we deduce some useful expressions, and to the Weyl–Titchmarsh m-function.

Preliminaries. Our starting point is the representation of the spectral measure of J as a weak-star limit of absolutely continuous measures. The following proposition joins Proposition 2.1 and Corollary 2.2 of [155] (see also [200]).

Proposition 4.5. *Let $y_n = y_n(z)$ and $v_n = v_n(z)$ be solutions of the eigenvalue equation $(J - z\mathbb{I})u = 0$, $z \in \mathbb{C}$, satisfying the initial conditions $y_0(z) = 0$ and $y_1(z) = 1$ and $v_0(z) = 1$ and $v_1(z) = 0$ ((4.1a) with Dirichlet and Newmann boundary condition: $y_n = u_n^{\phi=0}$ and $v_n = u_n^{\phi=\pi/2}$), respectively, and let*

$$M_N(z) = \frac{v_N(z) - w(z)v_{N+1}(z)}{y_N(z) - w(z)y_{N+1}(z)}, \qquad (4.43a)$$

$N \in \mathbb{N}$, *where*

$$w(z) = z/2 + i\sqrt{1 - z^2/4}.$$

Then, as N goes to ∞, $M_N(z)$ converges uniformly, in any compact subset of the upper half-plane $\mathbb{H}$, to the Weyl–Titchmarsh m-function

$$m(z) = \left(e_1, (J - z\mathbb{I})^{-1}e_1\right) = \int \frac{1}{\lambda - z} d\rho(\lambda); \qquad (4.43b)$$

$d\rho$ being the spectral measure of J associated with the state e_1 (see Proposition B.14).
In addition,

$$\int f(\lambda)d\rho(\lambda) = \lim_{N \to \infty} \frac{1}{\pi} \int_{-2}^{2} f(\lambda) \frac{\Im(w(\lambda))}{|y_N(\lambda) - w(\lambda)y_{N+1}(\lambda)|^2} d\lambda \qquad (4.43c)$$

holds for every continuous function $f : (-2, 2) \longrightarrow \mathbb{C}$.

Proof. We refer to Propositions B.1, B.2 and B.3 for analogous treatment. Write

$$f_n(z) = v_n(z) - M_N(z)y_n(z) \qquad (4.44a)$$

and note that f satisfies the eigenvalues equation $(J - z\mathbb{I})f = 0$ with 0-Dirichlet boundary condition at 0 and the following boundary condition at $N + 1$:

$$f_N(z) - w(z)f_{N+1}(z) = 0. \tag{4.44b}$$

We restrict the values of z to the upper half-plane $\mathbb{H} = \{z \in \mathbb{C} : \Im\mathrm{m}(z) > 0\}$.

Suppose that N is at least two unit distant from any scattering point a_k: $|N - a_k| > 1$, for all $k \in \mathbb{N}$, so $p_N = p_{N\pm1} = 1$. Since f satisfies $(J - z\mathbb{I}) f = 0$, we have

$$
\begin{aligned}
\|f\|_N^2 &= \frac{1}{z - \bar{z}}((f, Jf)_N - (Jf, f)_N) \\
&= \frac{1}{z - \bar{z}}(-W_N[f, \bar{f}] + W_0[f, \bar{f}]) \\
&= \frac{1}{z - \bar{z}}(\bar{f}_N f_{N+1} - \bar{f}_{N+1} f_N + M_N - \bar{M}_N)
\end{aligned}
$$

by the Green's identity and the boundary condition at 0 which, together with (4.44b), implies

$$\frac{\Im M_N(z)}{\Im z} = \|f\|_N^2 + \frac{\Im w(z)}{\Im z}|f_{N+1}(z)|^2 \geq \|f\|_N^2 > 0 \tag{4.44c}$$

for every $z \in \mathbb{H}$. Note that $w(z)$ maps $\mathbb{H}$ into itself and $\Im w(z)/\Im z > 0$.

The inequality (4.44c) is precisely the condition for M_N to lie inside a Weyl circle K_N:

$$|c_N - m|^2 = r_n^2 \tag{4.44d}$$

of center $c_N = W_N[v, \bar{y}]/W_N[y, \bar{y}]$ and radius $r_N = |W_N[y, \bar{y}]|^{-1}$ (see Proposition B.3).

The two statements: M_N lies inside the Weyl circle K_N together with the limit point case (see Proposition B.1), prove in particular that $M_N(z)$ converges, as $N \to \infty$, to the Weyl–Titchmarsh m-function $m(z)$, uniformly in compact sets of $\mathbb{H}$.

Since $M_N(z)$ is Herglotz, $\lim_{\nu\downarrow0} M_N(\lambda + i\nu)$ exists for every $N \in \mathbb{N}$ and a.e. $\lambda \in (-2, 2)$; the Borel measure $d\mu_N$ of its integral representation is absolutely continuous w.r.t. the Lebesgue measure (see, e.g., [137, Lemma I, Chap. II]):

$$d\mu_N(\lambda) = \lim_{\nu\downarrow0} \frac{1}{\pi}\Im M_N(\lambda + i\nu)d\lambda = \frac{1}{\pi}\frac{\Im w(\lambda)}{|y_N(\lambda) - w(\lambda)y_{N+1}(\lambda)|^2}d\lambda$$

in view of (4.43a) and $\lim_{\nu\downarrow0}(v_N\bar{y}_{N+1} - v_{N+1}\bar{y}_N)|_{z=\lambda+i\nu} = -W_N[\bar{y}(\lambda), v(\lambda)]$ $= -W_N[\bar{y}(\lambda), v(\lambda)] = 1$. This concludes the proof. $\square$

Remark 4.8. The conclusion (4.43c) of Proposition 4.5 would be true for any function $w(z)$ belonging to the Pick class. See [155, 200] and references therein. See also [201].

Problem 4.1. *Show that* $w(z) = z/2 + i\sqrt{1 - z^2/4}$ *is the Weyl–Titchmarsh m-function for the free Jacobi matrix* J^0, *given by* (4.4a).

We shall apply this formula with $f(\lambda) = |f(\lambda)|^2 e^{it\lambda}$, which becomes more transparent in terms of the Prüfer variables $(R_j, \theta_j)_{j \geq 1}$ (see Definition B.4). For $\lambda = 2\cos\varphi$, $\varphi \in (0, \pi)$, we have $w(\lambda) = e^{i\varphi}$ and the denominator of the integrand in (4.43c) can be written as (see (B.40a))

$$|y_N(\lambda) - w(\lambda)y_{N+1}(\lambda)|^2 = \left| UT(a_j + 1; \lambda) \begin{pmatrix} 1 \\ 0 \end{pmatrix} \right| = R_j^2 \qquad (4.45a)$$

for any N, $j \in \mathbb{N}$ satisfying $a_j < N < a_{j+1}$.

It follows from Proposition B.10, and respective Corollaries B.3 and B.5, that $y_n(\lambda)$ is a subordinate solution of $(J - \lambda\mathbb{I})u$ and this fact has implication to the measure (4.43c). According to Gilbert–Pearson theory (see Proposition B.17), the singular part $d\rho_s$ of spectral measure $d\rho$ is supported on a set Σ_s of λ's for which the Radon–Nikodym derivative $d\rho/d\lambda$ diverges and u obey the phase boundary ϕ. The measure on the r.h.s. of (4.43c) is a.c. with respect to Lebesgue measure for any finite N and concentrates in a set Σ of Lebesgue measure zero as N goes to ∞. By Corollary B.3, there exists a strong subordinate solution $u_n^{\phi^*}$ of $(J - \lambda\mathbb{I})u = 0$. Since the Prüfer angles are uniformly distributed by Theorem 4.3, the spectral norm $t_j = \|T(a_j + 1, \lambda)\|$ behaves, as a function of j, as $t_j^2 = O(r^j)$, for almost every λ, with the exceptional λ's in a countable set $[-2, 2] \cap 2\cos\pi\mathbb{Q}$. By Corollary B.5, there is a convergent sequence $(\phi_j)_{j \geq 1}$, defined by (B.51c), such that $\|T(a_j + 1, \lambda)\mathbf{v}_{\phi_j}\|^2 = O(r^{-j})$ for large j. So, λ is in the essential support $\Sigma_{\mathrm{ess}}^\phi$ of the spectral measure of J iff $\phi^* = \lim_{j \to \infty} \phi_j = \phi$. Since $y = u^{\phi=0}$,

$$|y_N(\lambda) - w(\lambda)y_{N+1}(\lambda)|^2 = \|UT(N, \lambda)\mathbf{v}_0\|^2 = O(r^{-j}) \qquad (4.45b)$$

implies that

$$\lim_{N \to \infty} \frac{1}{|y_N(\lambda) - w(\lambda)y_{N+1}(\lambda)|^2} = \begin{cases} \infty & \text{if } \lambda \in \Sigma_{\mathrm{ess}}^0, \\ 0 & \text{otherwise,} \end{cases}$$

and the limit $N \to \infty$ in (4.43c) cannot be interchanged with the integral, exactly as for the uniform measure in the ternary Cantor set (see [200,

Eq. (1)] and its following up for the general ideas). The generalization of the Cantor measure introduced in [200, Theorem 1] is hereby called *Pearson's fractal measure*.

Our next ingredient is related with the following identity

$$\frac{1}{R_j^2} = \frac{1}{R_{j^*-1}^2} + \sum_{k=j^*-1}^{j-1} \frac{1}{R_k^2}\left(\frac{R_k^2}{R_{k+1}^2} - 1\right) \tag{4.46a}$$

for some conveniently chosen $j^* \leq j$, where

$$\frac{R_k^2}{R_{k+1}^2} - 1 = F(\varphi, \theta_{k+1}) - 1 \equiv H(\varphi, \theta_{k+1}) \tag{4.46b}$$

satisfies $\bar{H} = \frac{1}{\pi}\int_0^\pi H(\varphi, \theta)d\theta = 0$, by (B.26), and

$$\frac{1}{R_k^2} = \prod_{l=1}^{k} F(\varphi, \theta_l) \frac{1}{R_0} \tag{4.46c}$$

by applying (B.24b) successively.

Plugging (4.45a), (4.46a) and (4.46b) into (4.43c), yields

$$\widehat{|f|^2 d\rho}(t) = \frac{1}{\pi}\int_0^\pi |f(2\cos\varphi)|^2 \frac{\sin^2\varphi}{R_{j^*-1}^2} e^{2it\cos\varphi}d\varphi$$

$$+ \sum_{k=j^*-1}^{\infty} \frac{1}{\pi}\int_0^\pi |f(2\cos\varphi)|^2 \frac{\sin^2\varphi}{R_k^2} H(\varphi, \theta_{k+1})e^{2it\cos\varphi}d\varphi.$$

$$\tag{4.47}$$

The last preparation step is the Fourier expansion of $H(\varphi, \theta)$:

$$H(\varphi, \theta) = \sum_{n=1}^{\infty}(A(\varphi)^n e^{2in\theta} + \bar{A}(\varphi)^n e^{-2in\theta})$$

$$A(\varphi) = \sqrt{1 - r^{-1}}e^{i(\delta+\pi)}, \tag{4.48}$$

with r given by (B.32) and $\tan\delta = c/b$. Since $r \geq 1$,

$$|A(\varphi)| \leq a < 1 \tag{4.49}$$

uniformly in each compact set K of $(0, \pi)$ and the series is uniformly and absolutely convergent. Equation (4.48) is similar to an expansion used in [155, Theorem 2.3].

Problem 4.2. *Show that* $H(\varphi, \theta)$ *satisfies* (4.48).

Hint: Use

$$H(\varphi, \theta) + 1 = \left(\frac{1 + |A|^2}{1 - |A|^2} + \frac{2|A| \cos \delta}{1 - |A|^2} \cos 2\theta + \frac{2|A| \sin \delta}{1 - |A|^2} \sin 2\theta\right)^{-1}$$

$$= F(\varphi, \theta).$$

The method of stationary phase. Equation (4.47) together with (4.48) can be written as

$$I(t) = I_{k,0}(t) + \sum_{k \geq j^* - 1} \sum_{n \geq 1} (I_{k,n}(t) + \bar{I}_{k,n}(-t)), \qquad (4.50)$$

where each integral involved in the sum (4.50) is of the form:

$$I_{k,n}(t) = \int_{-\infty}^{\infty} e^{ith_k(\varphi)} d(G_k \circ \lambda)(\varphi) \qquad (4.51a)$$

for some $\mathcal{C}_0^\infty$ function $G'_k \equiv dG_k/d\lambda$ with support $2\cos[\varphi_-, \varphi_+] \subset (-2, 2)$, and h_k of the type:

$$h_k(t, n; \varphi) = 2\left(\cos \varphi + \frac{n}{t}\theta_k(\varphi)\right), \quad n \in \mathbb{Z}. \qquad (4.51b)$$

The basic facts of an integral

$$\hat{I}(\lambda) = \int_a^b \exp(i\lambda \Phi(x)) \Psi(x) dx$$

depend on three principles: localization, scaling and asymptotics. We refer to Propositions 2.1, 2.2 and 2.3 for an exposition of the method of stationary phase. Localization means that, if Φ has compact support in (a, b), the asymptotic behavior of $\hat{I}(\lambda)$ is determined by the points where $\Phi'(x) = 0$, i.e., the points of *stationary phase*. Suppose we know only that

$$|\Phi^{(l)}(x)| \geq \rho > 0 \qquad (4.52)$$

for some *fixed* l, and we wish to obtain an estimate of $I(\lambda)$ for $\Psi(x) \equiv 1$, which is *independent* of a and b. The change of variable (scaling) $x \rightarrow \lambda^{-1/l} x'$ shows that the only possible estimate for the integral is $O(\lambda^{-1/l})$. The third principle, asymptotics, describes the full asymptotic development of $I(\lambda)$, using both localization and scaling.

Coming back to Eq. (4.50), the main contribution of $I(t)$ comes from the stationary phase:

$$h'_k(n/t; \varphi) = 2\left(-\sin \varphi + \frac{n}{t}\theta'_k(\varphi)\right) = 0, \quad n \in \mathbb{N}, \ k \geq j^* - 1.$$

The φ's that satisfy this equation will be called *resonant* or *critical values*. We cannot simply apply Proposition 2.1 to $I(t)$ since the equivalent Φ and Ψ functions for each integral $I_{k,n}(t)$ in the sum (4.50) depends on the sparseness parameter β_k, which increases superexponentially in k. The principles of localization and scale are still crucial for each k, but their interplay are more subtle as we need to keep track of the β_k dependence.

For $t \in \mathbb{R}$ fixed, let $j^* = j^*(t)$ be such that

$$\beta_{j^*} \leq |t| < \beta_{j^*+1} \tag{4.53}$$

and let us suppose there is no resonances inside the interval $[\varphi_-, \varphi_+]$. Partial integration m times, with m fixed number but as large as one wishes, yields (see [155, Lemmas 4.1 and 4.2])

$$|I_{k,n}(t)| \leq \begin{cases} a^n \tilde{C}_m D^k (\beta_{k-1}/\beta_k)^m & \text{if } k > j^*, \\ a^n \tilde{C}_m D^k (\beta_{k-1}/|t|)^m & \text{if } k = j^* - 1, \end{cases} \tag{4.54}$$

for some constants $\tilde{C}_m$, D and a as in (4.49). When $k = j^*$, the first inequality of (4.54) still holds provided $|t|$ stays away from the extremities of the interval (4.53), i.e., for any $\varepsilon > 0$ if (see [155, Theorem 4.3])

$$\beta_{j^*} (\log \beta_{j^*})^{1+\varepsilon} \leq |t| < \frac{1 - \varepsilon}{\max_{\varphi \in [\varphi_-, \varphi_+]} \sin \varphi} \beta_{j^*+1} \tag{4.55}$$

is not an empty set, then $|I_{j_*,n}(t)| \leq a^n \tilde{C}_m D^{j^*} (\beta_{j^*-1}/\beta_{j^*})^m$ holds for any fixed number m.

Regarding the resonant contributions of $I_{k,n}(t)$, the method employed by Krutikov–Remling in [155] differs very much from ours and, in certain aspects, both treatments complements each other. As explained in the introduction to this subsection, our aim is to search for the least possible sparseness so that the F.S. transform of μ_Ψ decays as (4.42). This is achieved by extending Krutikov–Remling's method in two directions. For each integrals $I_{k,n}$ in which (4.54) holds, we fix the pointwise decay t^{-1} and apply integration by parts as many times as necessary to obtain the least possible sparseness condition. We establish a Gevrey-type estimate for derivatives of the Prüfer angles $\theta_k(\varphi)$ in order to prove that $\tilde{C}_m$ in (4.54) has the following dependence on m

$$\tilde{C}_m = \frac{1}{R_0^2} C_m \delta^{-m} m! \tag{4.56}$$

for $C_m \leq 0.16/m^2$. This is our first improvement.

The second improvement concerns the integrals $I_{j^*,n}$, $n \in \mathbb{N}$, in which there are resonant values inside the interval $[\varphi_-, \varphi_+]$. We apply Lemma 4.1 below in which the scaling principle is embodied. Following the ideas in the proof of Theorem 10.12 of Zygmund's book on trigonometric series, we use Plancherel identity together with Proposition 2.2 in order to obtain the main contribution of the stationary phase, up to a logarithmic correction, provided the integral with the phase $e^{ith(\varphi)}$ replaced by $e^{it\lambda(\varphi)}$ decays as t^{-1}.

Full asymptotic development of $I_{k,n}(t)$ is deferred to a further contribution.

Statement of the main theorem. For J given by (4.1a) let f be a C_0^∞ function with support contained into the essential spectrum $[-2, 2]$ of J. We observe that, if $0 < p < 1$, there exists δ satisfying $0 < \delta < \min_{\varphi^- \le \varphi \le \varphi^+}(a/p - \sqrt{(a/p)^2 - 1})$, by (B.25), such that

$$\sup_{\varphi \in \text{supp} f \circ \lambda} \sup_\theta \frac{p^2}{a + b\cos 2\theta + c\sin 2\theta} < \frac{1}{\delta} \tag{4.57}$$

and

$$\frac{\delta^j}{R_j^2(\varphi)} \longrightarrow 0$$

exponentially fast, as j goes to infinity, uniformly in $\text{supp} f \circ \lambda$. We now state our result.

Theorem 4.7 ([64, Theorem 3.1]). *Let ρ be the spectral measure of J associated with the state e_1 localized at 1 and let f be a smooth function with compact support inside $(-2, 2)$ and such that $0 \notin \text{supp} f$. Suppose that the sequence $(a_j)_{j \ge 1}$, given by (4.1c), satisfies (4.41) with δ as in (4.57) and*

$$\beta_j = \frac{1}{\delta^j} \exp(cj(\ln j)^2) \tag{4.58a}$$

for some $c > 1/2$, as j tends to infinity. Then, there exists a constant C, depending on f and p, such that

$$|\widehat{|f|^2 \, d\rho}(t)| \le C|t|^{-1/2} \, \Omega(|t|) \tag{4.58b}$$

holds for $|t| \ge 2$, where $\Omega(t)$ increases less rapidly than any positive power of t.

Moreover, if the sparseness increments $(\beta_j)_{j \ge 1}$ are chosen to be

$$\beta_j = \frac{1}{\delta^j} \exp(\varepsilon^{-1} j \ln j) \tag{4.58c}$$

for some $\varepsilon > 0$ small enough as j tends to infinity, then the conclusion (4.58b) *holds with the upper bound replaced by $C(1 + |t|)^{-1/2+\varepsilon}$.*

The proof of Theorem 4.7 uses a classical result on the decay of Fourier–Stieltjes coefficients $c_n(dG)$ of a monotone increasing singular continuous function G originated from a Riesz product (see [280, Chap. XII, Theorem 10.12]).

Basic lemma. The following result reduces the resonant estimate to a nonresonant one, studied in Sec. 4.3.2.

Lemma 4.1. *Suppose $G : \mathbb{R} \longrightarrow \mathbb{R}$ is a monotone increasing continuous function, with dG supported in some closed interval $[a, b] \subset (-2, 0) \cup (0, 2)$, whose Fourier–Stieltjes transform satisfies*

$$\widehat{dG}(t) = \int_{-\infty}^{\infty} e^{it\lambda} dG(\lambda) \leq \frac{C}{1 + |t|} \tag{4.59a}$$

for some constant $C < \infty$ and every $t \in \mathbb{R}$. Let $\gamma : \mathbb{R} \longrightarrow \mathbb{C}$ be defined by

$$\gamma(t) = \int_{-\infty}^{\infty} e^{itx(t,\lambda)} dG(\lambda)$$

where

$$x(t, \lambda) = \lambda + \frac{\kappa}{\pi t} \cos^{-1} \frac{\lambda}{2}. \tag{4.59b}$$

If $\kappa = \kappa(t) = O(|t|)$, then

$$|\gamma(t)| \leq \frac{B}{|t|^{1/2}} \ln |t|$$

holds for some $B < \infty$ and every $t \in \mathbb{R}$ with $|t| \geq 2$.

Proof. Denoting by χ the characteristic function of the interval $[a, b]$: $\chi(\lambda) = 1$ if $a \leq \lambda \leq b$ and $= 0$ otherwise, by the Plancherel theorem

$$\gamma(t) = \int_{-\infty}^{\infty} e^{itx(t,\lambda)} \chi(\lambda) dG(\lambda) = \int_{-\infty}^{\infty} \Lambda(t, \tau) \widehat{dG}(\tau) d\tau,$$

where

$$\Lambda(t, \tau) = \frac{1}{2\pi} \int_{a}^{b} e^{i(tx(t,\lambda) + \tau\lambda)} d\lambda. \tag{4.60a}$$

We apply van der Corput estimates in order to obtain the asymptotic behavior of $\Lambda(t,\tau)$ for large t and τ. The integral (4.60a) is of the form

$$2\pi\Lambda(t,\tau) = \int_a^b e^{2\pi i f(\lambda)}d\lambda,$$

$$2\pi f(\lambda) = (t+\tau)\lambda + \frac{\kappa}{\pi}\cos^{-1}\frac{\lambda}{2},$$

where the second derivative f'' of f is strictly negative (under the hypothesis $0 \notin [a,b]$, $\cos^{-1}\lambda/2$ is strictly concave for every $\lambda \in [a,b]$) and proportional to $\kappa = O(t)$, uniformly in τ: $|f''(\lambda)| \geq \rho|\kappa|$ where $2\pi^2\rho = \sup_{\lambda\in[a,b]}\lambda/(4-\lambda^2)^{3/2}$. By Proposition 2.2

$$|\Lambda(t,\tau)| \leq \frac{4}{\sqrt{\rho\kappa}} \leq \frac{K}{\sqrt{1+|t|}} \tag{4.60b}$$

holds for a constant K, independent of t and τ. On the other hand, for $|\tau|$ large compared with $|t|$, let us say $|\tau| > \Delta|t|$ for $\Delta = 1 + |\kappa/t|\sup_{\lambda\in[a,b]}1/(\pi\sqrt{4-\lambda^2})$, the derivative f' of f is of order τ and Proposition 2.2 gives

$$|\Lambda(t,\tau)| \leq \frac{K'}{|\tau|} \tag{4.60c}$$

for some constant K', independent of t and τ.

Estimates (4.60b) and (4.60c), together with (4.59a), yield

$$|\gamma(t)| \leq \int_{|\tau|\leq\Delta|t|} |\Lambda(t,\tau)||\widehat{dG}(\tau)|d\tau + \int_{|\tau|>\Delta|t|} |\Lambda(t,\tau)||\widehat{dG}(\tau)|d\tau$$

$$\leq \frac{K}{\sqrt{1+|t|}}\int_{|\tau|\leq\Delta|t|}\frac{C}{1+|\tau|}d\tau + \int_{|\tau|>\Delta|t|}\frac{K'}{|\tau|}\frac{C}{1+|\tau|}d\tau$$

$$\leq B'|t|^{-1/2}\ln|t| + B''|t|^{-1} \tag{4.60d}$$

for $|t| \geq 2$ and some finite constants B' and B'', concluding the proof of the lemma. □

Remark 4.9. If the pointwise behavior of (4.59a) is replaced by $C/(1+|t|)^{1-\varepsilon}$ for some $\varepsilon > 0$, then the logarithmic correction $\ln|t|$ in the conclusion of lemma has to be replaced by $|t|^\varepsilon$ (see the last inequality of (4.60d)).

Remark 4.10. The function $x = x(\lambda)$ in [280, Chap. XII, Theorem 10.12], whose proof has suggested us Lemma 4.1, is a one-to-one mapping of $[-\pi,\pi]$

onto itself (it does not depend on t, as in our case; see (4.59b)) and, thereby, $\gamma(t) = \widehat{dF}(t)$, $F(x)$ is an increasing function with $F \circ x(\lambda) = \sqrt{2\pi}G(\lambda)$. These facts are not necessary for the conclusion of Lemma 4.1. The hypothesis on $\widehat{dG}(t)$ in that theorem is, in addition, stronger than ours. Because $\widehat{dG}(t)$ decays only "on the average", it is necessary an improved estimate K'/τ^2 instead of a simpler one (4.60c).

4.3.2. *Gevrey-type estimates*

To bring β_j (for large j) down from $\exp(\exp cj)$ to $(j!)^{1/\varepsilon}$ for some $\varepsilon > 0$, we need to establish (4.56) for the coefficients $\tilde{C}_m$ in (4.54).

Derivatives of Prüfer angles. The following propositions gather various crucial estimates.

Proposition 4.6. *Let $\theta_k = \theta_k(\varphi)$, $k = 0, 1, \ldots$, be the sequence of Prüfer angles starting from θ_0. Then, for every $m \geq 1$,*

$$|\theta'_m(\varphi)| \leq C_1 \delta \eta \beta_m \qquad (4.61a)$$

and for every $m > 1$ and $n > 1$

$$\frac{1}{n!}|\theta_m^{(n)}(\varphi)| \leq C_n \eta^n \beta_{m-1}^n \qquad (4.61b)$$

hold uniformly in compact subsets of $(0, \pi)$ with $C_n = K/n^2$, $K \leq 3/(2\pi^2) = 0.151981\ldots$ and

$$\eta = \frac{1 + \Delta}{\delta K},$$

where $\Delta < 1$ ($\Delta = O(\delta)$) is a constant satisfying $\theta'_m(\varphi) \leq (1+\Delta)\beta_m$, which is computable from [180, Proposition 5.2], and δ is suitably small.

Remark 4.11. Proposition 4.6 replaces an unspecified constant C_j appearing in [155, Lemma 3.1] by $K\eta^j j!/j^2$. The precise growth of $\theta_m^{(j)}(\varphi)$ in both j and m is essential for our method.

To prove Proposition 4.6 we need the following lemma.

Lemma 4.2. *Let $\mathbf{C} * \mathbf{D}$ denote the convolution product in $\mathbb{R}^{\mathbb{Z}_+}$:*

$$(\mathbf{C} * \mathbf{D})_n = \sum_{i=0}^{n} C_i D_{n-i} \qquad (4.62a)$$

for $n \geq 0$. If $\mathbf{C}$ has components given by $C_i = K/i^2$, $i \geq 1$, with $C_0 = K$, then

$$\underbrace{\mathbf{C} * \mathbf{C} * \cdots * \mathbf{C}}_{k\text{-}factors} \leq \mathbf{C}$$

holds for every $k \geq 1$ provided $K \leq 1/(2 + 2\pi^2/3)$. If $C_0 = 0$, then same result holds with $K \leq 3/(2\pi^2)$.

Remark 4.12. By definition (4.62a), the conclusion of the so-called *taming convolution lemma*, Lemma 4.2, can be written as

$$\sum_{\substack{i_1, \ldots, i_k \geq 0 \\ i_1 + \cdots + i_k = n}} C_{i_1} \cdots C_{i_k} \leq C_n. \tag{4.62b}$$

Remark 4.13. We have stated Lemma 4.2 for sequences $\mathbf{C} = (C_0, C_1, \ldots)$ $\in \mathbb{R}^{\mathbb{Z}+}$ with the zeroth component $C_0 = 0$ and $C_0 = K$ since both cases will appear in the following (see Proposition 4.8 below for the case $C_0 \neq 0$). Lemma 4.2 plays a key role in estimates involving higher-order chain rule.

Proof of Lemma 4.2. By definition,

$$\frac{1}{C_n} \sum_{i=0}^{n} C_i C_{n-i} = K \left(2 + \sum_{i=1}^{n-1} \frac{n^2}{i^2(n-i)^2} \right). \tag{4.63a}$$

For any real numbers a and b, we have $0 \leq (a-b)^2 = 2(a^2 + b^2) - (a+b)^2$ and this inequality with $a = 1/i$ and $b = 1/(n-i)$ implies

$$\frac{n^2}{i^2(n-i)^2} = \left(\frac{1}{i} + \frac{1}{n-i} \right)^2 \leq 2 \left(\frac{1}{i^2} + \frac{1}{(n-i)^2} \right). \tag{4.63b}$$

Replacing (4.63b) into (4.63a), yields

$$\frac{1}{C_n} \sum_{i=0}^{n} C_i C_{n-i} \leq 2K \left(1 + \frac{\pi^2}{3} \right) \leq 1 \tag{4.63c}$$

provided $K \leq 1/(2 + 2\pi^2/3)$. When $C_0 = 0$, the terms with $i = 0$ and n do not contribute to the sum and the inequality (4.63c) holds provided $K \leq 3/(2\pi^2)$. Once we have $\mathbf{C} * \mathbf{C} \leq \mathbf{C}$, Lemma 4.2 is proved by induction. $\square$

Proof of Proposition 4.6. The proof uses the recursive relation introduced in (B.24b):

$$\theta_m(\varphi) = g \circ \theta_{m-1}(\varphi) + \beta_m\varphi, \tag{4.64a}$$

where $g = g(\varphi, \theta)$ is given by (B.24c), together with the Scott's formula for higher-order chain rule (see, e.g., [86])

$$(g \circ f)^{[n]} = \sum_{k=1}^{n} g^{[k]} \circ f \sum_{\substack{i_1,\dots,i_k \geq 1 \\ i_1+\cdots+i_k=n}} f^{[i_1]}\dots f^{[i_k]} \tag{4.64b}$$

where, from here on, $h^{[n]}$ stands for $h^{(n)}/n!$, the nth derivative of h divided by $n!$.

Upper and lower bounds for the first derivative has been provided in [180]:

$$(1 - \Delta)\beta_m \leq \theta'_m(\varphi) \leq (1 + \Delta)\beta_m \tag{4.64c}$$

with $\Delta < 1$ a constant. Now, choosing $\eta = \frac{1+\Delta}{\delta K}$, (4.64c) establishes (4.61a) for every $m \geq 1$.

Since $(\beta_j)_{j\geq 1}$ is a fast increasing sequence, we apply the Scott's formula to g in (B.24c) as it were a function of a single variable θ. This really gives the main contribution to the derivatives. Now g as a function of $z = e^{i\theta}$ may be analytically continued to the complement of a disc of radius strictly less than $1 - e/\xi < 1$. The derivatives of $q(e^{i\theta}) = g(\theta)$ with respect to θ may be estimate by Cauchy formula:

$$|g^{[k]}(\theta)| \leq c_1\xi^k \tag{4.64d}$$

holds for $k \geq 1$ with c_1 as small as one wishes, by increasing ξ accordingly ((4.64d) can be bounded, e.g., by $\varepsilon(c_1\xi/\varepsilon)^k = \varepsilon\bar{\xi}^k$, for any $\varepsilon > 0$). Replacing f by θ_{m-1} in (4.64b) gives

$$\theta_m^{[n]}(\varphi) = \sum_{k=1}^{n} g^{[k]} \circ \theta_{m-1}(\varphi) \sum_{\substack{i_1,\dots,i_k \geq 1 \\ i_1+\cdots+i_k=n}} \theta_{m-1}^{[i_1]}(\varphi)\cdots\theta_{m-1}^{[i_k]}(\varphi). \tag{4.64e}$$

We prove (4.61b) by induction in n. Consider the case $n = 2$, for any $m > 1$. By Eq. (4.64a), together with (4.64d) and (4.61a), we have

$$\theta_m'' = g'' \circ \theta_{m-1} \cdot (\theta'_{m-1})^2 + g' \circ \theta_{m-1} \cdot \theta''_{m-1}$$
$$\leq c_1\xi^2 C_1^2 \delta^2 \eta^2 \beta_{m-1}^2 + g' \circ \theta_{m-1} \cdot \theta''_{m-1}.$$

The iteration of this relation, together with (4.64d) and (4.57), yields

$$\theta''_m \leq c_1 \xi^2 C_1^2 \delta^2 \eta^2 \beta_{m-1}^2 \sum_{j=1}^{m-1} (c_1 \xi)^{j-1} \frac{\beta_{m-j}^2}{\beta_{m-1}^2}$$

$$\leq \frac{c_1 \xi^2 \delta^2}{1 - c_1 \xi \delta^2} C_1^2 \eta^2 \beta_{m-1}^2$$

$$\leq 2 C_2 \eta^2 \beta_{m-1}^2$$

provided δ is chosen so small that $c_1 \xi \delta^2 < 1$ and

$$2K \frac{c_1 \xi^2 \delta^2}{1 - c_1 \xi \delta^2} \leq 1$$

are both satisfied, establishing (4.61b) for $n = 2$.

Now, suppose

$$\theta_m^{[j]}(\varphi) \leq C_j \eta^j \beta_{m-1}^j$$

holds for $m > 1$ and $j = 2, \ldots, n-1$ and we shall establish the inequality for n. By this assumption together with (4.57), we have

$$\theta_{m-1}^{[j]}(\varphi) \leq C_j \eta^j \beta_{m-2}^j = C_j \eta^j \left(\frac{\beta_{m-2}}{\beta_{m-1}} \right)^j \beta_{m-1}^j \leq C_j (\delta \eta)^j \beta_{m-1}^j. \quad (4.64f)$$

Plugging (4.61a), (4.64d) and (4.64f) into (4.64e), together with (4.62b), yields

$$\theta_m^{[n]}(\varphi) \leq c_1 (\delta \eta)^n \beta_{m-1}^n \sum_{k=2}^{n} \xi^k \sum_{\substack{i_1, \ldots, i_k \geq 1 \\ i_1 + \cdots + i_k = n}} C_{i_1} \cdots C_{i_k} + g' \circ \theta_{m-1} \cdot \theta_{m-1}^{[n]}$$

$$\leq c_1 \frac{\xi}{\xi - 1} C_n (\delta \xi \eta)^n \beta_{m-1}^n + g' \circ \theta_{m-1} \cdot \theta_{m-1}^{[n]}.$$

Here, we have separated the term with $k = 1$ which applies n derivatives on θ_{m-1}. Note that, for all the other terms with $k \geq 2$, we have $i_1, \ldots, i_k \geq 1$ and the derivatives applied on the θ_{m-1} are of order strictly smaller than n. The iteration of this relation gives

$$\theta_m^{[n]}(\varphi) \leq c_1 \frac{\xi}{\xi - 1} C_n (\delta \xi \eta)^n \beta_{m-1}^n \sum_{j=1}^{m-1} (c_1 \xi)^{j-1} \frac{\beta_{m-j}^n}{\beta_{m-1}^n}$$

$$\leq C_n \eta^n \beta_{m-1}^n.$$

provided

$$c_1 \frac{\xi}{\xi - 1} (\xi \delta)^n \frac{1}{1 - c_1 \xi \delta^n} \leq 1 \qquad (4.64\text{g})$$

holds for every $n > 2$. We pick δ satisfying both (4.57) and (4.64g) for $n \geq 2$, concluding the proof of Proposition 4.6. $\qquad \square$

Remark 4.14. The well-known formula for higher derivative of composite functions, Faà di Bruno's formula, cannot be used recursively since the constant η in (4.61b) deteriorates each time it is applied (see [20, Sec. 6.2, (6.10)]). The proof of (4.61b) using Scott's formula, which is reminiscent of KAM theory, was based on yet unpublished manuscript "Global formal series solution of a singularly perturbed first order partial differential equation coming from a problem in Statistical Mechanics" by D. H. U. Marchetti and W. R. P. Conti.

Miscellaneous collection of estimates. To obtain the t^{-1} decay from the summation in (4.47), it is necessary to apply an arbitrarily large number of the integration by parts for integrals of the type (4.50):

$$\int_0^\pi f_0 e^{ith} d\varphi = i \int_0^\pi \left(\frac{1}{th'} f_0 \right)' e^{ith} d\varphi,$$

where $\operatorname{supp} f_0 = [\varphi_-, \varphi_+] \subset (0, \pi)$. The following propositions gather tools to implement the estimate.

Proposition 4.7. *Let $f_0(\varphi)$ and $\varrho(\varphi)$ be, respectively, C^∞ complex and real-valued functions on $[0, \pi)$ and let $L = \frac{d}{d\varphi} \varrho(\varphi)$ be an operator defined in this space. If*

$$f_n = \frac{1}{n!} L^n f_0 = \frac{1}{n!} \frac{d}{d\varphi} \varrho \frac{d}{d\varphi} \varrho \cdots \frac{d}{d\varphi} \varrho f_0$$

denotes the nth application of L over f_0, divided by factorial of n, $n = 0, 1, \ldots$, then

$$f_n = \sum_{\substack{k_1, \ldots, k_n, p_n \geq 0 \\ k_1 + \cdots + k_n + p_n = n}} \varrho^{[k_1]} \cdots \varrho^{[k_n]} f_0^{[p_n]}. \qquad (4.65)$$

See [64, Proposition 3.12] for a proof. In our application, $\varrho = \frac{1}{th'_{m+1}(\varphi)}$ and $f_0 = |f(2 \cos \varphi)|^2 \frac{\sin^2 \varphi}{R_m^2(\varphi)} A^n(\varphi)$ (or its complex conjugate), where $h_k(\varphi)$, $R_k(\varphi)$ and $A(\varphi)$ are defined by (4.51b), (4.46c) and (4.48), respectively. The main contribution for $m \geq j^* + 1$, where $j^* = j^*(t)$ is defined by (4.53), comes from the derivatives of the Prüfer angles $\theta_k(\varphi)$ and in this case it is thus sufficient to consider

$$\varrho = \frac{1}{\theta'_{m+1}(\varphi)} \equiv s \circ \theta'_{m+1}(\varphi). \tag{4.66}$$

For $m < j^*$, we have

$$th'_{m+1}(\varphi) = -2t \sin \varphi + n\theta'_{m+1} = -2t \sin \varphi(1 + O(1)) \tag{4.67a}$$

and the derivatives of higher order

$$th^{(k)}_{m+1}(\varphi) - n\theta^{(k)}_{m+1} = \begin{cases} 2t(-1)^{(k+1)/2} \sin \varphi & \text{if } k \text{ is odd,} \\ 2t(-1)^{k/2} \cos \varphi & \text{if } k \text{ is even} \end{cases}$$

satisfy, in view of (4.61b),

$$th^{(k)}_{m+1}(\varphi) \leq 2|t| + nC_k \eta^k \beta_m^k k!. \tag{4.67b}$$

It is thus sufficient to consider in both cases

$$f_0 = \frac{1}{R_m^2(\varphi)} = \frac{1}{R_0^2} \prod_{j=1}^{m} F \circ \theta_j(\varphi)$$

by (4.46c) and (B.24d), where $F(\theta)$ satisfies, by direct computation,

$$F^{[k]} \leq \frac{1 - \delta/\zeta}{\delta} \left(\frac{\zeta}{\delta}\right)^k \tag{4.68}$$

for $k \geq 0$ and some positive number ζ. We also need

$$s^{[k]}(x) = \frac{(-1)^k}{x^{k+1}}, \quad k = 1, \ldots, \tag{4.69}$$

where $s(x) = 1/x$ has been introduced in (4.66).

Proposition 4.8. *Let $\theta_k = \theta_k(\varphi)$, $k = 0, 1, \ldots,$ be the sequence of Prüfer angles and let η, δ and $\{C_n\}$ be the constants that appear in Proposition 4.6. Then, there exist positive numbers d and $\hat{\eta}$, which can be expressed in terms* *of the previous constants, such that (with $\varrho = \varrho^{[0]} \le d/\beta_{m+1}$)*

$$\varrho^{[n]} = (s \circ \theta'_{m+1})^{[n]} \le \frac{d}{\beta_{m+1}} C_n \hat{\eta}^n \beta_m^n \qquad (4.70\text{a})$$

as well as ($f_0 = f_0^{[0]} \le R_0^{-2} \delta^{-m}$)

$$f_0^{[n]} = \frac{1}{R_0^2} \left(\prod_{j=1}^m F \circ \theta_j \right)^{[n]} \le \frac{1}{R_0^2} C_n (\zeta\eta)^n \frac{1}{\delta^m} \beta_m^n \qquad (4.70\text{b})$$

hold for every non-negative integer n, with ζ as in (4.68).

Proof. These inequalities are established as in Proposition 4.6, by using the Scott's formula. We begin with (4.70a). If $\tilde{\eta}$ is the smallest constant such that $(i-1)^2 \eta^i / i \le \tilde{\eta}^i$ holds for every $i \ge 1$, by (4.61b), (4.64f) and (4.64b) with g and f replaced by s and θ'_m, we have

$$\varrho^{[n]} = \sum_{k=1}^n s^{[k]} \circ \theta'_{m+1} \sum_{\substack{i_1,\ldots,i_k \ge 1 \\ i_1 + \cdots + i_k = n}} (i_1 + 1)\theta_{m+1}^{[i_1+1]} \cdots (i_k + 1)\theta_{m+1}^{[i_k+1]}$$

$$\le \tilde{\eta}^n \beta_m^n \sum_{k=1}^n \frac{1}{(\theta'_{m+1})^{k+1}} \tilde{\eta}^k \beta_m^k \sum_{\substack{i_1,\ldots,i_k \ge 1 \\ i_1 + \cdots + i_k = n}} C_{i_1} \cdots C_{i_k}$$

$$\le dC_n \hat{\eta}^n \frac{\beta_m^n}{\beta_{m+1}},$$

where $\hat{\eta} = \delta\tilde{\eta}^2 / (1 - \Delta)$ and $d = \delta\tilde{\eta}/(\delta\tilde{\eta} + \Delta - 1)$. In the third inequality we have used the lower bound (4.64c) for θ'_m and (4.62b). Note $\delta\tilde{\eta} > \delta\eta > 1$, by definition of η in Proposition 4.6.

For (4.70b), we start with the Scott's formula (4.64b) with g and f replaced by F and θ_j which, together with (4.61b), (4.64f) and (4.68), gives

$$(F \circ \theta_j)^{[n]} = \sum_{k=1}^n F^{[k]} \circ \theta_j \sum_{\substack{i_1,\ldots,i_k \ge 1 \\ i_1 + \cdots + i_k = n}} \theta_j^{[i_1]} \cdots \theta_j^{[i_k]}$$

$$\le C_n (\delta\eta)^n \beta_j^n \frac{1 - \delta/\zeta}{\delta} \sum_{k=1}^n \left(\frac{\zeta}{\delta} \right)^k$$

$$\le \frac{1}{\delta} C_n (\zeta\eta)^n \beta_j^n.$$

Now we take the nth derivative of the product. For this, we use the variation of Lemma 4.2 mentioned in Remark 4.13:

$$\left(\prod_{j=1}^{m} F \circ \theta_j(\varphi) \right)^{[n]} = \sum_{\substack{n_1,\ldots,n_m \geq 0 \\ n_1 + \cdots + n_m = n}} (F \circ \theta_1)^{[n_1]} \cdots (F \circ \theta_m)^{[n_m]}$$

$$\leq C_n (\zeta \eta)^n \frac{1}{\delta^m} \beta_m^n \tag{4.71}$$

concluding the proof of this proposition.   □

Remark 4.15. An estimate of (4.70a) with $\varrho = 1/t h'_{m+1}(\varphi)$ for $m < j^*$ is analogously given by

$$\varrho^{[k]} = (s \circ t h'_{m+1})^{[k]} \leq \frac{d_n}{|t|} C_k \hat{\eta}^k \beta_m^k \tag{4.72a}$$

with $d_n = 2dn/c$ where $c = \min_{\varphi \in \text{supp} f \circ \lambda} 2 \sin \varphi$. Note that, for any fixed m and t satisfying (4.53), the second term of the l.h.s. of (4.67b) rapidly overcomes t. On the other hand, an estimate for k in which t still dominates (4.67b) is, by the Scott's formula (4.64b), much better than (4.72a):

$$\varrho^{[k]} = \frac{1}{t} \sum_{l=1}^{k} s^{[l]} \circ h'_{m+1} \sum_{\substack{i_1,\ldots,i_l \geq 1 \\ i_1 + \cdots + i_l = k}} \frac{1}{i_1!} h_{m+1}^{(i_1+1)} \cdots \frac{1}{i_l!} h_{m+1}^{(i_l+1)}$$

$$\leq \frac{2^k}{|t|} \sum_{l=1}^{k} \frac{2^l}{c^{l+1}} \sum_{\substack{i_1,\ldots,i_l \geq 1 \\ i_1 + \cdots + i_l = k}} \frac{1}{i_1!} \cdots \frac{1}{i_l!}$$

$$\leq \frac{2^k}{|t|} \frac{1}{k!} \sum_{l=1}^{k} \frac{2^l l^k}{c^{l+1}}. \tag{4.72b}$$

Let us put all together. Plugging (4.70a) and (4.70b) into (4.65), deduced in Proposition 4.7 by applying n times integration by parts $n! f_n = L^n f_0$ to the integrand f_0 of (4.50), we arrive at the following estimate:

$$n! |f_n| \leq n! \sum_{\substack{k_1,\ldots,k_n,p_n \geq 0 \\ k_1 + \cdots + k_n + p_n = n}} |\varrho^{[k_1]} \cdots \varrho^{[k_n]} f_0^{[p_n]}|$$

$$\leq \frac{1}{R_0^2} D^n \frac{1}{\delta^m} \left(\frac{\beta_m}{\beta_{m+1}} \right)^n n! \sum_{\substack{k_1,\ldots,k_n,p_n \geq 0 \\ k_1+\cdots+k_n+p_n=n}} C_{k_1} \cdots C_{k_n} C_{p_n}$$

$$\leq \frac{1}{R_0^2} C_n D^n \frac{1}{\delta^m} \left(\frac{\beta_m}{\beta_{m+1}} \right)^n n!, \tag{4.73}$$

where $D = d \cdot \max(\hat{\eta}, \zeta\eta)$. Estimate (4.73) will be used to get an upper bound for all nonresonant integrals of (4.47).

Remark 4.16. As $C_n = K/n^2$, $n \geq 1$, $(C_0 = K)$ with $K \leq 1/(2 + 2\pi^2/3)$ are bounded constants, (4.73) makes explicit the dependence on the number n of times that integration by parts is applied to integral of type (4.50). Explicit dependence of n was not necessary in [155], since n is an arbitrarily large but fixed number. Apart this, (4.73) agrees with the estimate used in [155, p. 522].

4.3.3. *Proof of Theorem 4.7*

Let t be a fixed number. We assume t positive but the negative value can be dealt similarly. Let f be a C^∞ function with compact support in $(0, 2)$ and let $I_f = [\varphi_-, \varphi_+]$ be smallest closed interval that contains supp $f \circ \lambda$, $\lambda(\varphi) = 2\cos\varphi$. Since the spectral measure is symmetric, $d\rho(-\lambda) = d\rho(\lambda)$, we need only to consider f supported in one-half of the essential spectrum. We have excluded the origin to avoid that the curvature of $\cos^{-1}\lambda/2$ in (4.59b) vanishes (see observation right before (4.60b)).

For $j^* = j^*(t)$ defined by (4.53), let $n^* = n^*(t)$ be given by

$$(n^* - 1)\beta_{j^*} \leq t < n^*\beta_{j^*}. \tag{4.74a}$$

Since $\beta_{j^*}/t > 1/n^*$ there are at most n^* points $\varphi_1, \ldots, \varphi_{n^*}$ in the support of $f \circ \lambda$ satisfying

$$-\sin\varphi_l + \frac{l\theta'_{j^*}}{t} = 0. \tag{4.74b}$$

Observe that $1 \leq n^*(t) \leq \beta_{j^*+1}/\beta_{j^*} + 1$, by (4.53) and (4.74a). For the sparseness increment β_j in (4.58a), we have

$$j^*(t) = \frac{\ln t}{c \ln^2 \ln t} \left(1 + O\left(\frac{\ln \ln \ln t}{\ln \ln t} \right) \right)$$

and, consequently, the number of resonant values n^* is a monotone nondecreasing function of t in each interval $(\beta_{j^*}, \beta_{j^*+1}]$. Let $L(t)$ denote the continuous interpolation of $n^*(t)$. It follows from these observations that L is a piecewise linear function with inclination $1/\beta_j$ satisfying

$$1 < L(t) < Ee^{c\ln^2 \ln t} \equiv \Omega^2(t), \quad \beta_j < t \le \beta_{j+1} \qquad (4.74c)$$

for some constant E, independent of t. The inequality (4.74c) will be used at the end of this section.

By (4.47), the Fourier–Stieltjes transform of ρ can be written as

$$\widehat{|f|^2 d\rho}(t) = I_{j^*-1,0}(t) + \sum_{j=j^*-1}^{\infty} \sum_{n=1}^{\infty} (I_{j,n}(t) + \bar{I}_{j,n}(-t))$$

where

$$I_{j,n}(t) = \frac{1}{\pi} \int_0^\pi |f(2\cos\varphi)|^2 \frac{\sin^2 \varphi}{R_j^2} A^n(\varphi) e^{2it(\cos\varphi + n\theta_{j+1}/t)} d\varphi. \qquad (4.75a)$$

We apply integration by parts to all terms of this sum not satisfying the resonant condition (4.74b). Since the support of f is compact, every boundary term vanishes. Integration by parts may be repeated N_j times depending on the index j of the sum. Propositions 4.7 and 4.8, together with its combined estimate (4.73), can be used to get an upper bound for each integral (4.75a) with $j \ge j^*$. This yields

$$\sum_{j=j^*}^{\infty} \sum_{n=1}^{\infty} \left(|I_{j,n}| + |\bar{I}_{j,n}(-t)| \right)$$

$$\le 2 \sum_{j=j^*}^{\infty} \left(\sum_{n=1}^{\infty} \frac{1}{n} a^n \right) \frac{1}{R_0^2} C_{N_j} D^{N_j} \frac{1}{\delta^j} \left(\frac{\beta_j}{\beta_{j+1}} \right)^{N_j} N_j! \qquad (4.75b)$$

where $a = \sup_{\varphi \in I_f} |A(\varphi)| < 1$. We need the following lemma.

Lemma 4.3. *Let $(\beta_j)_j$ be a sequence given by (4.58a) with $c > 1/2$. If j^* is sufficiently large, then*

$$D^{N_j} \frac{1}{\delta^j} \left(\frac{\beta_j}{\beta_{j+1}} \right)^{N_j} N_j! \le \frac{1}{\beta_{j+1}} \qquad (4.76)$$

holds for $N_j = j + 1$ and $j \ge j^*$.

Proof. Applying the Stirling formula to

$$\Upsilon_j := \beta_{j+1} \left(\frac{\beta_j}{\beta_{j+1}} \right)^{j+1} (j+1)!,$$

where β_j is given by (4.58a) with $c > 1/2$, yields

$$\Upsilon_j = \frac{\delta^j}{\sqrt{2\pi(j+1)}} \left(\frac{j+1}{e} \right)^{j+1} \exp(-2c(j+1)\ln(j+1)) \left(1 + O\left(\frac{\ln j}{j} \right) \right)$$

$$(4.77)$$

which goes to 0 as j tends to ∞ faster than exponential and implies (4.76) provided j is large enough. $\qquad \square$

Problem 4.3. *Prove* (4.77).

Problem 4.4. *For* $(\beta_j)_{j\geq 1}$ *given by* (4.58c), *prove that*

$$\beta_{j+1}^{1-\varepsilon} \left(\frac{\beta_j}{\beta_{j+1}} \right)^{j+1} = \delta^{\varepsilon(j+1)} \exp(-(j+1)\ln(j+1) - \varepsilon^{-1}(j+1))$$

$$< \frac{1}{N_j!} \delta^j D^{-j-1}$$

and (4.76) *holds with* $1/\beta_{j+1}$ *replaced by* $1/\beta_{j+1}^{1-\varepsilon}$, *provided* $1 + 1/\varepsilon > \ln D - (1-\varepsilon)\ln\delta$ *and* j *is large enough.*

Together with Remark 4.9, the proof of $t^{-1+\varepsilon}$ decaying for the case of $(\beta_j)_{j\geq 1}$ given by (4.58c) may continued exactly as for decaying t^{-1}. We shall consider only the latter case.

If the sequences $(\beta_j)_j$ and $(N_j)_j$ are chosen as in Lemma 4.3, then the series in (4.75b) converges uniformly in t. By (4.57) and (4.53), we have

$$\frac{1}{\beta_{j+1}} = \frac{1}{\beta_{j^*+1}} \frac{\beta_{j^*+1}}{\beta_{j^*+2}} \cdots \frac{\beta_j}{\beta_{j+1}} \leq \delta^{j-j^*} \frac{1}{\beta_{j^*+1}}.$$

and $t/\beta_{j^*+1} < 1$. Consequently,

$$\sum_{j=j^*}^{\infty} \sum_{n=1}^{\infty} (|I_{j,n}| + |\bar{I}_{j,n}(-t)|) \leq \frac{2a}{1-a} \frac{1}{R_0^2} \sum_{j=j^*}^{\infty} C_{N_j} \frac{1}{\beta_{j+1}}$$

$$\leq \frac{2a}{1-a} \frac{1}{R_0^2} \frac{t}{\beta_{j^*+1}} \frac{K}{t} \sum_{l=0}^{\infty} \delta^l \leq \frac{C}{t} \qquad (4.78)$$

holds with $C < \infty$ independent of t.

For $j < j^*$, (4.67a) holds and we need to replace the estimate (4.70a) by (4.72a) and t occupies now the place of β_{m+1} in (4.73). Applying successive integration by parts to $I_{j^*-1,0}$ gives, analogously

$$|I_{j^*-1,0}| \le \frac{1}{R_0^2} C_{N_{j^*-1}} D^{N_{j^*-1}} \frac{1}{\delta^{j^*-1}} \left(\frac{\beta_{j^*-1}}{t}\right)^{N_{j^*-1}} N_{j^*-1}! \le \frac{C'}{t} \quad (4.79a)$$

for some constant C'. Note that, by (4.53),

$$\left(\frac{\beta_{j^*-1}}{t}\right)^{N_{j^*-1}} \le \frac{1}{t} \frac{\beta_{j^*-1}^{N_{j^*-1}}}{\beta_{j^*}^{N_{j^*-1}-1}}$$

and by (4.76)

$$D^{N_k} \frac{1}{\delta^k} \frac{\beta_k^{N_k}}{\beta_{k+1}^{N_k-1}} N_k! \le 1 \quad (4.79b)$$

for k large enough.

It remains to estimate the sum $S_{j^*}(t) = \sum_{n=1}^{\infty}(I_{j^*-1,n}(t) + \bar{I}_{j^*-1,n}(-t))$ which contains the most significant terms responsible for $t^{-1/2}$ decaying behavior. To extract this decay we write

$$I_{j^*-1,n}(t) = \frac{1}{\pi} \int_0^\pi |f(2\cos\varphi)|^2 \frac{\sin^2\varphi}{R_{j^*-1}^2} B_{j^*-1}^n(\varphi) e^{2it(\cos\varphi + n\beta_{j^*}\varphi/t)} d\varphi$$

$$(4.79c)$$

(analogously for $\bar{I}_{j^*-1,n}(t)$) where, by (4.64a), B_k is a function of the Prüfer angles $\theta_k(\varphi)$ such that $|B_k| = |A|$ and

$$\arg B_k = \arg A + \theta_{k+1}(\varphi) - \beta_{k+1}\varphi$$
$$= \arg A + g \circ \theta_k(\varphi)$$

with g given by (B.24c). We then apply Lemma 4.1 to (4.79c) with

$$d(G_n \circ \lambda)(\varphi) = \frac{1}{\pi} |f(2\cos\varphi)|^2 \frac{\sin^2\varphi}{R_{j^*-1}^2} B_{j^*-1}^n(\varphi) d\varphi$$

and

$$tx(t,\lambda) = t\lambda + 2n\beta_{j^*} \cos^{-1}\frac{\lambda}{2}.$$

Note that, by (4.53), $\kappa = 2\pi n\beta_{j^*}$ and $n^*\beta_{j^*} = O(t)$. In order to fulfill all assumptions of Lemma 4.6 it remains to show that $\widehat{dG_n}(t)$ decays as $|t|^{-1}$ (see (4.59a)).

We estimate the Fourier–Stieltjes transform

$$\widehat{dG_n}(t) = \frac{1}{\pi} \int_0^\pi |f(2\cos\varphi)|^2 \frac{\sin^2\varphi}{R_{j^*-1}^2} B_{j^*-1}^n(\varphi) e^{2it\cos\varphi} d\varphi \qquad (4.80)$$

as the nonresonant integrals (4.75a) with $j < j^*$ (see Remark 4.15). The estimate (4.70a) is replaced by (4.72b) and

$$f_0 = |f(2\cos\varphi)|^2 \frac{\sin^2\varphi}{R_{j^*-1}^2(\varphi)} A^n(\varphi) \exp(ing \circ \theta_{j^*-1}(\varphi))$$

includes now an extra exponential term depending on $\theta_{j^*-1}(\varphi)$.

To deal with this new term we need some more estimates. We refer to [64], (4.12) and the following equations, for details. Since the required estimates did not change significantly, we integrate (4.80) by parts N_{j^*-1} times and use (4.73) with (4.72b) in the place of (4.70a) to get, exactly as for (4.79a),

$$|\widehat{dG_n}(t)| \le \frac{C''}{t} d^k \tilde{a}^n, \quad \beta_k \le t < \beta_{k+1}$$

for some constants $C'' < \infty$, $d < 1$ and $\tilde{a} = e^{c_1} \sup_{\varphi \in I_f} |A(\varphi)| < 1$, as c_1 is arbitrarily small by the observation after (4.64d). Here, we have used the fact that (4.79b) holds with D replaced by D/d, for any $d > 1$, provided k is large enough. This immediately implies, by a slight modification of Lemma 4.1 (see Remark 4.9),

$$|I_{j^*-1,n}(t)| \le \int_{|\tau| \le \Delta|t|} |\Lambda(t,\tau)| |\widehat{dG}(\tau)| d\tau + O(1/t)$$

$$\le \frac{2K}{\sqrt{\kappa}} \tilde{a}^n \sum_{k=0}^{j^*} d^k \int_{\beta_k}^{\beta_{k+1}} \frac{C''}{\tau} d\tau + O(1/t)$$

$$\le \frac{2KC''}{\sqrt{n^*\kappa}} n^* \tilde{a}^n \sum_{k=0}^{\infty} d^k \ln\frac{\beta_{k+1}}{\beta_k} + O(1/t)$$

and by (4.74a) and the fact that $\ln\beta_{k+1}/\beta_k = O(\ln^2 k)$, we have

$$|S_{j^*}(t)| \le \sum_{n=1}^{\infty} (|I_{j^*-1,n}(t)| + |\bar{I}_{j^*-1,n}(-t)|)$$

$$\le \frac{C}{\sqrt{|t|}} \Omega(|t|),$$

where $\Omega(t)$ is defined in (4.74c).

Now, we show that $\Omega(t)$ increases slower than t^ε, for any $\varepsilon > 0$. Suppose, by contradiction, that $\lim_{t\to\infty} \Omega(t)/t^\varepsilon = k > 0$ holds for some $\varepsilon > 0$. Then, by L'Hospital,

$$\lim_{t\to\infty} \frac{\Omega(t)}{t^\varepsilon} = \lim_{t\to\infty} \frac{\Omega'(t)}{\varepsilon t^{\varepsilon-1}} = \frac{c}{\varepsilon} \lim_{t\to\infty} \frac{\Omega(t)}{t^\varepsilon} \cdot \lim_{t\to\infty} \frac{\ln\ln t}{\ln t} = 0,$$

concluding the proof of Theorem 4.7. $\square$

Chapter 5

Resonances and Quasi-exponential Decay

5.1. Introduction

As remarked by Harrell in his splendid review [110], "quantum theory makes a sharp distinction between bound states and scattering states, the former associated with point spectrum and the latter with continuous spectrum. Resonances, associated with quasi-stationary states, bridge this distinction, and have posed mathematical challenges since the beginning of Schrödinger theory."

The well-known approach to resonances through dilation-analyticity has met many of the above-mentioned challenges. Since, however, several excellent and comprehensive textbook treatments of this approach exist [54, 210], we shall not mention it further in this book.

On the other hand, in the dilation-analyticity approach, the analytic continuation which is necessary to define a resonance depends crucially on very restrictive conditions on the tail of the potential. As remarked by Lavine [165, 166], such conditions do not seem to be physically relevant, and it is not clear whether they are necessary for quasi-exponential decay in time.

As remarked by Harrell [110], Lavine's approach [165, 166] is conceptually distinct because it takes quasi-stationary time evolution of the wave-function as the starting point: it is based on the notion of sojourn time (see (3.85a) et seq., as well as the forthcoming (5.37a) et seq.).

Another approach, which also takes quasi-stationary time-evolution as the starting point, but from a different point of view, that of Gamow vectors, has been proposed in mathematical physics about 25 years ago, by Skibsted [249], and from the point of view of rigged Hilbert spaces by several authors (see [66] and references given there). Recently, Gamow vectors have been revisited in mathematical physics from a very appealing standpoint, see [76, 91]. Even more recently, a natural representation of the wave-function in terms of its transseries has been defined, in the context of Borel summability [50]. The dispersive part of the transseries is an asymptotic series in powers of $t^{-1/2}$, of which the (dispersive part of) the wave-function is the Borel sum, and the exponential part defines Gamow vectors nonperturbatively in a physically relevant way. The theory has been applied to model systems of atoms or ions in external periodic or quasi-periodic fields from [121], and this chapter presents a (hopefully comprehensive) outline of their work.

The notion of Gamow vectors was introduced by Gamow [90] in order to explain alpha decay, i.e., the process whereby a parent nucleus A decays spontaneously into an alpha particle and a (daughter) nucleus D, $A \rightarrow \alpha + D$. Treating the alpha particle as a single particle and the daughter nucleus as fixed, one may approximate the (radial) potential between the daughter nucleus and the alpha particle by

$$V(r) = \begin{cases} -Z_\alpha Z e^2 / R & \text{if } 0 \leq r \leq R, \\ Z_\alpha Z e^2 / r & \text{if } r \geq R, \end{cases}$$

where Z is the charge of the daughter nucleus. For an alpha particle of energy $0 < E_0 < \frac{Z_\alpha Z e^2}{R}$, we have a "barrier potential" whose treatment is qualitatively similar to the forthcoming (5.33), see also [33, Fig. 2]. With Gamow's Ansatz for the solution of the Schrödinger equation

$$u(r) = \exp(-ikr) - S \exp(ikr) \quad \text{with}$$

$$k = \frac{\sqrt{2mE}}{\hbar} \quad \text{and} \quad \Re e\, k > 0 \qquad \text{for } r \rightarrow \infty,$$

it follows that for $E = z = E_0 - i\Gamma/2$, S is supposed to have a pole, and u is a solution without incoming wave, i.e., satisfying the boundary condition (b.c.) $u(0) = 0$, $u(r) \rightarrow \exp(ikr)$ as $r \rightarrow \infty$. If one sets in the above potential the "nuclear radius" $R = 1.5 \times 10^{-15} A^{1/3} m$ as derived from Rutherford's experiments, where A is the number of nucleons in the nucleus, together with Gamow's Ansatz, values for the binding energy E_0 and lifetime Γ^{-1} for alpha decay of atomic nuclei A are obtained with reasonable agreement with experiment: we refer to [33, p. 214] et seq. for a more detailed discussion. In particle physics, the notion of resonance is particularly significant: except for a few (electron, u, d, photon, ...) most particles are transient. No satisfactory theory of resonances in relativistic quantum field theory (rqft) exists.

5.2. The Model System

Consider a one-dimensional particle described by the Hamiltonian

$$H = -\partial_x^2 + V(x), \tag{5.1a}$$

where $\Psi(x,0)$ has compact support, and the potential V is also assumed to be of compact support: it is supposed to model barriers or wells. The Schrödinger equation for H is

$$i\frac{\partial\Psi}{\partial t}(x,t) = -\frac{\partial^2\Psi}{\partial x^2}(x,t) + V(x)\Psi(x,t), \tag{5.1b}$$

where

$$V \in C^2 \text{ on its support.} \tag{5.1c}$$

$$\Psi_0 \text{ is of compact support, } C^2 \text{ on its support.} \tag{5.1d}$$

We normalize (5.1b) by

$$i\frac{\partial\Psi}{\partial t} = -\frac{\partial^2\Psi}{\partial x^2} + V(x)\Psi(x,t)$$
$$= (H\Psi)(x,t) \quad \text{with supp} V \in [-1,1]. \tag{5.1e}$$

We assume for simplicity that V possesses no bound states, although under the above assumption they are at most finite in number and their effect may be taken into account trivially.

5.3. Generalities on Laplace–Borel Transform and Asymptotic Expansions

Definition 5.1. The *asymptotic expansion of a function* f at a point t_0 is a formal series (i.e., there are no convergence requirements) of functions f_k

$$\tilde{f}(t) = \sum_{k=0}^{\infty} f_k(t) \tag{5.2a}$$

such that

$$f_{k+1}(t) = o(f_k(t)) \quad \text{as } t \to t_0^+ \tag{5.2b}$$

which is equivalent to the condition

$$\lim_{t \to t_0^+} \frac{f_{k+1}(t)}{f_k(t)} = 0. \tag{5.2c}$$

Note that the zero series is *not* an asymptotic expansion, since no $f_k = 0$ as follows from the above definition. There may exist a relation — which we denote by $f \sim \tilde{f}$ — between a function f and a formal expansion.

Definition 5.2. The function f is *asymptotic* to the formal series $\tilde{f}$ as $t \to t_0^+$ if

$$f(t) - \sum_{k=0}^{N} \tilde{f}_k(t) = f(t) - \tilde{f}^{[N]}(t)$$

$$= o(\tilde{f}_N(t)); \quad \text{or}$$

$$= O(\tilde{f}_{N+1}(t)); \quad \text{or}$$

$$= \tilde{f}_{N+1}(t)(1 + o(1)), \tag{5.3}$$

for all integer N.

The three options in Definition 5.2 are easily shown to be equivalent (see [49, Proposition 1.18]). The notation O means $g(t) = O(h(t)) \iff \limsup_{t \to t_0^+} \frac{|g(t)|}{|h(t)|} < \infty$. Simple examples (e.g., [49, Remark 1.25, p. 5]) show that asymptotic series cannot be added in general, but for a *power series* we have the following definition.

$$\tilde{S} = \sum_{k=0}^{\infty} c_k z^k \quad \text{as } z \to 0+. \tag{5.4a}$$

Definition 5.3. A function is asymptotic to a series as $z \to 0$ in the sense of power series iff

$$f(z) - \sum_{k=0}^{N} c_k z^k = O(z^{N+1}) \quad \forall N \in \mathbb{N} \text{ as } z \to 0. \tag{5.4b}$$

As an example, consider

$$f(z) = \sin z + \exp(-1/z) \sim z - \frac{z^3}{6} + \cdots + (-1)^n \frac{z^{2n+1}}{(2n+1)!} + \cdots \tag{5.5}$$

as $z \to 0+$. The series on the r.h.s. of (5.5) converges for any $z \in \mathbb{C}$ to $\sin z$, and *not* to $f(z)$: this shows that different functions may have the same asymptotic expansion, even if the latter is convergent. Nevertheless, if $\lim_{z \to 0+} z^{-N} [f(z) - \sum_{k=0}^{N} a_k z^k] = 0$, it follows that $\lim_{z \to 0+} z^{-k} [f(z) - \sum_{p=0}^{k} a_p z^p] = 0 \ \forall k = 0, 1, \ldots, N-1$, hence $\lim_{z \to 0+} f(z) = a_0 = f(0+)$,

$\lim_{z\to 0+} (f(z) - f(0+))z^{-1} = a_1 = f^{(1)}(0+)$, etc., where a suffix k denotes a derivative of order k.

Problem 5.1. *Show that, if f is asymptotic to a series $\sum_{k=0}^{\infty} c_k z^k$ as $z \to 0+$, it is C^{∞} at $z = 0+$ and*

$$c_k = \frac{1}{k!} f^{(k)}(0+) \quad for \ k = 0, 1, \ldots. \tag{5.6}$$

Example (5.5) is consistent with the above problem: the function $f(z) = \sin z + \exp(-1/z)$ is C^{∞} at $0+$ and has the same asymptotic power series as $\sin z$, which is entire analytic in $z \in \mathbb{C}$. Of course, as an everywhere convergent power series the r.h.s. of (5.5) defines a unique entire analytic function.

Laplace transforms. The method of the Laplace transform (see [231, Chap. VI, pp. 215–241]) has a long history of applications to the solution of the Cauchy problem for the heat (and Schrödinger's) equation (see, e.g., [231, p. 298]). Let $F \in L^1(\mathbb{R}_+)$; then the *Laplace transform*

$$(\mathcal{L}F)(p) \equiv \int_0^{\infty} \exp(-pt)F(t)dt \tag{5.7a}$$

is analytic in $p \in \mathbb{H} \equiv \{p \in \mathbb{C} : \Re e\, p > 0$ and continuous in $\overline{H}\}$.

Proposition 5.1.

(i) *If $F \in L^1(\mathbb{R}_+)$, then $\mathcal{L}F$ is analytic in $\mathbb{H}$ and continuous on the imaginary axis $\partial \mathbb{H}$;*

(ii) *$(\mathcal{L}F)(p) \to 0$ as $p \to \infty$ along any ray $\{p : \arg(p) = \theta\}$ if $\theta \leq \pi/2$.*

The proof is an easy consequence of the dominated convergence theorem (check!); for $\theta = \pi/2$ it follows from Lemma 3.8.

Extending F to $\mathbb{R}_-$ by zero and using (i) of Proposition 5.1,

$$(\mathcal{L}F)(ip) = \int_{-\infty}^{\infty} \exp(-ipt)F(t)dt = (\mathcal{F}F)(p) \tag{5.7b}$$

which is then the analytic continuation of the Fourier transform restricted to functions which are zero on $\mathbb{R}_-$. Let $\mathcal{H}$ denote the space of functions analytic in $\mathbb{H}$. We thus have the following proposition.

Proposition 5.2.

(i) $\mathcal{L} : L^1(\mathbb{R}_+) \to \mathcal{H}$; $\|\mathcal{L}F\|_\infty \leq \|F\|_1$;

(ii) $\mathcal{L} : L^1 \to \mathcal{L}(L^1)$ *is invertible, and the inverse is given by*

$$F(t) = \mathcal{F}^{-1}((\mathcal{L}F)(i\cdot))(t) \tag{5.7c}$$

for all $t \in \mathbb{R}_+$.

Lemma 5.1 (Uniqueness). *Let $F \in L^1(\mathbb{R}_+)$ and $(\mathcal{L}F)(p) = 0$ for all p in some set with accumulation point in $\mathbb{H}$. Then $F = 0$ a.e.*

Proof. By analyticity $\mathcal{L}F = 0$ in $\mathbb{H}$. Then $F = 0$ as an element of $L^1(\mathbb{R}_+)$, and hence $F = 0$ a.e. $\qquad\qquad\qquad\qquad\qquad\qquad\qquad\qquad\square$

The list of problems whose solution admits Laplace transform representations has been significantly expanded in the last decade, due to progress in showing the wide applicability of Borel–Ecalle (BE) summability, see [49]: linear or nonlinear ODEs, PDEs, Navier–Stokes equation, quantum field theory, KAM theory, among others. In this book we shall only touch on those concepts which are necessary, or useful, to understand the paper [50], and must refer to [49] for a comprehensive treatment. An essential tool of asymptotic analysis is provided by Watson's lemma, which yields the asymptotic series at ∞ of

$$(\mathcal{L}F)(t) = \int_0^\infty \exp(-tp)F(p)dp \tag{5.8}$$

in terms of the asymptotic series of F at zero:

Lemma 5.2 (Watson's lemma). *Let $F \in L^1(\mathbb{R}_+)$, and assume*

$$F(p) \sim \sum_{k=0}^\infty c_k p^{k\beta_1 + \beta_2 - 1} \quad as \; p \to 0+ \tag{5.9a}$$

for some constants β_i with $\Re\, \beta_i > 0$ for $i = 1, 2$. Then, for $a \leq \infty$,

$$f(t) = \int_0^a \exp(-tp)F(p)dp \sim \sum_{k=0}^\infty c_k \Gamma(k\beta_1 + \beta_2)t^{-k\beta_1 - \beta_2} \tag{5.9b}$$

along any ray in $\mathbb{H}$.

Proof. (see [49, p. 31]) Write $F = \sum_{k=0}^{N} c_k p^{k\beta_1 + \beta_2 - 1} + F_N(p)$. For the finite sum, use the fact that

$$\int_0^a p^q \exp(-tp)dp = \int_0^\infty p^q \exp(-tp)dp + O(\exp(-at))$$

$$= \Gamma(q+1)t^{-q-1} + O(\exp(-at)).$$

The remainder may be estimated by the following problem. □

Problem 5.2. *Show that if $F \in L^1(\mathbb{R}_+)$, $t = \rho \exp(i\phi)$, $\rho > 0$, $\phi \in (-\pi/2, \pi/2)$, and $F(p) \sim p^\beta$ as $p \to 0+$, with $\Re \beta > -1$, then*

$$\int_0^\infty F(p) \exp(-tp)dp \sim \Gamma(\beta+1)t^{-\beta-1} \quad as \ \rho \to \infty. \tag{5.10}$$

Hint: Divide the interval on the l.h.s. of (5.10) into $\int_0^a + \int_a^\infty$, with a such that $|F(p)| \leq C|p|^\beta$ for $p \in [0, a]$ and use dominated convergence.

The series on the r.h.s. of (5.9b) is a (in general divergent) asymptotic series in powers of $t^{-\beta}$ (for some $\Re \beta > 0$). Comparing (5.9a) and (5.9b), we are led to define the *Borel transform* $\mathcal{B}$ of a formal (in general divergent) series

$$g(p) \sim \sum_{k=0}^\infty d_k p^{k\beta_1 + \beta_2 - 1} \quad \text{with } \Re \beta_i > 0 \tag{5.11a}$$

by

$$(\mathcal{B}g)(p) \sim \sum_{k=0}^\infty \frac{d_k}{\Gamma(k\beta_1 + \beta_2)} p^{k\beta_1 + \beta_2 - 1} \tag{5.11b}$$

and the *Laplace–Borel transform* of g by

$$(\mathcal{LB}g)(t^{-1}) \equiv t \int_0^\infty \exp(-tp)(\mathcal{B}g)(p)dp$$

$$= \int_0^\infty (\mathcal{B}g)(p/t) \exp(-p)dp. \tag{5.11c}$$

Under certain conditions of analyticity for g, the Borel transform, which is represented by (5.11b) as a formal series, actually converges in some region of the complex plane (due to the division of the coefficients of g in (5.11a) by $\Gamma(k\beta_1 + \beta_2)$), and the l.h.s. of (5.11c) actually yields the function $g(t^{-1})$ for t^{-1} in some subsector of the analyticity domain of g in the complex plane (see [210, Theorem XII.21]). This method of obtaining a "sum" for a divergent series (5.11a) is an example of a *summability method* (see [210] for a thorough discussion of this and other methods of summability), the *Borel summability method*, and the Laplace transform defined by (5.11a) is (by the reasons explained above) called the inverse Borel transform.

5.4. Decay for a Class of Model Systems After Costin and Huang: Gamow Vectors and Dispersive Part

In [50], the authors prove, among other results for H, given by (5.1a) (under the assumption of no bound states, as previously remarked), the following theorem.

Theorem 5.1. *For all $t > 0$ we have under the assumptions* (5.1c)–(5.1e):

$$\Psi(x,t) = \Psi_1(x,t) + \sum_{k=1}^{\infty} g_k \Gamma_k(x) \exp(-\gamma_k t). \tag{5.12}$$

The infinite sum in (5.12) *is uniformly convergent on compact sets in x (rapidly so if t is large), and Ψ_1 is the* dispersive part, *which may be written*

$$\Psi_1(x,t) = (\mathcal{L}\mathcal{B}\tilde{\Psi})(x,t), \tag{5.13a}$$

where

$$\tilde{\Psi}(x,t) = \sum_{k=0}^{\infty} c_k(x) \Gamma(k\beta_1 + \beta_2) t^{-k\beta_1 - \beta_2} \tag{5.13b}$$

is a Borel summable formal series in the variable $t^{-1/2}$ which may be of two kinds: (A) and (B). In case (A), which is generic, we have:

$$\tilde{\Psi}(x,t) = c_0(x)\Gamma(3/2)t^{-3/2} + c_1(x)\Gamma(5/2)t^{-5/2}$$
$$+ c_2(x)\Gamma(7/2)t^{-7/2} + \cdots . \tag{5.13c}$$

In case (B), which is that of a zero-energy resonance,

$$\tilde{\Psi}(x,t) = c_0(x)\Gamma(1/2)t^{-1/2} + c_1(x)\Gamma(3/2)t^{-3/2}$$
$$+ c_2(x)\Gamma(5/2)t^{-5/2} + \cdots . \tag{5.13d}$$

In both cases (A) and (B)

$$\Psi_1(x,t) \text{ is analytic in the region } \Re t > 0. \tag{5.13e}$$

We have, in addition:

(i) Ψ_1 and Γ_k are twice differentiable in x;
(ii) *for potentials V satisfying $V(1-)V(-1+) \neq 0$, we have:*

$$\gamma_k \sim \text{const. } k \log k + \frac{i}{4} k^2 \pi^2 \quad as \ k \to \infty \tag{5.14}$$

(two other conditions on V yield results analogous to (5.14), see [50]).

Remark 5.1. We denote by $\Gamma_k(x)$ the *Gamow vectors* of H, and $i\gamma_k$ the associated *resonances.*

Remark 5.2. Equation (5.12) defines rigorously resonances and Gamow vectors in a nonperturbative regime.

Remark 5.3. The condition defining the (perturbative) region of parameters

$$|g_1\Gamma_1(x)|(\exp(-\Re e\,\gamma_1 t) \gg |c_0(x)|\Gamma(3/2)t^{-3/2} \tag{5.15a}$$

for which exponential decay is a good approximation in case (A) (with analogous condition in case (B)) cannot hold for two large times, as shown in the forthcoming Theorem 6.1 of Chap. 6. It will be satisfied for sufficiently small $\Re e\,\gamma_1$ for intermediate times — neither too short nor too large. The upper limit for the validity of the exponential law arises from the spreading effect studied in Chap. 2: the tail of the scattered wave-packet contains the "slow" components of the initial wave-packet, i.e., the long-wavelength components. Indeed, the components of wavelength much larger than the range of the potential are hardly affected by it, yielding free-time asymptotic decay (5.13c) and (5.13d) for very large times. One therefore speaks of *quasi-exponential decay*. As we shall prove in Theorem 6.1, strictly exponential decay of the return probability is impossible whenever the Hamiltonian is bounded from below (semibounded).

Remark 5.4. Equation (5.12) is an asymptotic expansion called (complex) *transseries*, exponential power series of type

$$\tilde{f}(t) = \sum_{k=0}^{\infty} \exp(-\gamma_k t)t^{\alpha_k} \tilde{f}_k(t), \tag{5.15b}$$

where $\tilde{f}_k$ are formal power (integer or non-integer) series in $1/t$, where, for disambiguation purposes, the real part of the leading power of $1/t$

in $\tilde{f}_k(t)$ is chosen to be one, no $\tilde{f}_k$ is zero and the pairs (γ_k, α_k) are all distinct. With these conventions, the important result of *uniqueness* of the transseries representation may be proved — see [50, Proposition 20]. Thus, under certain conventions, the representation (5.12), (5.13a) and (5.13b) is unique.

We now present an outline of the proof of Theorem 5.1. Our main objective is to explain the main concepts and methods, making it easier for the reader to study the original reference [50].

Outline of the proof of Theorem 5.1. The function H satisfies the assumptions of [208, Theorem X.71, p. 290]. Thus, for any t, $\Psi(\cdot, t)$ is in the domain of $-\frac{d^2}{dx^2}$, which implies continuity in x of $\Psi(x, t)$. It also follows that the unitary propagator $U(t)$ is strongly differentiable in t, which implies existence of the Laplace transform

$$\hat{\Psi}(x, p) = \int_0^\infty \exp(-pt)\Psi(x, t)dt$$

$$= \left(\int_0^\infty \exp(-pt)U(t)dt \right) \Psi_0(x), \quad \text{for } \Re e\, p > 0 \qquad (5.16)$$

where Ψ_0 is the initial condition. Taking the Laplace transform of (5.16), we obtain

$$ip\hat{\Psi}(x, p) - i\Psi_0(x) = -\frac{\partial^2 \hat{\Psi}}{\partial x^2}(x, p) + V(x)\hat{\Psi}(x, p). \qquad (5.17a)$$

We see that the PDE (5.1b) has been converted into an ODE (5.17a). This is one of the basic advantages of the Laplace transform. Treating p as a parameter, let $y(x; p) \equiv y(x) = \hat{\Psi}(x, p)$. Rewrite (5.17a) as

$$y''(x) - (V(x) - ip)y(x) = i\Psi_0(x) \qquad (5.17b)$$

where primes denote differentiation, and $y \in L^2(\mathbb{R})$. The associated homogeneous equation is

$$y''(x) = (V(x) - ip)y(x). \qquad (5.17c)$$

If $y_\pm(\cdot)$ are two linearly independent (l.i.) solutions of (5.17c) with additional restrictions (and the usual branch of the square root)

$$y_+(x) = \exp(-\sqrt{-ip}x), \quad \text{when } x > 1 \qquad (5.17d)$$

$$y_-(x) = \exp(\sqrt{-ip}x), \quad \text{when } x < -1 \qquad (5.17e)$$

then, for $\Re e\, p > 0$, the L^2 solution of (5.17b) is

$$\hat{\Psi}(x,p) = \frac{i}{W_p[y_+,y_-]}$$
$$\times \left(y_-(x) \int_\infty^x y_+(s)\Psi_0(s)ds - y_+(x) \int_{-\infty}^x y_-(s)\Psi_0(s)ds \right),$$
(5.18a)

where

$$W_p[y_+,y_-] \equiv y_+(x)y'_-(x) - y_-(x)y'_+(x)$$
(5.18b)

is the *Wronskian*: it is easily seen (check!) to be nonzero and independent of x.

Problem 5.3. *Derive* (5.18a).

Hint: Let G be the Green's function (distribution) (or fundamental solution) [231] satisfying (in the sense of distributions)

$$G''(x,x') - (V(x) - ip)G(x,x') = \delta(x - x'),$$
(5.19a)

where δ is the Dirac measure at x', as well as the b.c. for x' fixed:

$$G(x,x') = \begin{cases} O(\exp(\sqrt{-ip}x)) & \text{for } x \to -\infty, \\ O(\exp(-\sqrt{-ip}x)) & \text{for } x \to \infty. \end{cases}$$
(5.19b)

The Ansatz

$$G(x,x') = \begin{cases} A(x')y_-(x) & \text{if } -\infty < x < x', \\ B(x')y_+(x) & \text{if } x' < x < \infty \end{cases},$$
(5.19c)

satisfies (5.19b); imposing continuity of G as $x \to x'$, and unit discontinuity of the right-derivative of G when $x \to x'+$, in the sense of distributions, both conditions following from (5.19a), one obtains (5.18a). We shall need to analyze the analyticity and decay properties of $\hat{\Psi}(x,p)$: it will be seen that $\hat{\Psi}$ is meromorphic in p except for a possible branch point at zero, and has subexponential bounds in the left half (cut) plane when not close to poles, so that it can be written in the form of the *Bromwich integral*

$$\Psi(x,t) = \frac{1}{2\pi i} \int_{a_0-i\infty}^{a_0+i\infty} \hat{\Psi}(x,p) \exp(pt)dp$$
(5.20)

with some $a_0 > 0$. The authors of [50] show that the contour of integration may be pushed through the left half-plane: collecting the contributions from poles and branch points, the decomposition in Theorem 5.1 follows.

A key result in the proof of Theorem 5.1 is the following proposition.

Proposition 5.3. *The function $\hat{\Psi}(x,p)$ is meromorphic in p on the Riemann surface of the square root at zero, which we denote by $\mathbb{C}_{1/2;0}$, and zero is a possible square root branch point.*

Proof. Continuity of y and y' imply the matching conditions

$$y_+(1) = \exp(-\sqrt{-ip}),$$
$$y'_+(1) = -\sqrt{-ip}\exp(-\sqrt{-ip}),$$
$$y_-(-1) = \exp(-\sqrt{-ip}),$$
$$y'_-(-1) = \sqrt{-ip}\exp(-\sqrt{-ip}).$$

Consider now the solutions f_1 and f_2 of (5.17c) with initial conditions $f_1(-1) = 1$, $f'_1(-1) = 0$, and $f_2(-1) = 0$, $f'_2(-1) = 1$. By standard results on analytic dependence on an external parameter of solutions of ODE (see [113, Theorem 2.8.5]), f_1 and f_2 are defined on $\mathbb{R}$ and, for fixed x, are entire in p. Note that, by construction, the Wronskian $W_p[f_1, f_2]$, defined as in (5.18b), is one. Then

$$y_+(x) = C_1 f_1(x) + C_2 f_2(x) \tag{5.21a}$$

and

$$y_-(x) = C_3 f_1(x) + C_4 f_2(x), \tag{5.21b}$$

where

$$C_1 = \exp(-\sqrt{-ip})(\sqrt{-ip}f_2(1) + f'_2(1)), \tag{5.21c}$$
$$C_2 = -\exp(-\sqrt{-ip})(\sqrt{-ip}f_1(1) + f'_1(1)), \tag{5.21d}$$
$$C_3 = \exp(-\sqrt{-ip}), \tag{5.21e}$$
$$C_4 = \sqrt{-ip}\exp(-\sqrt{-ip}). \tag{5.21f}$$

Problem 5.4. *Derive (5.21c)–(5.21f) and prove that*

$$W_p[y_+, y_-] = -\exp(-2\sqrt{-ip})(ip(f_2(1) - f'_2(1)) - \sqrt{-ip}(f_1(1) + f'_2(1))). \tag{5.22}$$

By (5.21a)–(5.21f), (5.22), $y_\pm$ and $W_p[y_+, y_-]$ are analytic in $\mathbb{C}_{1/2;0}$ with a possible branch point at zero. By (5.21a)–(5.21f), (5.22) and (5.18a), written in the form

$$\hat{\Psi}(x,p) = \frac{i}{W_p[y_+, y_-]}$$

$$\times \left(y_-(x) \left(\int_1^x y_+(s)\Psi_0(s)ds + \int_\infty^1 \exp\left(-\sqrt{-ips}\right) \Psi_0(s)ds \right) \right.$$

$$\left. - y_+(x) \left(\int_{-1}^x y_-(s)\Psi_0(s)ds + \int_{-\infty}^{-1} \exp\left(\sqrt{-ips}\right) \Psi_0(s)ds \right) \right),$$

$$(5.23)$$

the same follows for $\hat{\Psi}$, with the exception of the existence of a (possibly infinite) number of isolated poles arising from the zeros of $W_p[y_+, y_-]$. □

In order to calculate the asymptotic position of the poles as $p \to \infty$ in the left half-plane, one needs a more convenient choice of the pair of independent solutions than the choice f_1, f_2 of (5.17c) — a choice for which the asymptotic behavior as $p \to \infty$ is manifest. These are the WKB solutions (see [50, Proposition 5]). Expressing $y_\pm$ as linear combinations of these solutions (in a manner analogous to Problem 5.4), it is possible to prove that $W_p[y_+, y_-]$ has infinitely many simple zeros $\{p_k\}_{k=1,2,\ldots}$ in the left half-plane, of which (5.14) is an example (see [50, Proposition 6]). The residua at the poles $\{p_k\}$ may also be proved to be of polynomial growth in p, for large $|p|$. Together with the subexponential bounds in p on $y_\pm$ obtained through the properties of the WKB solutions (see [50, Lemma 9]), and (5.18a), (5.18b) subexponential bounds on $\hat{\Psi}$ are obtained, as well as the existence of a set of curves $p_k(s)$, parameterized by $s \in [0,1]$, with $p_k(0)$ on the negative imaginary axis, $p_k(1)$ on the negative real axis, such that $|p_k(s)| \geq k$, and $1/W_p[y_+, y_-]$ is bounded uniformly in k by a polynomial in p along these curves (see [50, Lemma 11]). The subexponential bounds and (5.18a) yield estimates of type

$$|\hat{\Psi}(x,p)| = O(\exp(2|\Re\mathfrak{e}(\sqrt{-ip})|(|x| + M + 2))), \qquad (5.24)$$

where we assume supp $\Psi_0 \subseteq [-M, M]$, for $|p|$ sufficiently large along these curves (see [50, Lemma 12]). In order to justify pushing the contour into the left, the authors rewrite (5.17b) as an integral equation

$$y = \mathcal{T}(Vy + i\Psi_0), \qquad (5.25a)$$

where

$$(\mathcal{T}f)(x) \equiv \frac{-1}{2\sqrt{-ip}} \int_{-\infty}^{\infty} \exp(-\sqrt{-ip}|s - x|)f(s)ds. \tag{5.25b}$$

Writing

$$y(x) = (\mathcal{T}(i\Psi_0))(x) + p^{-3/2}h(x) \tag{5.25c}$$

the integral equation (5.25a) and (5.25b) may be rewritten

$$h = p^{3/2}\mathcal{T}(V\mathcal{T}(i\Psi_0)) + \mathcal{T}(Vh). \tag{5.25d}$$

Problem 5.5. *Prove* (5.25d).

What is crucial here is the fact that $\exp(-\sqrt{-ip})$ is bounded in the region

$$\Omega \equiv \{p \in \mathbb{C} : -\pi/2 \le \arg p \le \pi\} \cup \{p \in \mathbb{C} : -\Im m\, p > \text{const.}\, (\Re p)^2\} \tag{5.25e}$$

(They choose the constant $1/9$.) Then it may be shown (see [50, Lemma 13]) that (5.25c) and (5.25d) is contractive in the (Banach) space of functions analytic in $p \in \Omega_0$, where

$$\Omega_0 \equiv \Omega \cap \{p \in \mathbb{C} : |p| > p_v\} \tag{5.25f}$$

where

$$p_v = 9\left(\sup_{x \in [-1,1]} |V(x)| + \mu + 1 \right)^2 \tag{5.25g}$$

and

$$\mu = \sup_{p \in \Omega} |\exp(-\sqrt{-ip})| \tag{5.25h}$$

equipped with the sup norm

$$\|f\| = \sup_{p \in \Omega_0, x \in [-x_1, x_1]} |f(x, p)|$$

within a ball of size

$$2\mu \sup_{x \in [-1,1]} |V(x)| + 2M \sup |\Psi_0''(x)| \sup_{s \in [0, M+|x_1|]} |\exp(-\sqrt{-ips})|.$$

In particular, the solution h of (5.25d) is bounded as $x_1 \to \infty$ if $\Re p > 0$. Note that analyticity in p is preserved by $\mathcal{T}$ and convergence in the sup norm.

The above has as consequence that the solution $\hat{\Psi}$ has the decomposition

$$\hat{\Psi}(x,p) = \frac{\Psi_0(x)}{p} + \frac{G_2(x,p)}{p^{3/2}} \qquad (5.26a)$$

with $G_2(x,p)$ bounded in $p \in \Omega_0$, $x \in [-x_1, x_1]$ with $x_1 > 0$ arbitrary: the general solution to (5.17b) may be written in the form

$$y_{\text{gen}}(x,p) = \hat{\Psi}(x,p) + c_1(p)y_+(x,p) + c_2(p)y_-(x,p)$$

and, therefore,

$$y(x,p) = \hat{\Psi}(x,p) + c_1(p)y_+(x,p) + c_2(p)y_-(x,p)$$

where y is the solution of (5.25a) mentioned above. The contractivity region is depicted in Fig. 5.1 (courtesy of the authors of [50]).

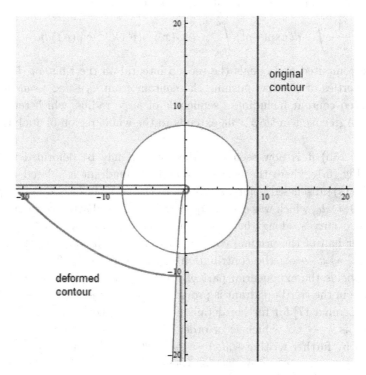

Fig. 5.1.   The contour with shaded area indicating contractivity region.

Since in the region $\{\Re e\, p > p_v,\; x < -1\}$, y_+ is unbounded and y_- is bounded, and in $\{\Re e\, p > p_v, x > 1\}$ y_+ is bounded and y_- is unbounded, while both y and $\hat{\Psi}$ are bounded (the latter property following from Problem 5.6 below), we must have $c_1 = c_2 = 0$ in $\Re e\, p > p_v$. Thus $\hat{\Psi}$ and y coincide in $\Re e\, p > p_v$, and also in Ω_0 by the uniqueness of analytic continuation. From (5.26a) we have

$$\Psi(x,t) = \Psi_0(x) + \frac{1}{2\pi i}\int_{a_0-i\infty}^{a_0+i\infty} G_2(x,p)p^{-3/2}\exp(pt)dp \qquad (5.26b)$$

where $a_0 > 0$ is a constant. The integral in formula (5.26b) needs justification.

Problem 5.6. *Prove* (5.26b).

Hint: Use the boundedness of $G(x,p)$ (uniformly in $x \in \mathbb{R}$) in Ω_0, defined by (5.25e) and (5.25f). Since $\{p \in \mathbb{C} : \Re e\, p > 0$ and $|p| > p_v\} \subset \Omega$, the second integral in (5.26b) is well-defined. Use now Fubini's theorem and Cauchy's formula to show that, if $\Re e\, p > a_0$, the double integral

$$\frac{1}{2\pi i}\int_0^\infty dt\, \exp(-pt)\int_{a_0-i\infty}^{a_0+i\infty} G(x,p')(p')^{-3/2}\exp(p't)dp'$$

may be computed and equals the second integral on the r.h.s. of (5.26b): the properties of G allow pushing the contour from $a_0 - i\infty$ to $a_0 + i\infty$ to a closed contour including a semicircle of large radius, which tends to infinity. Agreement in $\Re e\, p > a_0$ extends to the whole region of analyticity of $\hat{\Psi}$.

In (5.26b) it is now seen that the contour may be deformed to the one in Fig. 5.1, where the curve in the third quadrant is related to the curve $\{p_k(s)\}$ discussed before, and joins to the curve in $\{p \in \mathbb{C} : -\Im m\, p > (\Re e\, p)^2/9\} \cap \Omega_0$, which was chosen to be a vertical line downward to infinity. These two curves, along the one from below the real axis to the origin and lower half of the original contour surround all the poles in the third quadrant, as $k \to \infty$: the contribution of all the poles (5.13a)–(5.13d) in (5.26b) yields the exponential part of the transseries (5.12). Decay along the curve in the third quadrant is provided by estimates of the type (5.24): see [50, Lemma 17] for further details, in particular concerning the length of $p_k(s)$ as $k \to \infty$, which is of order k^2. The proof of Theorem 5.1 is achieved by further writing $\Psi_0(x) = \frac{1}{2\pi i}\int_C \Psi_0(x)p^{-1}dp$ and combining it with $\frac{1}{2\pi i}\int_C G_2(x,p)p^{-3/2}\exp(pt)dp$, where the contour C is the horizontal part around the negative real axis in Fig. 5.1.

Problem 5.7. *Derive* (5.14), *by the following considerations:*

(i) *The final integral* (5.26b) *may be written*

$$\frac{1}{2\pi i}\int_{C_1} G_2(x,p)p^{-3/2}\exp(pt)dp + \frac{1}{2\pi i}\int_{C_1}\Psi_0 p^{-1}dp$$

where C_1 is the part of the contour C of Fig. 5.1 consisting of the horizontal line from $-\infty$ to $-\epsilon$ below the cut, a small counterclockwise circle surrounding the origin, and a horizontal line from $-\epsilon$ to $-\infty$ above the cut. The only part which contributes is the Laplace transform of the non-analytic part of $\hat{\Psi}(x,p)$, for $p \in \mathbb{R}_-$, which may be written in the form

$$\hat{\Psi}(x,p) = \sum_{k=0}^{\infty} c_k p^{k\beta_1+\beta_2-1} \tag{5.27}$$

i.e., an analytic function of the variable $\sqrt{p}$ with a possible branch cut at zero (by Proposition 5.3).

(ii) *In order to compute the exponents β_1 and β_2 in* (5.27), *consider* (5.18a), *with $W_p[y_+, y_-]$ given by* (5.22). *The non-analytic part in* (5.18a) *has two sources: the Wronskian W_p and the numerator of* (5.18a). *The latter is bounded as $p \to 0$ since Ψ_0 has compact support, and thus the only singularities in $\hat{\Psi}(x,p)$ are due to W_p.*

(iii) *Show that if $f_1'(1) \neq 0$ (general case) $W_p = f_1'(1)(1 + o(1))$, yielding $\sqrt{p}(\text{const.} + o(1))$ as leading singularity of $\hat{\Psi}(x,p)$, and if $f_1'(1) = 0$ (zero energy resonance)*

$$W_p = [f_1(1) + f_2'(1)]\sqrt{-ip}(1 + o(1))$$

(note that $W_p[f_1, f_2] = 1$ by construction), yielding a leading singularity $\frac{\text{const.}}{\sqrt{p}}(1 + o(1))$ in this case.

(iv) *Identify from* (iii) *the exponents $\beta_1 = 1$, $\beta_2 = 1$ in the generic case and in* (5.27), *$\beta_1 = 1$, $\beta_2 = 3/2$ in the zero resonance case. Using Watson's lemma and* (5.11a)–(5.11c), *show* (5.13a), (5.13b) *with* (5.13c), (5.13d) *as the special cases above.*

Finally, analyticity of $\hat{\Psi}_1(x,t)$, (5.13e), in the region $\Re\, t > 0$, follows from the fact that $\Psi_1(x,t) - \Psi_0(x)$ is by (5.26b) a Laplace transform of a function $G_2(x,p)p^{-3/2}$, uniformly bounded by const. $|p|^{-3/2}$ for large $|p|$, $p \in \mathbb{R}_-$, since $G_2(x,p)$ is uniformly bounded in Ω, which, by (5.25e), includes $\mathbb{R}_-$. This concludes the outline of the proof of Theorem 5.1. □

By (5.13b), the exponential part in (5.12) of the transseries (5.15b) is analytic in the region $\{t \in \mathbb{C} : \Im t < 0\}$, because the term $\exp(-\gamma_k t) \sim \exp(-ik^2(\pi)^2 t/4) \sim \exp(k^2(\pi)^2 \Im m\, t)$ for large k. Additional estimates show (see the proof of [50, Proposition 3, p. 862]), however, that $\mathbb{R}_+$ is a natural boundary, i.e., the exponential part of the transseries is *not* analytic at any point $t_0 \in \mathbb{R}_+$! This surprising phenomenon occurs, as the authors remark, even in simple examples such as square wells: $\Psi(x, t)$ is C^∞ in t but nowhere analytic in $\mathbb{R}_+$.

5.5. The Role of Gamow Vectors

Consider generalized solutions $\Psi(x, k)$ (i.e., bounded uniformly in x, but not square-integrable) to the stationary Schrödinger equation corresponding to (5.1b), i.e.,

$$(H\Psi)(x, k) = \left(-\frac{d^2}{dx^2} + V(x) \right) \Psi(x, k) = k^2 \Psi(x, k). \tag{5.28}$$

If $k \in \mathbb{R}_+ \cup \{0\}$, we refer to $\Psi(x, k)$ as *generalized eigenfunctions*. If $k^2 = z^2$, with $z \in \mathbb{C}$, $\Re e\, z \geq 0$ and Ψ satisfies the b.c.

$$\lim_{x \to \pm\infty} (\Gamma(x, z) - \exp(\pm izx)) = 0, \tag{5.29a}$$

we refer to *Gamow functions or vectors* (see Remark 5.1). Coming back to (5.17c), with $z^2 = ip = -(-ip)$ we see that, with the usual branch of the square root, (5.29a) is equivalent to

$$\Gamma(x, z) \sim \exp(\mp\sqrt{-ip}\,x), \quad \text{as } x \to \pm\infty. \tag{5.29b}$$

Thus, after a rescaling, the Gamow function equals $y_+(x)$ for $x > 1$ and a nonzero constant times $y_-(x)$ for $x < -1$, with $y_\pm(x)$ defined by (5.17d) and (5.17e). The existence of the Gamow function is therefore equivalent to the linear dependence of y_+ and y_-, because the Wronskian $W_p[\Gamma, y_\pm]$ is independent of x, and this in turn is equivalent to the vanishing of the Wronskian $W_p[y_+, y_-]$. It follows that $-i\gamma_k$, found from the poles of $\hat{\Psi}$, are resonances corresponding to the Gamow vectors $\Gamma_k(x)$ which are, then, constant multiples of y_+ (see again Remark 5.1). The latter are multiples of the residues of $\hat{\Psi}$ by (5.18a), which may be rewritten in the form (with the assumption of l.d. of $y_+ = y_+(x, p)$ and $y_- = y_-(x, p)$):

$$-iW_p\hat{\Psi}(x, p) = -c\left(\int_{-M}^{M} y_+(s, p)\Psi_0(s)ds \right) y_+(x, p). \tag{5.30}$$

From (5.30), we see that

$$\Gamma_k(x) = dy_+(x, p_k), \tag{5.31a}$$

where d is a constant and p_k are the poles related to the γ_k in (5.14), i.e.,

$$p_k = -\gamma_k = -\frac{\pi^2 i k^2}{4} - \pi k \log k + a_v k + o(k)$$

in the case corresponding to (5.14), with similar expressions in the other cases. Since, for $\Re \sqrt{-ip_k} > 0$, which is generally true for the poles p_k, it follows that $y_-(x, p_k) = O(\exp(\Re \sqrt{-ip_k}x))$ for $x > 1$ by the constancy of the Wronskian $W_{p_k}[y_+, y_-]$, and $y_+(x, p_k) = O(\exp(-\Re \sqrt{-ip_k}x))$ for $x < -1$ for the same reason. As $y_+(x, p_k)$ and $y_-(x, p_k)$ are proportional, it follows from (5.31a) that

$$\Gamma_k(x) = O(\exp(c_k|x|)), \quad \text{as } |x| \to \infty \tag{5.31b}$$

with $c_k > 0$ for all $k \in \mathbb{N}$.

By (5.30) and Proposition 5.3, we see that the Laurent series of $\hat{\Psi}(x, p)$ about the pole (resonance) p_k may be written

$$\hat{\Psi}(x, p) = \eta(p)\Gamma_k(x) + a_0(x) + a_1(x)(p - p_k) + o(p - p_k), \tag{5.32a}$$

where

$$\eta(p) = \frac{c}{p - p_k} \tag{5.32b}$$

with c a constant. Equation (5.32a) is obtained by writing $W_p = W_p'(p_k)(p - p_k)(1 + o(1))$, with $W_p'(p_k) \neq 0$, and the Taylor series for the entire function $y_+(x, p)$ around $p = p_k$. Since $\hat{\Psi}(x, p)$ is, for $p \neq p_k, k \in \mathbb{N}$, uniformly bounded in x, and $\Gamma_k(x)$ increases exponentially by (5.31b), the remainder in (5.32a) must compensate the exponential tails of $\Gamma_k(x)$ for large $|x|$.

The above discussion, adapted from [76] to the present framework, shows clearly why the pole term in (5.32a) dominates only *locally*, i.e., on bounded intervals.

A different approach, due to [91], illuminates still different physical and mathematical aspects of the theory. Consider a potential plateau

$$V(x) = -\Delta E(\Theta(x - a) + \Theta(-x - a)), \tag{5.33}$$

where Θ denotes the Heaviside function. A particle starting on the plateau with energy $E > 0$ could remain there for a very long time, i.e., much longer than the maximal time $\tau_{\text{cl}} = a\sqrt{(2m/E)}$ a classical particle would remain there. This was proved in [91, Theorem 1]: during an arbitrarily long time interval $[0, t_0]$, with arbitrarily small error $\epsilon > 0$, $\Psi_t^{\Delta E}$ stays concentrated in the plateau region $[-a, a]$, that is,

$$\int_{-a}^{a} \left| \Psi_t^{\Delta E}(x) \right|^2 dx > 1 - \epsilon, \quad \forall t \in [0, t_0] \tag{5.34}$$

provided ΔE is large enough, $\Delta E \geq \Delta E_0(\Psi_0, t_0, \epsilon)$, where Ψ_0 is the initial normalized wave-function, equal to zero for $|x| > a$. Above, $\Psi_t^{\Delta E} = \exp(-iHt/\hbar)\Psi_0$, with $H = -\frac{\hbar^2}{2m}\frac{\partial^2}{\partial x^2} + V$, and V is given by (5.33).

The above result corresponds to the particle having a large "lifetime" $\Gamma_{qu} = \tau_{qu}^{-1}$ with

$$\tau_{qu} = a\frac{\sqrt{2m\Delta E}}{4E} = \frac{\sqrt{\Delta E}\tau_{\mathrm{cl}}}{4\sqrt{E}} \tag{5.35}$$

which may be obtained semiclassically by a simple argument [91]: imagine a particle traveling along the plateau with speed $\sqrt{2E/m}$, getting reflected at the edge with probability

$$R = \left(\frac{\sqrt{E + \Delta E} - \sqrt{E})}{\sqrt{E + \Delta E} + \sqrt{E}} \right)^2 \tag{5.36a}$$

corresponding to a potential step, traveling back with the same speed, getting reflected at the other edge with the same probability R, and so on. The transmission probability

$$T = 1 - R = 4\frac{\sqrt{E}}{\sqrt{\Delta E}} + \text{higher powers of } \frac{E}{\Delta E} \tag{5.36b}$$

is of the order of $4\sqrt{E}/\sqrt{\Delta E}$ for $E/(\Delta E)$ small, corresponding to a decay rate T/τ_{cl}, and hence to a decay time (5.35). The assumption implies a *large reflection coefficient*, having as consequence that the potential plateau has *metastable states*, which remain on the plateau for a long time — of the order of (5.35). The above-mentioned metastable states are related to the previous Gamow functions, as proved in [91], i.e., behaving in time as the second term on the r.h.s. of (5.12), written in a slightly different form:

$$\Gamma_k(x, t) = \exp(-iz_k t)\Gamma_k(x) \tag{5.37a}$$

with $(\gamma_k = iz_k)$

$$z_k = E_k - i\Gamma_k \quad \text{with } \Gamma_k > 0 \tag{5.37b}$$

and

$$\Gamma_k = 1/\tau_k \tag{5.37c}$$

(with the above choices, the resonances are placed in the fourth quadrant, rather than the third quadrant). More generally, Theorem 3 of [91], suggests the following. Near an $E > 0$ (with the conventions of the forthcoming (5.38a)–(5.38c)), there exists, immersed in the a.c. spectrum of H, suitably localized $\Psi \in \mathcal{H}$ (i.e., square integrable), such that the survival probability $\tau_H(\Psi)$ of the particle, initially in the state Ψ, defined by (3.85a), is "very large" . This statement may be quantified in the following way. Let

$$\delta_\epsilon(\lambda - E_k) \equiv \frac{\epsilon}{\pi}[(\lambda - E_k)^2 + \epsilon^2]^{-1} \tag{5.38a}$$

be the Lorentzian, and define the energy width about E_k for a unit vector Φ to be [165]:

$$\Delta E(\Phi, E_k) = \inf\{\epsilon > 0 : \pi\epsilon(\Phi, \delta_\epsilon(H - E_k)\Phi) \geq 1/2\}. \tag{5.38b}$$

The quantity $\Delta E(\Phi, E_k)$ is an integral of the values of the energy distribution (spectral) measure $d(\Phi, E_\lambda\Phi)$, where E_λ is the spectral family associated to H, near E_k: the smaller ΔE, the more is the spectral measure concentrated near E_k. By *Lavine's form of the time–energy uncertainty principle,*

$$\Delta E(\Phi, E_k)\tau_H(\Phi) \geq 1/2. \tag{5.38c}$$

This result is specially nice because the Fourier transform of (5.38a), with $\epsilon = \Gamma_k$, corresponds precisely to the exponential decay $\exp(-\gamma_k t)$ found in (5.12) or, writing $\gamma_k = iz_k$ as before, $\exp(-iz_k t)$; thus $\Re\gamma_k$ is the width Γ_k of the Lorentzian. By (5.15a), we expect that $\Gamma_k \ll 1$ in the perturbative regime — this joins well with the heuristics after (3.85d) in Chap. 3.

Coming back to the plateau example (5.33), the physical wave-function $\Psi(x,t)$ is such that the amount of $|\Psi|^2$ in the plateau region continuously shrinks due to a flow of $|\Psi|^2$ away from the plateau. On any large but *finite* interval $[-b, b]$ containing the plateau $[-a, a]$, $\Psi(x, t)$ approaches $\Gamma_k(x, t)$, and thus becomes a quasi-steady state, i.e., stationary up to an exponential shrinking due to the outward flux through $x = \pm b$.

A simple model to show in a physically clear way why (5.31b) occurs was also provided in [91]. Let P_t denote the probability that the particle is in the plateau region at time t, $P_t = \int_{-a}^{a} \rho_t(x)dx$. One should think of $\rho_t(x)$

as $\Im\mathrm{m}\,(\Psi^*\Psi')$, which satisfies a continuity equation also for complex z_k in (5.37a), shrinking with time due to (5.37b). Suppose $P_0 = 1$ initially, and that the particle leaves the plateau at a rate $\tau = \tau_{\mathrm{qu}}$, and $P_t = \exp(-t/\tau_{\mathrm{qu}})$, corresponding to (5.37a). For simplicity, assume that the distribution in the plateau region is flat, $\rho_t(x) = P_t/2a$, for $-a < x < a$. After leaving the plateau, the particle should move away from it, say at speed v. Then $\rho_t(x) = 0$ for $|x| > a + vt$, because such x cannot be reached by time t, and the amount of probability between x and $x + dx$ (with $a < x < a + vt$) at time t, $\rho_t(x)dx$, is what flowed off at $x = a$ between $\tilde{t} = t - (x - a)/v$ and $\tilde{t} - d\tilde{t} = t - (x + dx - a)/v$, which is half the decrease in P_t between $\tilde{t} - d\tilde{t}$ and $\tilde{t}$ (half because the other half was lost at $x = -a$), or

$$\rho_t(x)dx = \frac{1}{2}\left|\frac{dP_{\tilde{t}}}{d\tilde{t}}\right|d\tilde{t}$$

$$= \frac{1}{2\tau_{\mathrm{qu}}}\exp(-\tilde{t}/\tau_{\mathrm{qu}})$$

$$= \frac{1}{2v\tau_{\mathrm{qu}}}\exp(-t/\tau_{\mathrm{qu}})\exp((x - a)/(v\tau_{\mathrm{qu}})).$$

Likewise, for $-a - vt < x < -a$,

$$\rho_t(x) = \frac{1}{2v\tau_{\mathrm{qu}}}\exp(-t/\tau_{\mathrm{qu}})\exp((x - a)/(v\tau_{\mathrm{qu}})).$$

Although, as the authors of [91] remark, this model is oversimplified, it explains part of the reason for the exponential growth in (5.31b), visible in the above formulae for ρ: (primitive) causality and conservation of probability. One way to make these ideas precise is to introduce a cutoff factor, which removes the "exponential catastrophe", in the initial-value problem for the Green's function for the time-dependent Schrödinger equation (5.1b), of the type

$$G_k(x, x', 0) = a_j\theta(-x - x')\exp(i\omega_j(x + x')).$$

This is a cutoff exponential wave-packet with "complex wave-number" ω_j and amplitude a_j. The step function introduces a sharp cutoff at the wave-front, corresponding to the excitation at the definite instant $t = 0$. This approach is due to Nussenzveig [195, pp. 168–170], who applied it to the problem of a Schrödinger particle interacting with a penetrable spherical shell: we refer to that excellent book for further details and references.

A very important application of resonances arises in atomic physics, when an atom is subject to interaction with the quantized electromagnetic field. Every excited state of the isolated atom becomes immersed in the continuum and is expected to give rise to a resonance, while the ground state remains stable. This picture has been established by Bach, Fröhlich and Sigal in a remarkable series of papers; we refer to [17] for the aspects associated to the property of return to equilibrium defined in Chap. 6, and additional references, and to [105], Chaps. 16 and 17 for a textbook treatment. An ultraviolet cutoff is used in the quantized electro magnetic field, and the authors employ the method of dilation analyticity coupled to several other important techniques (Feshbach map, renormalization group) also expounded in [105].

From the point of view of the present monograph, one of the very few nontrivial models of atoms in interaction with radiation for which resonances may be proved, without relying on analyticity, is due to King [151], see also the paper by Sewell [234].

In the next section, we adopt the approximation of treating the radiation field as an external field, i.e., not quantized. A first important phenomenon which is related to quantum stability (Sec. 3.4) is that of ionization, to which we now turn.

5.6. A First Example of Quantum Instability: Ionization

Consider a many-body Hamiltonian for N nonrelativistic electrons and M nuclei interacting through Coulomb forces, as well as with an external static electromagnetic field; Z_{tot} will denote the total nuclear charge, and let $E_0(N)$ denote the ground-state energy of the system (for relativistic versions of the result mentioned below, see [170, Chap. 12]). One says (see [170, Chap. 12]) that the N electrons can be *bound* if

$$E_0(N, M) < E_0(N - 1, M) \qquad (5.39a)$$

i.e., it is necessary to use some energy to move an electron to infinity. A remarkable theorem was proved, which expresses a bound on the maximum number of bound electrons in the system: by [170, Theorem 12.1, p. 223], if (5.39a) holds, then

$$N < 2Z_{\text{tot}} + M. \qquad (5.39b)$$

The authors of [170] remark that this bound is far from optimal. It means that, if $N > 2Z_{\text{tot}} + M$, the excess electrons will flow away to infinity (with no cost of energy), i.e., the probability of *ionization* or *detachment* of electrons (the word ionization is used here in a sense different from [170]) is one.

We now consider the problem of ionization of stable atoms or ions, subject to external *time-dependent* electric fields, which has attracted great theoretical and experimental interest (see [44, 154]), particularly in relation to the Bayfield–Koch experiment described in Chap. 3. According to Sec. 3.4, ionization is a special example of quantum instability, corresponding to delocalization of the wave-function as $t \to \infty$. We say that a stable atom, e.g., the hydrogen atom, *ionizes completely* if the probability of finding the electron in any bounded spatial region $S \subset \mathbb{R}^n$ ($n \geq 1$) tends to zero for large times:

$$\lim_{t \to \infty} \int_S |\Psi(x,t)|^2 dx = 0. \tag{5.40}$$

The time-dependent external field continuously imparts energy to the atom, so that it eventually becomes larger than the energy of the initial state, and thus (5.40) is expected to occur in great generality. The phenomenon has been rigorously analyzed for a large class of Hamiltonians of the type

$$H = H(t) = H + V_1(x,t), \tag{5.41a}$$

where

$$H = -\triangle + V_0(x) \tag{5.41b}$$

on $\mathcal{H} = L^2(\mathbb{R}^n)$, $n \geq 1$, with V_0 a binding potential having both bound and continuum states, and

$$V_1(x,t) = \sum_{j=1}^{\infty} (\Omega_j(x) \exp(ij\omega t) + \text{c.c.}) \tag{5.41c}$$

is a time-periodic electric field of zero average. This extensive and elegant work is due to Costin, Lebowitz, Stucchio, Tanveer, Costin and Rokhlenko, and is excellently reviewed in [53], to which we refer for the original references: it includes the important case of the Coulomb potential $V_0(x) = -b|x|^{-1}$ for $n = 3$, with $\Omega_j(\cdot)$ compactly supported and equal to zero for j different from -1 or 1.

On the other hand, much of the rich behavior observed in (multiphoton) ionization is displayed by a relatively simple model described in the next section.

5.7. Ionization: Study of a Simple Model

In this section we consider two versions of (5.41a)–(5.41c), in which $n = 1$,

$$V_0(x) = -\alpha\delta(x) \tag{5.42a}$$

and

$$\Omega_j(x) = \begin{cases} -rx/2 & \text{if } j \in \{-1, 1\}, \\ 0 & \text{otherwise.} \end{cases} \tag{5.42b}$$

Equation (5.42a) describes a one-dimensional "electron" in an "atom" or "ion" with short-range potential, in an external sinusoidal electric field

$$V_1(x, t) = E(t)x = -r\cos(\omega t)x. \tag{5.42c}$$

From the point of view of the Bayfield–Koch experiment, this model was treated in [229], to which we refer for additional literature. Let

$$H_0 = -\frac{\partial^2}{\partial x^2} - \alpha\delta(x) \tag{5.43a}$$

denote the Hamiltonian of the isolated ion, and

$$H(t) = H_0 + V_1(x, t) \tag{5.43b}$$

the full Hamiltonian corresponding to (5.42c). It may also be written as

$$H(t) = K_0(t) - \alpha\delta(x) \tag{5.43c}$$

with

$$K_0(t) = -\frac{\partial^2}{\partial x^2} + E(t)x. \tag{5.43d}$$

The propagator (Green's function) $U_0(t)$ corresponding to $K_0(t)$ is explicitly known (see [54, Hunziker's Remark 3, p. 140]). Taking Ψ_0 the initial state, to be the unique normalized bound state of H_0,

$$\Psi_0(x) = \exp(-|x|/2)/\sqrt{2} \tag{5.44a}$$

with energy

$$E_0 = -\omega_0 = -1/4 \tag{5.44b}$$

and defining

$$b(t) = \int_0^t E(s)ds,$$

$$a(t) = \int_0^t b(s)^2 ds, \tag{5.44c}$$

$$c(t) = -2 \int_0^t b(s)ds,$$

$$\Phi(t) \equiv \sqrt{2}\Psi(0,t)\exp(ia(t)), \quad \text{with } \Phi(0) = 1 \tag{5.44d}$$

one obtains formally the following (weakly) singular Volterra equation for Φ (see [229, p. 3252]):

$$\Phi(t) = \exp(ic(t)^2/4t)\overline{w}(t) + \left(\frac{i\omega_0}{\pi}\right)^{1/2} \int_0^t ds \frac{\Phi(s)}{\sqrt{t-s}} \exp\left(i\frac{c(t,s)^2}{4(t-s)}\right), \tag{5.44e}$$

where $\sqrt{i} \equiv \exp(i\pi/4)$,

$$\overline{w}(t) = 1/2[w(z_+(t)) + w(z_-(t))], \tag{5.44f}$$

$$z_\pm(t) = \sqrt{it}[i\omega_0 \pm c(t)/2t] \tag{5.44g}$$

and

$$w(z) = \exp(-z^2)\mathrm{erfc}(-iz) \tag{5.44h}$$

is the complex error function [1]. In order to avoid domain problems, it is simpler to regard (5.44d) and (5.44e) as *defining* the wave-function at $x = 0$ and time t corresponding to the time-dependent Hamiltonian $H(t)$ given by (5.43c) (in a similar way as it is best to regard (3.137c) as defining the monodromy matrix of the quantum kicked rotor). Let

$$p(t,r,\omega) \equiv |(\Psi_t, \Psi_0)|^2 \tag{5.45a}$$

i.e., the survival probability of the electron in the bound state Ψ_0, and

$$P(t,r,\omega) \equiv 1 - p(t,r,\omega) \tag{5.45b}$$

the ionization probability.

Powerful numerical methods are known for (5.44e), with rigorous bounds on the errors. Thus, remarkably precise results have been obtained in [229] for various quantities. Let

$$N_I \equiv \frac{\omega_0}{\omega} \qquad (5.46)$$

denote the "number of photons" energetically required for ionization, and

$$U_P \equiv \frac{r^2}{2\omega^2} \qquad (5.47a)$$

the mean kinetic energy of a classical electron in the field (5.42c) — the *ponderomotive energy*. One of the results of [229] is their Fig. 9, which displays the ionization probability as a function of $\frac{\omega}{\omega_0}$: it is seen to have peaks (maxima) at points given, approximately, by reciprocal integers $1/n$, with $n = 1, 2, 3, \ldots$. Their excellent precision allowed to show that $P(t, \mu, \omega)$ has equidistant maxima for not too small times, given by the condition

$$N_P = n - \frac{\omega_0}{\omega}, \qquad (5.47b)$$

where $N_P = U_P/\omega$. To some extent, this confirms a prediction of Eberly [77] attributing the phenomenon to the quivering motion of the electron, as well as tunneling: in the process, known as *multiphoton ionization*, an electron escapes from the nucleus by simultaneous absorption of several photons.

The model (5.43a)–(5.43d), at least for large r, is quite reasonable to describe an ion such as H^- (note that, amusingly, this saturates the inequality (5.39b) — for integer values of N and M, with $N = 2$, $Z_{\text{tot}} = 1$, $M = 1$). The loosely bound electron in H^- is almost completely shielded from the nucleus by the other one, so that it no longer captures the long range of the Coulomb potential, but rather a short range one, which we approximate by a delta potential. For high electric fields, the motion of the electron in the field direction is much more important than the other degrees of freedom, so that the one-dimensional model captures the most essential part of the dynamics. Comparison of the theoretical results with the experimental data for the thresholds of multiphoton ionization in H^-, with $-\omega_0 = -0,754\,\text{eV}$, its binding energy, was done in [229], with very good agreement.

In spite of the rigorous bounds on the errors, the reported results do not explain their physical/mathematical origin. This was remedied, in part, by the work of Susskind, Cowley and Valeo [253]; we refer to the article of Costin, Lebowitz and Rokhlenko [51] for a more complete list of references.

Most significant along the previous lines is the introduction of the Zak transform $\mathcal{Z}$ in [52],

$$\mathcal{Z}[\Psi](x, \sigma, t) \equiv \sum_{j \in \mathbb{Z}} \exp(i\sigma(t + 2\pi j/\omega))\Psi(x, t + (2\pi j/\omega)). \qquad (5.48)$$

The Zak transform commutes with the integral operator defined by the r.h.s. of (5.44e), and it turns out [52] that, after Zak transformation by (5.48), this operator becomes compact. Use of the analytic Fredholm alternative (see [207, Theorem VI.14, p. 201]) within a mathematically very elegant framework related to the generalized Floquet or quasi-energy operator (3.27h) yields the following.

Theorem 5.2 ([52]). *For the model* (5.43a)–(5.43d), *with any* $\Psi_0 \in L^2(\mathbb{R})$, *complete ionization* (*defined by* (5.40)) *occurs, i.e.,*

$$\lim_{t \to \infty} \int_{-L}^{L} dx |\Psi(x, t)|^2 = 0$$

for any $0 < L < \infty$. *If* $\Psi_0 \in L^1(\mathbb{R}) \cap L^2(\mathbb{R})$ (*such as the bound-state wavefunction* (5.44a)),

$$\|\Psi(\cdot, t)\|_{L^2(-L,L)}^2 = O(t^{-1}).$$

The proof in [52] includes cases more general than (5.42c), i.e., $\cos \omega t$ may be replaced by any trigonometric polynomial. Note that $\Psi(x, t)$ may be obtained from $\Psi(0, t)$: we shall come back to this in an equivalent form in the next section.

Although Theorem 5.2 clarifies several of the aforementioned physical/mathematical issues, it leaves open some of the most striking multiphoton-ionization aspects. For this, we now turn to a different model.

5.8. A Second Example of Multiphoton Ionization: The Work of M. Huang

We now turn to (5.43a) and (5.43b), replacing $V_1(x, t)$ by

$$V_1(x, t) = -2\alpha\delta(x)r \cos \omega t. \qquad (5.49)$$

The model (5.43a), (5.43b) and (5.49) has been studied by Huang [121] in connection to the theory of Gamow vectors of [50] outlined in Sec. 5.2.

After a normalization of the equation, using $\delta(Ax) = \delta(x)/A$, we obtain for the time-dependent Schrödinger equation for (5.43a) and (5.43b), (5.49):

$$i\frac{\partial\Psi}{\partial t}(x,t) = -\frac{\partial^2\Psi}{\partial x^2}(x,t) - 2\delta(x)(1 + r\cos(\omega t))\Psi(x,t). \tag{5.50}$$

Taking the Laplace transform of (5.50) we have, with the same notation as in Sec. 5.2,

$$ip\hat{\Psi}(x,p) - i\Psi_0(x)$$

$$= -\frac{\partial^2\hat{\Psi}}{\partial x^2} - 2\delta(x)[\hat{\Psi}(x,p) + r(\hat{\Psi}(x,p - i\omega) + \hat{\Psi}(x,p + i\omega))]. \tag{5.51a}$$

We rewrite (5.51a) as an integral equation by inverting the operator $\frac{\partial^2}{\partial x^2} + ip$, as in (5.25b):

$$\hat{\Psi}(x,p) = \frac{\sqrt{i}\exp(-i^{3/2}\sqrt{p}x)}{2\sqrt{p}}\int_\infty^x \exp(i^{3/2}\sqrt{p}s)g(s)ds$$

$$- \sqrt{i}\frac{\exp(i^{3/2}\sqrt{p}x)}{2\sqrt{p}}\int_{-\infty}^x \exp(-i^{3/2}\sqrt{p}s)g(s)ds \tag{5.51b}$$

with

$$g(x) = i\Psi_0(x) - 2\delta(x)\hat{\Psi}(x,p) - 2r\delta(x)[\hat{\Psi}(x,p - i\omega) + \hat{\Psi}(x,p + i\omega)] \tag{5.51c}$$

Equation (5.51b) yields an expression for $\hat{\Psi}(x,p)$ as a functional of $\hat{\Psi}(0,p)$ (using the delta function in (5.51c)); setting $x = 0$ in this expression, one obtains an integral equation for $\hat{\Psi}(0,p)$ whose *solution*, when substituted in (5.51b) and (5.51c), finally yields $\hat{\Psi}(x,p)$. It thus suffices to analyze $\hat{\Psi}(0,p)$ *setting* $x = 0$ in the integral equation (5.51b) and (5.51c).

Problem 5.8. *Show that the equation for* $\hat{\Psi}(0,p)$ *assumes the form*

$$(\sqrt{-i}\sqrt{p} - 1)\hat{\Psi}(0,p) = r\hat{\Psi}(0,p - i\omega) + r\hat{\Psi}(0,p + i\omega)$$

$$+ \sqrt{-i}\sqrt{(p)}f(0,p), \tag{5.52a}$$

where

$$f(0,p) \equiv \frac{i^{3/2}}{2\sqrt{p}}\int_\infty^x \exp(i^{3/2}\sqrt{p}x)\Psi_0(s)ds$$

$$- \frac{i^{3/2}}{2\sqrt{p}}\int_{-\infty}^x \exp(-i^{3/2}\sqrt{p}x)\Psi_0(s)ds. \tag{5.52b}$$

Let

$$\tilde{\Psi}(p) \equiv \hat{\Psi}(0,p) - f(0,p). \tag{5.52c}$$

From (5.52a) the recurrence relation for $\tilde{\Psi}$ is

$$(\sqrt{-i}\sqrt{p} - 1)\tilde{\Psi}(p) = r[\tilde{\Psi}(p - i\omega) + \tilde{\Psi}(p + i\omega)] + (1 + 2r)f(0,p). \tag{5.53a}$$

Denoting by $p = i + in\omega + z$, $y_n(z) = \tilde{\Psi}(i + in\omega + z)$, $f_n(z) = (1 + 2r)f(0, i + in\omega + z)$, we have from (5.53a),

$$(\sqrt{-i}\sqrt{i + in\omega + z} - 1)y_n(z) = r(y_{n-1}(z) + y_{n+1}(z) + f_n(z)). \tag{5.53b}$$

The homogeneous equation associated to (5.53b) is

$$(\sqrt{(-i)}\sqrt{(i + in\omega + z)} - 1)y_n(z) = r(y_{n-1}(z) + y_{n+1}(z)). \tag{5.53c}$$

Huang writes (5.53c) in the form

$$C\vec{y} = \vec{y}, \tag{5.53d}$$

where $\vec{y} \equiv \{y_n\}_{n\in\mathbb{Z}}$ is regarded as an element of the Hilbert space $\mathcal{H} \equiv \{\vec{y} : \sum_{n\in\mathbb{Z}}(1+|n|^{3/2})|y_n|^2 < \infty\}$, and C is (proved to be) a *compact* operator. By use of the analytic Fredholm alternative, it follows that for all $r \in \mathbb{C}$ there exists at most a finite set of $z = z_1, \ldots, z_{l_r}$ for which (5.53c) has a solution (in $\mathcal{H}$, as a consequence, $\sqrt{p}\tilde{\Psi}$ is meromorphic in p with square root branches at every $\{in\omega : n \in \mathbb{Z}\}$, and poles at $\{p_k + in\omega : k = 1, 2, \ldots, l_r, \ n \in \mathbb{Z}\}$, obtained by a contour deformation of the Bromwich integral, now consisting of infinitely many horizontal and vertical line segments surrounding the branch cuts). The residues at the poles $\{p_k + in\omega\}$ are $C_n \exp[(p_k + in\omega)t]$, while the (rapidly convergent) integral along the branch cut n is a function whose large t-behavior is $C_n \exp(in\omega t)t^{-3/2}$ (the t-power being due to the generic $(p - in\omega)^{1/2}$ behavior at the branch points, see Problem 5.2). One finally obtains ($\omega_0 = 1$)

$$(\Psi_0, \Psi_t) = \exp[-\gamma(r;\omega)t]F_\omega(t) + \sum_{n=-\infty}^{\infty} \exp[i(n\omega - 1)t]h_n(t) \tag{5.54a}$$

with F_ω periodic of period $2\pi/\omega$ and

$$h_n(t) \sim \sum_{j=0}^{\infty} c_{n,j} t^{-3/2-j} \tag{5.54b}$$

as $t \to \infty$, $\arg t \in (-\pi/2 - \epsilon, \pi/2 + \epsilon)$. The analysis of the Gamow resonances $\gamma(r; \omega)$ is quite involved, but Huang was able to prove the following.

Proposition 5.4 ([121, Proposition 4.1]). *For*

$$1/(m+1) < \omega < 1/m \tag{5.55a}$$

$$\Re\, \gamma(r; \omega) = O(r^{2m+2}) \tag{5.55b}$$

for small r. More precisely,

$$\Re\, \gamma(r; \omega) \approx \frac{2\sqrt{(m+1)\omega - 1}\, r^{2m+2}}{(m+1)\omega \prod_{k=1}^{m}(\sqrt{1 - k\omega} - 1)^2}. \tag{5.55c}$$

The content of Proposition 5.4 is best grasped by the figure below (courtesy of M. Huang).

As one sees from Proposition 5.4 and Fig. 5.2, there is a rapid change in the real part of the resonance when the photon frequency ω is near $1/m$, m integer; this is precisely the condition that the energy of $N_I = m$ photons is precisely equal to the ionization energy $= 1$ ($\omega_0 = -1$) — the electron escapes from the nucleus by absorbing simultaneously m photons. For these values of the photon frequency the bound state is *exceptionally stable* against ionization. They correspond to the approximate zeros in Fig. 5.2. The origin of this behavior is left to the next section. As r increases, these resonances shift in the direction of increasing frequency. For small r

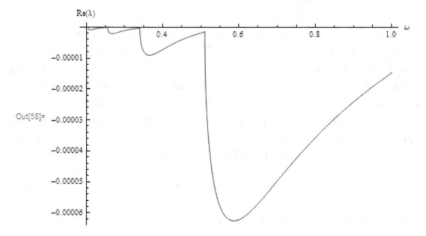

Fig. 5.2.   Real part of the resonance as function of photon frequency.

and $\omega = 1$, the shift in the position of the resonance is

$$\omega - 1 = \frac{r^2}{\sqrt{2}}. \tag{5.55d}$$

The last term reminds of (5.48), but it is of different form because of the difference between (5.42c) and the external field term in (5.50). In fact, the role of the ponderomotive energy (5.47a), which is only meaningful for the more realistic model with V_1 given by (5.42c), which was observed in [229], is not yet clear.

Between two minima in Fig. 5.2 there are pronounced *maxima*. For the model of the previous section, (5.47b) describes these maxima quite well, with a physical interpretation that the photon energy $n\omega$ equals the difference between the ponderomotive energy and the bound state energy.

The following properties of (5.54a) may be proved (see also [51]): not too close to resonances, i.e., when $|\omega - 1/n| > \epsilon$ for all integer n, $|F_\omega(t)| = 1 + O(r^2)$, and its Fourier coefficients decay faster than $r^{2|m|}|n|^{-|m|/2}$; the h_n functions decrease with n faster than $r^{|n|}$. When r gets larger, the ripples of $|F_\omega(t)|$ become visible, and the polynomial-oscillatory behavior starts sooner: since the sum in (5.54a) is $O(r^2 t^{-3/2})$ for large t and small r, increased r may yield a higher late-time survival probability. This phenomenon is called *adiabatic stabilization* and has been studied by Schrader and collaborators [88].

In the example which originated Gamow's analysis the (alpha) particle has to tunnel away, but when the potential has a bound state, an external field causes a level-shift in its energy, due to virtual transitions which are also associated with tunneling (see the next section). Thus, tunneling is always involved in the processes studied in this chapter, and we mentioned in Sec. 3.4 the tunneling instability. This tunneling instability may also be of *dynamical* nature. Since this phenomenon has a close analogy to adiabatic stabilization, we present a very simple illustration of it [99]: it may be considered one of the simplest models of *decoherence*.

Let $\Psi_{1,2} \in \mathcal{H} = \mathbb{C}^2$ be the normalized eigenvectors of a double-well Hamiltonian (see [109] for a mathematical treatment of double wells, and [185] for an explicitly soluble model), corresponding to eigenvalues $\lambda_{1,2}$, and

$$\Psi_\pm = \frac{1}{\sqrt{2}}(\Psi_1 \pm \Psi_2) \tag{5.56a}$$

be the single-well ground states, such that

$$(\Psi_\pm, \Psi_\pm) = 1 \quad \text{and} \quad (\Psi_\pm, \Psi_\mp) = 0 \tag{5.56b}$$

and

$$H\Psi_\pm = \Omega\Psi_\pm - \omega\Psi_\mp. \tag{5.56c}$$

We have Ψ_1 corresponding to the smallest eigenvalue $\lambda_1 = \Omega - \omega$, Ψ_2 corresponding to the highest $\lambda_2 = \Omega + \omega$. Let

$$W(t) = \epsilon v(\mu t) f_0 P_+ \tag{5.56d}$$

where $P_+\Psi = (\Psi_+, \Psi)\Psi$, i.e., a time-dependent perturbation with frequency μ, strength ϵf_0, acting on one side of the well only; for simplicity we adopt

$$v(t) = \cos t \tag{5.56e}$$

(for the quasi-periodic case, see [276]) and consider the Schrödinger equation for the perturbed system

$$i\frac{\partial\Psi}{\partial t} = (H + W(t))\Psi, \tag{5.56f}$$

where H is given by (5.56c), and W by (5.56d) and (5.56e). From (5.56f), with $A_\pm = (\Psi_\pm, \Psi)$,

$$iA'_+ = -\omega A_- + \epsilon\cos(\mu t) f_0 A_+ \tag{5.57a}$$

and

$$iA'_- = -\omega A_+. \tag{5.57b}$$

Let $\alpha_+(t) = A_+(t)\exp(iq(t))$, $\alpha_-(t) = A_-(t)$; then (5.57a) and (5.57b) becomes

$$\alpha'_\pm = i\omega\alpha_\mp \exp(\pm iq(t)) \tag{5.58a}$$

with

$$q(t) \equiv \epsilon(f_0\mu)\sin(\mu t). \tag{5.58b}$$

Assuming $\omega \ll \mu$, it follows from the theorem of averaging (see, e.g., [264, Theorem 11.1, p. 150]) that this equation has approximate solutions of the form

$$\alpha_0 \cos(\tilde{\omega}t - \phi_0) \tag{5.59a}$$

for some α_0 and ϕ_0, in the time scale

$$t \in \left[0, c\frac{\mu}{\omega}\right] \tag{5.59b}$$

for some constant c, where

$$\tilde{\omega} = \frac{\omega}{2\pi} \int_{-\pi}^{\pi} \exp[i\epsilon(f_0/\mu)\sin(\tau)]d\tau = \omega J_0\left(\frac{\epsilon f_0}{\mu}\right), \tag{5.59c}$$

where J_0 is the zeroth Bessel function, Hence, by (5.59a) and (5.59c) the beating period is given by

$$T_{\text{per}} = T\left(J_0\left(\frac{\epsilon f_0}{\mu}\right)\right)^{-1}, \tag{5.59d}$$

where $T = 2\pi/\omega$ is the beating period of the unperturbed double-well model. In the large perturbation limit $\frac{\epsilon}{\mu} \to \infty$, $T_{\text{per}} \approx T\sqrt{\frac{\epsilon f_0}{\mu}} \gg T$ and thus we have a *destruction of tunneling*, with the particle *localizing* at one of the single-well ground states; for intermediate values of ϵ, this effect also takes place when $\frac{\epsilon f_0}{\mu}$ coincides with one of the zeroes of the Bessel function.

The above result is, however, rigorous only for a limited time-interval (5.59b). Full results, valid for all times, have been obtained in [99, 276].

We finish this section with a remark: it is quite impressive that the ionization results for this simple model compares relatively well with actual experimental ones obtained for Rydberg atoms using an effective value of r (see [51, Figs. 4 and 5])!

5.9. The Reason for Enhanced Stability at Resonances: Connection with the Fermi Golden Rule

The reader might at this stage be asking for the real reason for the enhanced stability at (multiphoton) resonances, as well as the qualitative agreement of the resonance vs. frequency in the one-dimensional model ($n = 1$) and experiment ($n = 3$). Some of it relates to the famous *Fermi golden rule* (see [224, p. 39]), which asserts that the probability per unit time induced by absorption of a photon of energy ω by an atom initially in a state Ψ_l (corresponding to energy E_l) is

$$w_{d\Omega} = 2\pi|(\Psi_m, H'_I\Psi_l)|^2 \rho_{\omega,d\Omega} \tag{5.60a}$$

where $H_I(t) = H'_I \exp(-i\omega t)$, and Ψ_m is the state of energy

$$E_m = E_l - \omega. \tag{5.60b}$$

Above, H_I is the perturbation Hamiltonian corresponding to the quantized field (see [224, (2.102)]) or the equivalent formula for an external classical field — our case — and $\rho_{\omega,d\Omega}$ is the density of photon states, which equals (for $n = 3$)

$$\rho_{\omega,d\Omega} = |k|^2 \frac{d|k|}{d\omega} d\Omega \tag{5.60c}$$

(see [224, (2.116)]) in the continuum version; $d\Omega$ is the element of solid angle. In the one-dimensional case, $d\Omega = 2$ corresponding to the two possible directions, and

$$\rho_{\omega,d\Omega} = 2\frac{d|k|}{d\omega}. \tag{5.61}$$

The probability per unit time corresponds precisely to the inverse lifetime (see [224, p. 67, (2.236)])

$$\Gamma = 2\,\Re e\,\gamma(r;\omega) \tag{5.62a}$$

which, for $\omega > 1$ ($m = 0$ in (5.55c)), becomes

$$\Gamma = \frac{\sqrt{\omega - 1}}{\omega} r^2. \tag{5.62b}$$

The Fermi golden rule leads to the Wigner–Weisskopf theory of unstable states (see [224, p. 52]). The imaginary part $\Im m\,\gamma(r;\omega)$ is the *level shift* given in the same theory by [224, (2.235)], which is a sum over photons of all possible momenta and polarizations, unrestricted by energy conservation, such as (5.60b) — "virtual" processes; this is the main input (together with mass renormalization, see [224, p. 68]) in Bethe's treatment of the Lamb shift (see [224, p. 68]).

We emphasize that the above is (highly) nonrigorous: a rigorous version, using the concept of level-shift operator, is due to Merkli [184], to which we refer for references to the extensive rigorous work which has been done on the topic by several authors.

Problem 5.9. *Show that the Fermi golden rule and (5.61) imply (5.62b).*

Hint: The continuum generalized eigenfunctions of (5.43a) (with $\alpha = 2$), with energy k^2 ($2m = \hbar = 1$) are given by

$$u(k, x) = \frac{1}{\sqrt{2\pi}} \left(\exp(ikx) - \frac{\exp(i|kx|)}{1 + i|k|} \right). \qquad (5.62c)$$

We see from Problem 5.9 that for $\omega > 1$, Γ, given by (5.62a), is given by Fermi's golden rule. As remarked by [51, p. 6316], the origin of the enhanced stability at resonance is the vanishing of the matrix element in (5.60a) for $k = 0$ (with $\Psi_m(x) = \Psi_k(x) = u(k, x)$), which also holds for $n = 3$ due to the factor $|k|^2$ in (5.60c).

This chapter leaves a large number of open problems, among them a theory of Gamow vectors in higher dimensions, where Wronskian methods are not available: for radially symmetric potentials such methods do exist, however.

Our last chapter is concerned with exponential decay which, in the form (A-36) of decay of correlations, is an important characteristic of uniformly hyperbolic systems in classical mechanics (but there are many results outside this class — see App. A for a survey with references). What happens in quantum mechanics?

Chapter 6

Aspects of the Connection Between Quantum Mechanics and Classical Mechanics: Quantum Systems with Infinite Number of Degrees of Freedom

6.1. Introduction

In this chapter, we explore some aspects of the connection between quantum mechanics and classical mechanics (classical dynamical systems). As remarked in the preface, we speak here not of the singular and highly intricate (semi) classical limit of quantum mechanics (see [108] and references given there), but of results, ideas and methods in classical mechanics which have suggested analogies, albeit necessarily different in some essential features, in quantum mechanics. Some of these have already appeared in Chap. 3 in relation to the KAM perturbation theory and stability problems, as well as Howland's method in classical mechanics.

For clarity, we divide this chapter into several subsections, each of which is devoted to one of the aforementioned aspects. The first one is: Sec. 6.2 exponential decay and quantum Anosov systems. Related to Anosov systems (see App. A), a fundamental issue will be studied in Sec. 6.3: approach to equilibrium. In Sec. 6.3.2, we briefly explain the ideas of the classical theory: ergodicity, mixing and the Anosov property, the Gibbs entropy and the "coarse graining" involved in the second law; Sec. 6.3.3 is devoted to the Ehrenfest model, and Sec. 6.3.4 to the question of the initial state and macroscopic states: as mentioned in the preface, we provide a general discussion of the evolution of densities. Section 6.3.5 briefly discusses the (mostly open) analogous issues in quantum mechanics.

The concept of mixing in classical mechanics is naturally related to an analogous concept in quantum mechanics, but now for systems with an infinite number of degrees of freedom, due to the existence in the latter of an invariant state. The associated exponential decay now refers to *correlation functions* rather than the return probability (see (A.63)). This decay is not (necessarily) to zero, but to the expectation values (of certain observables) in an *equilibrium state*. We explain these concepts in an interlude (Sec. 6.4), and then come to Sec. 6.5 — quantum systems with infinite number of degrees of freedom — which is divided into three subsections: Secs. 6.5.1–6.5.3. Section 6.5.1 discusses quantum mixing and related concepts (return to equilibrium, weak asymptotic abelianness), as well as, rather briefly, for completeness, the important concept of stability (a counterpart to Sec. 3.4): dynamical stability. Sections 6.5.2 and 6.5.3 are concerned with special and important examples: Sec. 6.5.1 with quantum mixing: the vacuum and thermal states in relativistic quantum field theory (rqft), and Sec. 6.5.3 with the approach to equilibrium for quantum spin systems, where we discuss the Emch–Radin model and rates of decay, relating them to stability requirements.

6.2. Exponential Decay and Quantum Anosov Systems

6.2.1. *Generalities: Exponential decay in quantum and classical systems*

In Chap. 5, we mentioned that strict exponential decay is not possible in quantum theory. Consider the return probability (3.1a)

$$R_H^\Psi \equiv |(\Psi_0, \Psi(t))|^2, \tag{6.1a}$$

where

$$\Psi(t) = \exp(-itH)\Psi_0. \tag{6.1b}$$

We assume that H is a self-adjoint operator, whose spectrum is bounded from below, i.e., $\lambda \geq -c$ for some $0 < c < \infty$, for all $\lambda \in \sigma(H)$: we say that H is bounded from below or *semibounded*. Let μ_Ψ be the spectral measure of H associated to the vector $\Psi \in \mathcal{H}$ (see Theorem 1.4). We have the following theorem.

Theorem 6.1. *Let H be self-adjoint and*

$$R_H^{\Psi_0} \leq c \exp(-\beta|t|), \quad \text{for some } 0 < \beta < \infty, 0 < c < \infty. \tag{6.2}$$

Then, the spectrum $\sigma(H)$ is a.c. and

$$\sigma(H) = \mathbb{R}. \tag{6.3}$$

Proof. The amplitude

$$f_{\Psi_0}(t) \equiv (\Psi_0, \Psi(t)) = \int d\mu_{\Psi_0}(\lambda) \exp(-it\lambda) = O(\exp(-\beta|t|/2)) \quad (6.4)$$

by (6.2) is the F.S. transform of the spectral measure μ_{Ψ_0}, which is a.c. by (6.2) and Theorem 3.14. Thus, by the spectral theorem

$$g_{\Psi_0}(\lambda) \equiv \frac{d\mu_{\Psi_0}}{d\lambda} \quad (6.5a)$$

exists a.e. and defines a L^1-function. Hence, by the Fourier inversion formula for L^1 (see, e.g., [208, p. 10]),

$$g_{\Psi_0}(\lambda) = \frac{1}{2\pi} \int_{-\infty}^{\infty} \exp(-it\lambda) f_{\Psi_0}(t) dt. \quad (6.5b)$$

By (6.4) and (6.5a), the r.h.s. of (6.5b) may be extended to a function of a complex variable $z = \lambda + i\mu$, by the relation

$$g_{\Psi_0}(z) = \frac{1}{2\pi} \int_{-\infty}^{\infty} \exp(-itz) f_{\Psi_0}(t) dt. \quad (6.6)$$

The integral on the r.h.s. of (6.6) is defined for z in the strip $-\beta/2 < \mu < \beta/2$, and defines a function g_{Ψ_0} analytic in the strip, with boundary value $g_{\Psi_0}(\lambda)$ as $\mu \to 0$ a.e. in λ. Its essential support must therefore be the whole line, and (6.3) follows. $\qquad \square$

Theorem 6.1 has an immediate corollary.

Corollary 6.1. *If H is a semi-bounded self-adjoint operator, (6.2) does not hold, i.e., exponential decay of the survival probability is excluded.*

What is the quantity analogous to (6.1a) for classical dynamical systems? Intuitively, we expect that the analogous quantity is the "probability" $P(t)$ of return of a particle to a point x after a time t, and we are interested in this quantity for large t. We restrict ourselves to discrete time systems and take the hyperbolic automorphisms (Definition A.7) such as the cat map (A.62) as prototypes of Anosov systems, which describe irregular ("chaotic") motion, see App. A. Define, now,

$$N_n(T) \equiv |\{x : T^n x = x\}| \quad (6.7)$$

which is the number of points x returning to themselves after n iterations of the map T, and let

$$\omega(T) \equiv \limsup_{n \to \infty} \frac{\log N_n(T)}{n}. \tag{6.8}$$

Now $\omega(T)$ may be computed as follows: by definition (6.7),

$$N_n(T) = |\mathrm{Ker}(T^n - \mathbb{I})|. \tag{6.9}$$

We now have the following proposition.

Proposition 6.1 ([266, p. 180]). *If $B : \mathbb{T}^k \to \mathbb{T}^k$ is an endomorphism (see App. A) of $\mathbb{T}^k$ onto $\mathbb{T}^k$, with corresponding matrix $[B]$ so that $\det[B] \neq 0$ by the definition of endomorphism (before Definition A.2), the B is a $|\det[B]|$: to one map.*

We apply Proposition 6.1 to

$$B = T^n - \mathbb{I}, \tag{6.10}$$

where $[T]$ has no roots of unity as eigenvalues due to hyperbolicity (see Definition A.7); as a consequence of (6.9),

$$N_n(T) = |\det([T]^n - \mathbb{I})| = \left| \prod_i (\lambda_i^n - 1) \right|, \tag{6.11}$$

where λ_i are the eigenvalues of $[T]$.

Problem 6.1. *Show that (6.11) implies that*

$$\omega(T) = \sum_{i:|\lambda_i|>1} \log |\lambda_i|. \tag{6.12}$$

By Problem 6.1, $\omega(T) = \log 2$ for the Arnold cat map (A.62), if T is baker's map (the two-side shift on two symbols, see App. A), it is not a diffeomorphism, but it follows directly from (6.7) that $N_n(T) = 2^n$ so that $\omega(T) = \log 2$.

By (6.9) and (6.12),

$$N_n(T) = O(\exp(an)), \quad \text{for } a > 0 \quad \text{and} \quad n \to \infty. \tag{6.13a}$$

The probability $P(n)$ of return to x after n iterations is thus inversely proportional to $N_n(T)$ (see also the discussion in [59]):

$$P(n) = O(\exp(-an)), \quad \text{with } a > 0 \quad \text{and} \quad n \to \infty. \tag{6.13b}$$

Equation (6.13b) is a classical version of (6.2): it shows the exponential decay which is the "trademark" of classical chaotic systems. Here $P(n)$ is also the probability of having periodic orbits of large period n (see also the discussion in [59] related to the Ozorio–Hannay uniformity principle for periodic orbits of large period in chaotic systems) for the Arnold cat map and related Anosov systems. An independent check of (6.13b) is provided by the fact that, ignoring m-fold repeated orbits, the distribution of periods satisfies the prime-number theorem (for large prime numbers n), which is also consistent with (6.13b); but, for large n, $N_{n/m}(T) \ll N_n(T)$ by (6.13a): see, again, the discussion in [59], as well as Keating's [147] popular article "Physics and the queen of mathematics".

A different reason is, sometimes, given in the literature for the absence of "quantum chaos", viz. that the two limits $\hbar \to 0$ (the "classical limit") and $t \to \infty$, i.e., the limit of large times, do not commute. It has, however, been recognized by Berry and Goldberg [25], on the basis of the apparently trivial example of the free classical versus quantum rotor that this feature is a general one, independent of the classical dynamics being chaotic or not. Although their treatment is not rigorous, because no estimations for the remainders in their approximation are presented, a beautiful rigorous treatment due to Cellarosi and Marklof [181] confirms their conclusions.

Coming back to quantum systems, one may ask whether the conclusion of Theorem 6.1, viz. absence of exponential decay (6.2), is general for quantum systems. Semiboundedness of the Hamiltonian is a fundamental stability requirement, called *stability of the first kind* in [170]. It may be violated, however, in approximate theories. For instance, in treating systems such as oscillators or atoms, molecules or ions interacting with the quantized electromagnetic field, semiboundedness is lost, but only as a consequence of neglecting the back-reaction of the field. This approximation may, however, be excellent, e.g., considering the field as an external classical field, either constant (Stark effect), periodic or bichromatic — quasiperiodic — see Chaps. 3 and 5, and is widely employed in quantum optics; we refer to [194] for a careful physical discussion. In addition, several of the most important nonperturbative effects in rqft, such as the Unruh effect (see Sec. 6.5.2), are of the external field type, and are related to the forthcoming Anosov group [259].

Thus, although "quantum chaos" in the sense of (6.2) does not exist for general semibounded Hamiltonians, models such as the kicked rotor (3.23a) and (3.23b), which are considered to be prototypes of "quantum chaos" for reasons explained after (3.136), are of the external field type referred to above, and the corresponding Floquet or quasi-energy operator (3.24a)–(3.24c) is not semibounded. Therefore, exponential decay, either in the sense (6.13b) or, perhaps in a way more susceptible to analysis, of the autocorrelation function (3.144), is, in principle, possible. However, Duarte [73, 74] showed, roughly speaking, an "extraordinary abundance of elliptic islands" even in the region of "hard chaos" of the standard map (A.64), which is the classical system underlying the kicked rotor, see App. A. As briefly discussed in App. A, the standard or Chirikov–Taylor map is a prototype of a *mixed system*, whose behavior is essentially different from the uniformly hyperbolic systems, and much harder to analyze — see, again, App. A. For such systems, which are closer to physically more "realistic" systems, exponential decay seems doubtful: indeed, the decay of correlations in the standard map is an important open problem in the theory of classical dynamical systems. For a (nonrigorous) result on the decay of (classical and quantum) correlations in the standard map, see [238].

For mixed systems, it may be expected that "sticking" or trapping of trajectories by regular islands takes place [263], and that $P(t)$ decays algebraically:

$$P(t) = O(t^{-\gamma}) \quad \text{as } t \to \infty \tag{6.14}$$

for some $0 < \gamma < \infty$. The algebraic decay (6.14) was first observed by Channon and Lebowitz [41] in the study of stochastic motion between two KAM surfaces in the Hénon quadratic map. We refer to [263] for the several (nonrigorous, but supported by numerous analytic and numerical results) conjectures regarding (6.14), related, in particular, to the anomalous diffusion exponent associated to (3.138c) and (3.143). It is remarkable that asymptotic algebraic decay, which is the rule in quantum mechanics (Chaps. 2 and 5), is one of the most subtle and challenging problems in classical mechanics! Indeed, very few proofs of (6.14) exist — see [37] for a survey of open problems, and the kicked oscillator model of Lowenstein and Vivaldi [172] (see also Lowenstein's forthcoming book), which displays a surprisingly rich behavior.

6.2.2. *Quantum Anosov systems*

We now come back to the problem of exponential decay in quantum systems, which is related to the issue of "quantum chaos", and we now know is restricted to non-semibounded Hamiltonians or Floquet operators.

As remarked by Thirring in his review article [259], the intuition behind the sensitive dependence on initial conditions is that, next to any point on any trajectory, there is a point at a distance s such that, in the course of time, their distance grows exponentially:

$$\tau^t \circ \sigma^s = \sigma^{s \exp(-\lambda t)} \circ \tau^t. \tag{6.15}$$

Here we assume that we are dealing with a C^*-algebra $\mathcal{A}$ of (bounded) observables, and τ and σ are automorphisms or endomorphisms of $\mathcal{A}$, corresponding, respectively, to time translations and space translations: these notions are explained in the forthcoming interlude (Sec. 6.4.2). The quantity $\lambda > 0$ in (6.15) is defined to be the *quantum Lyapunov exponent* in analogy to the classical Lyapunov exponent, see Definition A.6. We shall come back to the interpretation of (6.15) in a slightly different context later. The multiplication law (6.15) may be abstracted to:

(i) $(t, s) \circ (t', s') = (t + t', s + s' \exp(-\lambda t))$;
(ii) $e = (0, 0)$;
(iii) $(t, s)^{-1} = (-t, -\exp(\lambda t)s)$.

Properties (i)–(iii) define (see [80, 259]), for (a) $(t, s) \in \mathbb{Z} \times \mathbb{R}_+$ the *Anosov semigroup*; (b) $(t, s) \in \mathbb{Z} \times \mathbb{R}$ the *Anosov group*; (c) $(t, s) \in \mathbb{R} \times \mathbb{R}$ the *continuous Anosov group*. An Anosov (quantum) (C^*) dynamical system is a realization of these groups by endomorphisms of $\mathcal{A}$.

According to the philosophy of this chapter, we first consider this definition in the context of classical dynamical systems. We have then an abelian C^*-algebra $\mathcal{A}$, which is a space of continuous functions, and any automorphism τ is induced by a homeomorphism τ_* of this underlying space: $\tau(f) = f \circ \tau_*^{-1}$ for all $f \in \mathcal{A}$. For the flows (see App. A) corresponding to τ_*^t and σ_*^s, we take the Hamiltonian flows generated by H and K (the latter denoting the generator of dilations), and, because of the relation $(\tau^t \circ \sigma^s)_* = \tau_*^t \circ \sigma_*^s$:

Problem 6.2. *Show that* (6.15) *has the differential version*

$$\{H, K\} = \lambda K, \tag{6.16a}$$

$$\tau^t K = \exp(-\lambda t)K, \tag{6.16b}$$

$$\sigma^s H = H + sK, \tag{6.16c}$$

where $\{\cdot, \cdot\}$ *denotes the Poisson bracket (see* [14]).

Hint:

$$\left.\frac{\partial \tau^t}{\partial t}(f)\right|_{t=0} = \{H, f\},$$

$$\left.\frac{\partial \sigma^s}{\partial s}(g)\right|_{s=0} = \{K, g\}$$

(see [14, Lemma 1, p. 209]).

Prototype of sensitive dependence of initial conditions are the repulsive harmonic forces, because the particle escapes to infinity with exponentially increasing velocity:

$$H = \frac{p^2}{2} - \lambda^2 \frac{x^2}{2} \quad \text{and} \quad K = p - \lambda x \tag{6.17}$$

with

$$\tau^t(x, p) = (\lambda x \cosh t + p \sinh t, p \cosh t + \lambda x \sinh t) \tag{6.18a}$$

and

$$\sigma^s(x, p) = (\lambda x + s, p + s). \tag{6.18b}$$

The reader may check that (6.17), (6.18a), (6.18b) satisfy (6.16a)–(6.16c). It is well-known that the flow lines of a relativistic particle with constant proper (i.e., relative to its own reference system) acceleration are hyperbolas (in usual phase space) (see [214, 3.7, p. 70; 3.8, p. 81]). Thirring [259] shows that in the extended phase corresponding to Howland's method in classical mechanics (after (3.23b)), a relativistic particle of charge e and mass m in a constant electric field with vector potential $A = 1/2(-Ex_1, Ex_0)$ (where $x_0 = t$, x_i, $i = 1, 2$ and 3, are the coordinates) (Stark effect) is, in suitable canonical coordinates, of the form (6.17), with $\lambda = eE/m$. Thus, the Unruh effect (see 6.5.2) is directly related to a quantum Anosov system (see also [259]). In this representation, the Lyapunov exponents are $\pm\lambda$. The systems we shall treat are *time-reversible*, i.e., there is an automorphism (or anti-automorphism in the quantum case) T such that

$$T^{-1} \circ \tau^t \circ T = \tau^{-t}. \tag{6.19a}$$

In this case,

$$\sigma^s_- \equiv T^{-1} \circ \sigma^s \circ T \tag{6.19b}$$

satisfies (6.15), with

$$\lambda \to -\lambda. \tag{6.19c}$$

Example (6.17) may be translated into quantum mechanics of one-particle, where x and $p = -i\partial/\partial x$ are operators on $\mathcal{H} = L^2(\mathbb{R})$. As elsewhere in this book, we do not use hats to denote quantum operators, but shall use the subscript c for classical variables, p_c and x_c. The restriction to one dimension is unimportant, see [228, Remark 4, p. 710] for the extension to the multidimensional case, even in the quasi-periodic case). In the prototypic case (6.17) and extensions to come, H and K are essentially self-adjoint on $C_0^\infty(\mathbb{R})$, and their self-adjoint closures, again denoted by the same symbols, define two families of unitaries $\exp(iHt)$, $\exp(iKs)$, satisfying

$$\exp(iKs\exp(-\lambda t))\exp(iHt) = \exp(iHt)\exp(iKs). \qquad (6.20a)$$

They generate, therefore, automorphisms of $\mathcal{B}(\mathcal{H})$ by

$$\tau^t(A) = \exp(iHt)A\exp(-iHt) \quad \text{and} \quad \sigma^s(A) = \exp(iKs)A\exp(-iKs) \qquad (6.20b)$$

which satisfy (6.15). This structure, introduced by Emch, Narnhofer, Sewell and Thirring [80] will be our departing idea, but we shall employ a modified description, introduced in [130] and further developed in [228]. We refer to the latter references for further important references to the literature, particularly the works of Majewski and Kuna and Vilela–Mendes.

Consider a quantum mechanical particle as above, with $\hbar = 1$. We define the self-adjoint operator

$$L_\alpha = \alpha_p p + \alpha_x x \quad \text{and} \quad \alpha = (\alpha_p, \alpha_x) \in \mathbb{R}^2. \qquad (6.21a)$$

It defines a derivation δ_α (see Sec. 6.4.2) acting on operators $A \in \mathcal{A}$, where $\mathcal{A}$ denotes the algebra of observables, by

$$\delta_\alpha A = [L_\alpha, A], \quad \forall A \in D(\delta_\alpha), \qquad (6.21b)$$

where the domain of the derivation $D(\delta_\alpha)$ is

$$D(\delta_\alpha) \equiv \{A \in \mathcal{A} : [L_\alpha, A] \in \mathcal{A}\}. \qquad (6.21c)$$

This derivation may be interpreted as a derivation in the direction of phase space determined by α, and is therefore naturally suggested by classical mechanics.

Let $U(t, t_0)$ denote the unitary propagator defining the dynamics, with initial time t_0. Experience with examples [259] suggests choosing as $\mathcal{A}$

the Weyl algebra $\mathcal{W}$, which consists of finite linear combinations of the operators

$$W(\beta,\gamma) = \exp[(i\beta x + \gamma p)], \quad \text{with } (\beta,\gamma) \in \mathbb{R}^2. \tag{6.22a}$$

We see that $\mathcal{W} \subset D(\delta_\alpha)$ because

$$[L_\alpha, W(\beta,\gamma)] = (\alpha_p\beta - \alpha_x\gamma)W(\beta,\gamma). \tag{6.22b}$$

We also assume that the dynamics defines an automorphism of $\mathcal{W}$, i.e., $U(t,t_0)AU(t,t_0) \in \mathcal{W}$ for all $A \in \mathcal{W}$ and for all $t, t_0 \in \mathbb{R}$. Under the above assumptions, we may formulate the following.

Definition 6.1. The *upper quantum Lyapunov exponent* is defined as

$$\overline{\lambda} = \sup_{\alpha \in \mathbb{R}^2} \overline{\lambda_\alpha}, \tag{6.23a}$$

where

$$\overline{\lambda_\alpha}(U, L_\alpha, A, t_0) \equiv \limsup_{t\to\infty} \frac{1}{t} \log \|[L_\alpha, A(t,t_0)]\| \tag{6.23b}$$

and

$$A(t,t_0) \equiv U^\dagger(t,t_0)AU(t,t_0) \tag{6.23c}$$

and $A \in \mathcal{W}$. The norm is chosen as usual, $\|A\| = \sup_{\Psi \in \mathcal{H}} \|A\Psi\|/\|\Psi\|$.

The above definition is adapted from the general formulation for cocycles described, e.g., in [145]. $\overline{\lambda}$ is expected to be independent of t_0 and of the choice of the observable A, under suitable conditions, e.g., $[L, A(t,t_0)] \neq 0$. If the limit in (6.23b) exists, then it is called the *quantum Lyapunov exponent* $\lambda_\alpha(U, L, A, t_0)$. Because of the unitarity of time evolution, the exponent $\overline{\lambda_\alpha}$ may also be expressed as

$$\overline{\lambda_\alpha}(U, L_\alpha, A, t_0) = \limsup_{t\to\infty} \frac{1}{t} \log \|[L_\alpha(t_0,t), A]\| \tag{6.23d}$$

with

$$L_\alpha(t_0,t) \equiv U^\dagger(t_0,t)L_\alpha U(t_0,t). \tag{6.23e}$$

6.2.3. *Examples of quantum Anosov systems and Weigert's configurational quantum cat map*

In order to show the usefulness of the above definition, we consider the parametric quantum oscillator, one of the simplest paradigms of the transition from regular to unstable behavior in classical mechanics [14, 15]. The Hamiltonian (we take the mass equal to one) is

$$H(t) = \frac{p^2}{2} + f(t)\frac{x^2}{2}, \qquad (6.24a)$$

where f is a periodic function of period T,

$$f(t + T) = f(t). \qquad (6.24b)$$

It is convenient to decompose f as $f(t) = E + f_{za}$ with $E = (1/T)\int_0^T dt\, f(t)$, and thus f_{za} satisfies $(1/T)\int_0^T dt\, f_{za}(t) = 0$. We shall analyze the one-parameter family of systems defined by varying the average $E \in \mathbb{R}$.

The classical equation corresponding to (6.24a) is Hill's equation [115]:

$$x_c'' + f(t)x_c = 0 \qquad (6.24c)$$

which is well-known (see [115, Chap. 4]) to have bands of stability regions S and instability regions I ("gaps") when the parameter E is varied.

For the quantum parametric oscillator, we shall prove the following result.

Theorem 6.2. *For any observable $A = W(\beta, \gamma)$ of the form (6.22a), in the stability region $E \in S$, one has*

$$\overline{\lambda_\alpha}(U, L_\alpha, A, t_0) = 0, \quad \forall\, \alpha, \ \forall\, t_0.$$

In the instability region $E \in I$, there is a stable direction α_s, which depends on t_0, for which

$$\overline{\lambda_{\alpha_s}}(U, L_{\alpha_s}, A, t_0) = -\lambda_r < 0$$

whereas for all other directions α

$$\overline{\lambda_\alpha}(U, L_\alpha, A, t_0) = \lambda_r > 0$$

where λ_r is the absolute value of the real part of the Floquet exponent of the corresponding classical oscillator defined below in (6.28). Thus, the upper *quantum Lyapunov exponent is*

$$\overline{\lambda} = \sup_{\alpha} \overline{\lambda_{\alpha}} = \lambda_r > 0.$$

There is therefore a transition in the upper quantum Lyapunov exponent $\overline{\lambda}$ as the parameter E ranges from the classical system's region of stability to the instability region.

Theorem 6.3. *If we consider the time evolution in the instability region* $E \in I$ *at discrete times* $t_n = n2T$ *given by integer multiples of the period* T, *there is an unstable eigendirection* $\alpha_+ = (\alpha_{p+}, \alpha_{x+})$ *such that the corresponding derivation* L_{α_+} *satisfies for all* t_n,

$$\exp(iL_{\alpha_+} s \exp(-\lambda_r t_n))U^\dagger(t_n, 0) = U^\dagger(t_n, 0)\exp(iL_{\alpha_+} s) \qquad (6.25a)$$

and a stable eigendirection $\alpha_- = (\alpha_{p-}, \alpha_{x-})$ *such that*

$$\exp(iL_{\alpha_-} s \exp(\lambda_r t_n))U^\dagger(t_n, 0) = U^\dagger(t_n, 0)\exp(iL_{\alpha_-} s) \qquad (6.25b)$$

where $\lambda_r > 0$ *is the real part of the classical Floquet exponent. Thus, the system satisfies the discrete Anosov relations proposed in* [80] *and corresponding to* (6.15).

Relation (6.25a) may be written equivalently, by changing $t_n \to -t_n$ as

$$\exp(iL_{\alpha_+} s \exp(\lambda_r t_n))U(t_n, 0) = U(t_n, 0)\exp(iL_{\alpha_+} s) \qquad (6.25c)$$

with a similar expression corresponding to (6.25b).

With the representation (6.25a) one may give an intuitive interpretation for the Lyapunov exponent in the Schrödinger picture: compare the forward time evolution ($t_{n+1} > t_n$) of two initial states Ψ and $\Psi + \delta\Psi$ related to each other by a translation in phase space in direction α_+ of size s. The time evolution $U(t_n, 0)[\Psi + \delta\Psi]$ to time t_n will yield a state $\Phi + \delta\Phi$ which can be related to $\Phi = U(t_n, 0)\Psi$ by a translation in phase space in the same direction α_+ but of a size s' which is exponentially amplified $s' = s\exp(\lambda_r t_n)$.

In order to determine the upper quantum Lyapunov exponent according to (6.23d), we must first calculate $L_\alpha(t_0, t)$.

Lemma 6.1. *We have*

$$L_\alpha(t_1, t_2) = \alpha_p(t_1, t_2)p + \alpha_x(t_1, t_2)x \qquad (6.26a)$$

with

$$\alpha_p(t_1, t_2) = \exp[(t_1 - t_2)\lambda_+]h_{p+}(t_1, t_2)$$
$$+ \exp[(t_1 - t_2)\lambda_-]h_{p-}(t_1, t_2) \qquad (6.26b)$$

and

$$\alpha_x(t_1, t_2) = \exp[(t_1 - t_2)\lambda_+]h_{x+}(t_1, t_2)$$
$$+ \exp[(t_1 - t_2)\lambda_-]h_{x-}(t_1, t_2) \qquad (6.26c)$$

and $h_{p\pm}, h_{x\pm}$ are periodic functions of the same period T of the function f, and $\lambda_\pm$ are the Floquet exponents associated with the classical dynamics (defined below in (6.28)). In the stability bands $E \in S$, $\lambda_\pm$ is purely imaginary, whereas in the instability regions $E \in I$, $\lambda_\pm$ has a nonvanishing real part, which we denote $\pm\lambda_r$.

Proof. The classical equation (6.24c) may be written as

$$\begin{pmatrix} \dfrac{dp_c}{dt} \\ \dfrac{dx_c}{dt} \end{pmatrix} = \begin{pmatrix} 0 & -f(t) \\ 1 & 0 \end{pmatrix} \begin{pmatrix} p_c \\ x_c \end{pmatrix}.$$

The propagator of the classical equation, defined by

$$\begin{pmatrix} p_c(t) \\ x_c(t) \end{pmatrix} = F(t, t_0) \begin{pmatrix} p_c(t_0) \\ x_c(t_0) \end{pmatrix} \qquad (6.27)$$

with $F(t, t) = 1 \; \forall t$, may be written, by Floquet's theorem [176] as the following.

Lemma 6.2. *We have*

$$F(t, t_0) = G(t) \exp[(t - t_0)B]G^{-1}(t_0),$$

where

$$G(t + T) = G(t)$$

is an invertible matrix, and B a constant traceless matrix.

We shall not prove Lemma 6.2 here, and refer to [176] or to [130, p. 4381]. It follows from Lemma 6.2 that B is one of the following three forms: (1) B has two complex eigenvalues

$$\lambda_\pm = \pm(\lambda_r + i\lambda_i) \neq 0 \qquad (6.28)$$

(we choose notation such that $\lambda_r \geq 0$). In this case it follows that B is diagonalizable and there are two cases: (1a) $\lambda_r = 0$ (stable case, band), or (1b) $\lambda_r \neq 0$ (unstable case, gap). Denote now by Λ the eigenvalue matrix:

$$\Lambda = \begin{pmatrix} \lambda_+ & 0 \\ 0 & \lambda_- \end{pmatrix}.$$

In these two cases above one may write

$$F(t, t_0) = G(t) S \exp[(t - t_0)\Lambda] S^{-1} G^{-1}(t_0),$$

where S is some invertible matrix. The second case is (2) $\lambda_\pm = 0$, in this case B is not diagonalizable, and has Jordan canonical form $\begin{pmatrix} 0 & 1 \\ 0 & 0 \end{pmatrix}$. It corresponds to a band edge, and was not discussed in [130]: its treatment is open.

The second crucial point is the fact that the Heisenberg equations of motion for the operators $x(t)$ and $p(t)$ have the same form as the classical equations of motion for the corresponding observables, i.e.,

$$\begin{pmatrix} U^\dagger(t, t_0) p U(t, t_0) \\ U^\dagger(t, t_0) x U(t, t_0) \end{pmatrix} = F(t, t_0) \begin{pmatrix} p \\ x \end{pmatrix}$$

$$= g(t) \begin{pmatrix} e^{(t-t_0)\lambda_+} & 0 \\ 0 & e^{(t-t_0)\lambda_-} \end{pmatrix} g^{-1}(t_0) \begin{pmatrix} p \\ x \end{pmatrix} \quad (6.29)$$

with $g(t) \equiv G(t)S$. Writing explicitly the matrix elements $g(t_1) \equiv [g_{ij}(t_1)]$ and $g(t_2) \equiv [g_{ij}(t_2)]$, we find that the functions in (6.26b) are given by the following.

Problem 6.3. *Show that* (6.26a) *and* (6.29) *imply that the functions in* (6.26b) *and* (6.26c) *are given by*

$$\begin{aligned} h_{p+}(t_1, t_2) &= (\alpha_p g_{11}(t_1) + \alpha_x g_{21}(t_1)) g_{11}^{-1}(t_2), \\ h_{p-}(t_1, t_2) &= (\alpha_p g_{12}(t_1) + \alpha_x g_{22}(t_1)) g_{21}^{-1}(t_2), \\ h_{x+}(t_1, t_2) &= (\alpha_p g_{11}(t_1) + \alpha_x g_{21}(t_1)) g_{21}^{-1}(t_2), \\ h_{x-}(t_1, t_2) &= (\alpha_p g_{12}(t_1) + \alpha_x g_{22}(t_1)) g_{22}^{-1}(t_2), \end{aligned} \quad (6.30)$$

which are periodic in t_1 and t_2 since $g(t)$ is periodic.

In order to see what is involved in Theorem 6.2, note that the Weyl algebra of observables is in the domain of the family of derivations defined as

$$\delta_\alpha^{t_0, t} A = [L_\alpha(t_0, t), A]$$

parameterized by the time variable,

$$\mathcal{W} \subset D(\delta_\alpha^{t_0,t}) \quad \forall t_0, t \in \mathbb{R}.$$

Indeed, by (6.22b) and (6.26a), if $A = \exp(\beta x + \gamma p)$, then

$$[L_\alpha(t_0, t), A] = (\alpha_p(t_0, t)\beta - \alpha_x(t_0, t)\gamma)A \in \mathcal{W}.$$

In order to determine the Lyapunov exponent, we calculate

$$\|[L_\alpha(t_0, t), A]\| = |\alpha_p(t_0, t)\beta - \alpha_x(t_0, t)\gamma|,$$

where we have used $\|A\| = 1$. By (6.26a)–(6.26c) and (6.30), we may write, for $t_0, t \in \mathbb{R}_+$,

$$\|[L_\alpha(t_0, t), A]\| = |Q + R|,$$
$$Q = e^{(t-t_0)\lambda_+}(\alpha_p g_{12}(t_0) + \alpha_x g_{22}(t_0))(\beta g_{21}^{-1}(t) - \gamma g_{22}^{-1}(t)),$$
$$R = e^{-(t-t_0)\lambda_+}(\alpha_p g_{11}(t_0) + \alpha_x g_{21}(t_0))(\beta g_{11}^{-1}(t) - \gamma g_{12}^{-1}(t)),$$

$$(6.31)$$

where we have used that $\lambda_- = -\lambda_+$. The stable direction is determined by the condition that the curly bracket in the first term vanishes:

$$\alpha_p g_{12}(t_0) + \alpha_x g_{22}(t_0) = 0, \quad (6.32a)$$

which leads to

$$\alpha_s \equiv (\alpha_{ps}, \alpha_{xs}) = \eta_s(-g_{22}(t_0), g_{12}(t_0)), \quad (6.32b)$$

where η_s is an arbitrary constant.

Problem 6.4. *Show that the second term in* (6.31) *is not identically zero.*

In the stable case we may write by (6.28), (6.31) and (6.32a),

$$\|[L_{\alpha_s}(t_0, t), A]\| = \exp(-t\lambda_r)S(t), \quad (6.33a)$$

where $S(t)$ is a periodic function of t. By (6.33a), there is thus a sequence of times $t_k \to \infty$ such that $S(t_k) > \delta > 0$, for some constant δ,

$$\overline{\lambda_{\alpha_s}}(U, L_{\alpha_s}, A, t_0) = -\lambda_r + \limsup_{t \to \infty} \frac{\log S(t)}{t} = -\lambda_r. \quad (6.33b)$$

By the same token, (6.31) implies the remaining assertions of Theorem 6.2.

$\square$

In the stability bands $E \in S$, $\lambda_\pm$ are imaginary and thus $\overline{\lambda_\alpha} = 0$ for all α. We refer the interested reader to [130] for the proof of Theorem 6.3.

It may be argued that Theorems 6.2 and 6.3 are "induced" by classical mechanics, in particular the apparently crucial fact (6.29) that the Heisenberg equations of motion for the observables are classical. We emphasize that the latter is *not* a general requirement: it is only important as a tool to obtain explicit results. It should not be forgotten that the space of *states* is, nevertheless, subject to the laws of quantum mechanics, without classical analogue even in the analyzed case. Thus, as remarked by [228], the Lyapunov exponents are defined in an intrinsically quantum mechanical way, and, if one were to add small perturbations to these models, it is expected that nonzero Lyapunov exponents will continue to exist, while the exact equivalence to classical systems in terms of observables will be destroyed.

A basic difference between (6.24a) and (6.17) is the nonglobal character of the Anosov structure of the former in the space of parameters E, i.e., if $E \in S$, it does not take place, i.e., in this case the upper quantum Lyapunov exponent is zero, a case included in our definition but not that of [80, 259]. This leads to a much richer structure than (6.17), with a transition from quantum regular to irregular motion, which even has a physical interpretation in a model of quadrupole radio-frequency traps (Paul-Penning traps), see [39, 45]. This complexity becomes even more pronounced in the case (6.24a) with f almost-periodic (Definition 3.6). A nontrivial extension of the previously described framework to systems in enlarged phase space, described in (3.24a) et seq. was made in [228], which allows for a unified treatment of all previously treated models (6.24a) with f periodic or almost-periodic, as well as a kicked rotor model due to Weigert, to which we now come.

Consider a charged particle of unit mass constrained to move in a unit square with periodic b.c. (period one), subject to external time-dependent electromagnetic fields. It was shown by Weigert [267, 268] that the external fields may be chosen in such a way that the configuration space of the particle is mapped periodically to itself according to Arnold's cat map (A.35): the monodromy matrix $U(T,0)$ is given on $\mathcal{H} = L^2([0,1] \times [0,1])$ by

$$U(T,0) = \exp(-iTp^2/2)\exp[-i(x^T V^T p + p^T V x)/2], \qquad (6.34)$$

where

$$p = \begin{pmatrix} p_1 \\ p_2 \end{pmatrix}$$

and

$$x = \begin{pmatrix} x_1 \\ x_2 \end{pmatrix}$$

and V is a matrix such that exp(V) is given by the matrix corresponding to T_A (A.62), i.e., the matrix associated to Arnold's cat map:

$$\exp(V) = \begin{pmatrix} 2 & 1 \\ 1 & 1 \end{pmatrix}.$$

Note the similarity with (3.137a)–(3.137c), the first unitary operator in (6.34) describing free propagation during a time interval T and the second one, the "kick" operator which acts during an "infinitesimally short" time interval. It was proved in [228] that Weigert's configurational quantum cat has a positive upper quantum Lyapunov exponent (6.23a). Of course, the profound difference between (6.34) and (3.137c) lies in the nature of the kick operator, reflecting the fact that the classical analogue of (3.137c) is, unlike (6.34), a mixed system, a major source of difficulty.

The importance of Anosov systems in classical statistical mechanics is that certain "billiards" (caricatures of the hard-sphere gas — itself a caricature of realistic classical interacting gases) are Anosov systems, and this property explains their approach to equilibrium — a subject to which we now turn.

6.3. Approach to Equilibrium

6.3.1. *A brief introductory motivation*

There has been great progress in nonequilibrium quantum statistical mechanics in the last decade, in particular as regards the dynamics of open systems, which started to flourish in the 1970s (see [58] and references given there). The work on NESS (nonequilibrium stationary systems) (see [126] and references given there), which still proceeds intensively (see the recent IAMP article 2011 by Jaksic and Pillet), introduced several new conceptual insights and ideas into the subject. On the other hand, the framework adopted in the quantum theory of open systems of a "small" quantum system, viewed as a system with finite number of degrees of freedom, interacting with a "large" bath or reservoir, depicted as a quantum system with infinite number of degrees of freedom, does not seem adequate — at least at the present time — to describe several of the most interesting quantum systems out of equilibrium, e.g., collective systems carrying a current, such as superconductors or superfluids, which are themselves systems with an infinite number of degrees of freedom. Ultimately, we have our Universe, the most extraordinary dynamical system, which is (by definition!) a closed system which, nevertheless, is not in equilibrium.

This state of affairs, coupled to the fact that there are several excellent textbook treatments and review articles on open systems, led us to restrict ourselves in this book to the theory of the approach to equilibrium of *closed* systems. This theory is, conceptually, very subtle, and only recently there has been significant progress to understand its quantum foundations (see [96, 97, 255]). We shall mostly concentrate on the classical theory in Secs. 6.3.2–6.3.5, and, after an interlude (Sec. 6.4), turn to quantum systems with an infinite number of degrees of freedom in Sec. 6.5.

6.3.2. *Approach to equilibrium in classical (statistical) mechanics* 1: *Ergodicity, mixing and the Anosov property. The Gibbs entropy*

In this section we assume that the reader is familiar with the content of App. A.

We consider the dynamical system generated by the motion of N matter points (gas molecules) in a fixed volume V, as in App. A ((A.37) et seq.). Corresponding to the flow equation

$$x_t = T_t(x_0) \tag{6.35a}$$

we have, for the measure describing the evolution of the system from an initial distribution μ_0, (A.38), the equation:

$$\mu_t(B) = \mu_0(T_{-t}(B)). \tag{6.35b}$$

Let ρ_0 denote the (Radon–Nikodym) derivative of μ_0 with respect to the (microcanonical Gibbs) measure (A.39), supposed to exist a.e. (see (A.38) et seq.). Then (6.35b) implies

$$\rho_t(x) = \rho_0(T_{-t}(x)) \tag{6.35c}$$

(Liouville's theorem). Let G be a dynamical variable, and let us assume that

$$G \in L^1(\Gamma, d\mu) \tag{6.35da}$$

and

$$\rho_0 \in L^\infty(\Gamma, d\mu). \tag{6.35db}$$

Then, the expectation value of G at time t is given by

$$\langle G \rangle_t = \int_\Gamma G(x)\rho_t(x)d\mu(x) \tag{6.35e}$$

with ρ_t given by (6.35c); under assumption (6.35da), $\langle G \rangle_t$ is finite. Approach to equilibrium in the sense of (A.43) means that $\langle G \rangle_t$ approaches a limit, which is the (microcanonical Gibbs) *equilibrium value* $\langle G \rangle_{\rm eq}$ of G:

$$\lim_{t \to \infty} \langle G \rangle_t = \langle G \rangle_{\rm eq}, \tag{6.35fa}$$

where

$$\langle G \rangle_{\rm eq} = \int_\Gamma G(x)d\mu. \tag{6.35fb}$$

Equation (6.35fa) follows just as (A.43): this result holds for *mixing systems*, as shown and explained in App. A.

Thus, we are now concerned with decay, not to zero, but to an equilibrium value: more generally, we have correlation functions, and one may inquire into their rate of decay, e.g., whether it is exponential, see (A.63).

There remains the important question of how to actually *prove* mixing in real physical systems. It turns out that this is a mathematically most profound and difficult issue, and we shall only indicate the main ideas involved. In App. A we showed how the exponential instabilities of Anosov systems, such as the cat map (A.62), lead to the mixing property. As remarked by Penrose, it was recognized implicitly by Gibbs and explicitly by Krylov that the kind of exponential instability of Anosov systems could also occur in systems such as the hard sphere gas, which are caricatures of "realistic" systems. This problem is already very difficult, but a caricature of its dynamical behavior is suggested by billiards, e.g., the motion of a billiard ball on a square table with a circular obstacle in the middle. It turns out that even this is an enormously difficult problem, solved by Sinai (see the references in [202]), but an idea of why exponential sensitivity takes place may be obtained thus (see [202, p. 1950]): consider a small frictionless billiard ball moving on a table on which there are some perfectly elastic circular obstacles of radius a. Denote the distance traveled by the ball before its first collision with an obstacle by l_1. Then, if the initial direction of motion is changed by a small angle ϵ, the first point of impact shifts by at least ϵl_1, so that the direction of the boundary there changes by at least $\epsilon l_1/a$. The direction of the ball after reflection changes therefore by at least $2\epsilon l_1/a$, in addition to the original ϵ, making $(1 + 2l_1/a)\epsilon$ in all. This magnification of the change in direction by the first reflection is repeated by further collisions: the second one yields $(1 + 2l_2/a)$, etc., where l_2 is the distance traveled between the first and second collisions, and so on, and if l denotes the mean free path, the angular change is magnified by $(1 + 2l/a)$ per collision — an exponential rate of increase — until the path deviations become of the order of the size of one of the obstacles. We refer to Penrose (see [202, p. 1951] et seq.) for early references and a still very readable and pedagogic discussion, to [24] for a colorful description of several types of billiard and of the "defocalizing shocks" (with pictures), and to [167] for a complementary discussion, with a selected list of references to the more recent literature.

The Gibbs entropy and the second law. The second law of thermodynamics tells us that there exists an additive thermodynamic function — the entropy — which is defined at equilibrium and which does not decrease in an adiabatic process, i.e., a process in which the system exchanges no heat with its surroundings.

Gibbs assumed that the "entropy" $S_f(\rho)$ associated to a phase space density ρ is given by

$$S_f(\rho) = \int_\Gamma f(\rho(x))dx, \qquad (6.36a)$$

where f is a function to be chosen. By Liouville's theorem, however

$$S_f(\rho_t) = S_f(\rho_0). \qquad (6.36b)$$

Choosing f to be a concave function (see the forthcoming (6.55a)), modeling the adiabatic process by a time-dependent Hamiltonian starting at $t = 0$ and ending at $t = 1$, after which time the Hamiltonian is assumed to remain constant and yield an ergodic and mixing motion, and defining the final "coarse grained" equilibrium density $\bar\rho$ by an ergodic average,

$$\bar\rho(x) \equiv \lim_{T\to\infty} \frac{1}{T} \int_1^{1+T} \rho_t(x)dt \qquad (6.36c)$$

it may be proved that the entropy which describes the final equilibrium is higher than the one describing the initial equilibrium (see [202]); this is *Gibbs' theorem*. Further requiring the property of additivity almost completely fixes the form of the function f, see [202] and references given there, and yields the *Gibbs entropy*:

$$S_G(\bar{\rho}) = k \int_\Gamma \bar{\rho}(x) \log\left(\frac{c}{\bar{\rho}(x)}\right) dx, \qquad (6.37a)$$

where k is Boltzmann's constant, and c is a constant which may differ from one system to another. In classical mechanics, the simplest choice — the "thickened microcanonical ensemble" — is

$$\rho(x) = \begin{cases} \text{const.} & \text{if } E - \Delta E < H(x) \leq E, \\ 0 & \text{otherwise,} \end{cases} \qquad (6.37b)$$

where ΔE is a positive number representing the accuracy ("tolerance") of energy measurements. This choice yields

$$S_G^{\text{micr}} = k \log W \quad \text{with } W = c \int_{E-\Delta E < H(x) \leq E} dx. \qquad (6.37c)$$

The quantum analogue of (6.37a), with ρ replaced by the density operator, and the integration over phase space by the trace, is the *von Neumann entropy*

$$S = -k \operatorname{Tr}(\rho \log \rho). \qquad (6.38)$$

The above approach to irreversibility has some drawbacks: in particular, one compares the Gibbs entropy of the initial density with that of a "coarse-grained density", but the ergodic "coarse graining" is meaningless for the initial distribution. The next model is a probabilistic model, which illuminates other important aspects of this subtle phenomenon.

6.3.3. *Approach to equilibrium in classical mechanics 2; The Ehrenfest model*

The ergodic average used to obtain the "coarse graining" also appears, but in a different way, in a probabilistic model proposed by Ehrenfest. We follow the exposition of Galves, Nogueira and Vares [89].

The following elementary concepts of Markov chains [43] are essential for the development.

Definition 6.2. Let S be a finite or countable set, $T \subseteq [0, \infty)$. A collection of random variables (r.v.) $\{X(t), t \in T\}$ in a probability space $(\Omega, \mathcal{B}, P)$ is a *Markov process* with values in S if $\forall\, t_1 < t_2 < \cdots < t_n < t$:

$$P(X(t) = x | X(t_1) = x_1, \ldots, X(t_n) = x_n) = P(X(t) = x | X(t_n) = x_n),$$
(6.39)

where $x_1, \ldots, x_n \in S$ and $P(A|B)$ is the conditional probability $P(A|B) = \frac{P(A \cap B)}{P(B)}$.

In the case $T = \{0, 1, 2, \ldots\}$ we have a *Markov chain* (M.C.) with discrete parameter. In addition:

(i) if $P(X(n+1) = y | X(n) = x)$ does not depend on n, but only on x, y, we say the M.C. is *homogeneous in time*. This function $P(x, y)$ is called *transition matrix* (t.m.);

(ii) $X(\cdot)$ is *stationary* if for all $n \geq 1$, for all $k_1, \ldots, k_n, k \in \mathbb{N}$, $(X(k_1), \ldots, X(k_n))$ and $(X(k_1 + k), \ldots, X(k_n + k))$ have the same distribution;

(iii) $X(\cdot)$ is *reversible* if, for all $k_1, \ldots, k_n \in \mathbb{N}$, if $k_i \leq k\ \forall\, i = 1, \ldots, n$, $(X(k_1), \ldots, X(k_n))$ and $(X(k - k_1), \ldots, X(k - k_n))$ have the same law;

(iv) A probability distribution $\pi(i), i \in S$ is *stationary* for the M.C. $X(\cdot)$ if $X(0)$ with distribution π implies $X(1)$ has distribution π:

$$\pi(j) = \sum_{i \in S} \pi(i) P(i, j) \quad \forall\, j \in S.$$
(6.40)

If π is stationary for the M.C. and $X(0)$ has distribution π, then $X(n)$ will be a stationary process.

(v) i is a *recurrent state* (respectively, *positive recurrent state*) if

$$P_i(T_i < \infty) = 1$$
(6.41)

(respectively, $\mathbb{E}_i(T_i) < \infty$) where

$$T_i = \begin{cases} \min\{n \geq 1 : X(n) = i\} & \text{if this set is not empty,} \\ +\infty & \text{otherwise,} \end{cases}$$
(6.42)

P_i denotes the distribution of the M.C. conditioned to $X(0) = i$, and $\mathbb{E}_i$ is the expectation relative to P_i.

If all states of a M.C. are recurrent (respectively, positive recurrent), we say that the M.C. is recurrent (respectively, positive recurrent). We have the following theorem.

Theorem 6.4 (Ergodic theorem for Markov chains). *Let $X(n)$ be a finite irreducible (i.e., all states "communicate with each other", see [43, Paragraph 8.4]) M.C. Then, for any initial distribution and for all $k \in S$,*

$$\lim_{n \to \infty} \frac{1}{n} \sum_{m=1}^{n} \chi(X(m) = k) = \frac{1}{\mu_k} \qquad (6.43a)$$

with probability one, where

$$\mu_k = \mathbb{E}_k(T_k) \qquad (6.43b)$$

is the average time of recurrence for the state k. Moreover,

$$\pi(k) = \frac{1}{\mu_k} \qquad (6.43c)$$

provides the unique stationary distribution for the chain.

For the proof, see [43, Theorems 10–12].

Coming back to Ehrenfest's model, let us suppose that there are M balls, numbered from 1 to M, which, at a given instant of time, are distributed among two urns A and B. At each previously chosen unit of time a number in the set $\{1, 2, \ldots, M\}$ is drawn at random (uniform distribution in $\{1, 2, \ldots, M\}$), each trial being independent of any other one. The ball with the chosen number is required to change urns: this defines the "dynamics". Let $X(0) = (x_1(0), \ldots, x_M(0))$ be the "initial state", where

$$x_i(0) = \begin{cases} 1 & \text{if the } i\text{th ball is in } A, \\ 0 & \text{otherwise}, \end{cases} \qquad (6.44a)$$

and let $X(n) = (x_1(n), \ldots, x_M(n))$ be the state of the system after the nth trial. We have then a homogeneous reversible M.C. $(X(n))_{n \geq 0}$ with

$$S = \{0, 1\}^M = \{x = (x_1, \ldots, x_M), \ x_i = 0 \text{ or } 1\} \qquad (6.44b)$$

as state space, and 2^M possible states, corresponding to a "microscopic description" of the system: it identifies the position — at the urn A or B — of each "particle" (ball) in the "gas" of M particles. The transition matrix is homogeneous and given by

$$P(x, y) = \begin{cases} \dfrac{1}{M} & \text{if } x, y \in S \text{ differ by precisely one coordinate}, \\ 0 & \text{otherwise}. \end{cases} \qquad (6.45a)$$

By (6.40) and the fact that, by definition, $\sum_{x \in S} \pi(x) = 1$, we find that the (unique) stationary distribution for this M.C. is

$$\pi(x) = 2^{-M} \quad \forall x \in S. \tag{6.45b}$$

Now comes the basic point: suppose, now, that instead of this microscopic distribution, we only have access to the *total* number of balls in A: i.e., after each trial, as well as for the initial state, the dice-throwing observer just communicates the new number of balls in A; our information is thus reduced to

$$Y(n) = \sum_{i=1}^{M} x_i(n) \quad \forall n \geq 0. \tag{6.46a}$$

For large M, this means an *enormous reduction of information*. The new process $(Y(n))_{n \geq 0}$ is now a M.C. with state space $S' = \{0, 1, \ldots, M\}$, with transition matrix $Q = Q(i, j)$, where

$$Q(i, i+1) = 1 - \frac{i}{M} \quad \text{and} \quad Q(i, i-1) = \frac{i}{M} \tag{6.46b}$$

(check!), for $i = 0, \ldots, M$, $Q(i, j) = 0$ otherwise. Obviously, it is always more likely that the balls corresponding to the chosen number belong to the urn with the largest number of balls, and it is therefore more likely that the difference in the number of balls in A and B tends to decrease. A direct consequence of the information reduction is thus a "restoring force" tending to equalize the number of balls in A and B: indeed, for i very small, $Q(i, i+1) \gg Q(i, i-1)$.

Problem 6.5. *Show that*

$$p(i) \equiv \binom{M}{i} 2^{-M}, \tag{6.46c}$$

where $i = 0, \ldots, M$, *verifies*

$$p(i) \geq 0, \quad \sum_{i=0}^{M} p(i) = 1 \tag{6.46d}$$

and

$$p(j) = \sum_{i=0}^{M} p(i)Q(i, j). \tag{6.46e}$$

Thus, $\{p(i)\}_{i=0,\ldots,M}$ is the unique stationary probability distribution for the M.C. $(Y(n))_{n=0,\ldots,M}$. Since the latter is a finite M.C., and thus positive recurrent, the chain with starting point i returns to i in finite time with probability one. We now use Theorem 6.4. By (6.43a), if $N_n(k)$ denotes the number of returns to the state k up to the instant n (note the similarity to (6.7)), $\lim_{n\to\infty} N_n(k)/n = 1/\mu_k$, but clearly $\sum_{k=0}^{M} N_n(k)/n = 1$, for all $n \geq 1$, which implies that $\sum_{k=0}^{M} 1/\mu_k = 1$ $(1/\mu_1 > 0)$, meaning that $a_k = 1/\mu_k$ defines a probability distribution on the space of states S'. This is a first indication of the statement in Theorem 6.4 that $\pi(k) = 1/\mu_k$ provides the unique stationary distribution for the chain; for the rest of the proof, which is elementary but nontrivial, see [43, Theorem 12, p. 278]. Equation (6.46c) now implies that

$$\mu_k = \frac{2^M}{\binom{M}{k}}. \tag{6.46f}$$

It follows from (6.46f) that the average recurrence time for the equilibrium state $k = M/2$ is

$$\mu_{M/2} = \frac{2^M}{\binom{M}{M/2}} \sim \sqrt{\pi M/2} \tag{6.46g}$$

time units, very much smaller (for large M) than the recurrence time for states far from equilibrium: e.g., if $k = 0$, or $k = M$,

$$\mu_0 = \mu_M = 2^M. \tag{6.46h}$$

It may be seen that, for large M, the curve for $\Delta_n \equiv |2Y(n) - M|$ as a function of n, for $Y(0) = M$, initially drops from M to zero, with various subsequent fluctuations close to zero, but then returns to M several times (with probability one): however, the *average time* spent near M or zero is *exceedingly small* in comparison with the same quantity at equilibrium, i.e., for $Y(n) = M/2$ (see (6.46g) and (6.46h)). This is true independently of *whether the "curve" for Δ_n is traversed from left to right or from right to left!* Thus, there is no conflict with reversibility!

Although the "dynamics" of the model is not physical, we may provide a "physical" interpretation of the Ehrenfest model by identifying the balls with particles, and assuming that the two urns represent boxes with the same volume containing an "ideal gas", separated by a permeable membrane. We may assume that the macroscopic variable of interest is the pressure, which is proportional to the number of particles, with the temperature fixed. At equilibrium, one expects equality of pressures, and hence of the particle numbers in each box. It is clear that no return of the gas to the initial state with all the particles in one box is expected or observed. However, the reversibility does not have this consequence, because traversing the above-mentioned "curve" in the direction of decreasing, instead of increasing n does not change the proportions of the average times spent near, and far from, equilibrium. That this is, however, not the whole story, and is illustrated by a remark due to Schrödinger [232], quoted in [167]:

"First, my good friend, you state that the two directions of your time-variable, from $-t$ to t, and from t to $-t$, are *a priori* equivalent. Then, by fine arguments appealing to common sense, you show that disorder (or 'entropy') must, with overwhelming probability, increase with time. Now, if you please, what do you mean by 'with time'? Do you mean in the direction $-t$ to t? But, if your inferences are sound, they are equally valid for the direction $+t$ to $-t$. If these two directions are equivalent *a priori*, then they remain so *a posteriori*. The conclusions can never invalidate the premise. Then your inference is valid for both directions, and that is a contradiction."

Thus, Ehrenfest's model, while illustrating well part of the mechanism of the issue "arrow of time" versus reversibility, viz. the "restoring force" due to the loss of information in a large system, does not account for the process of entropy growth before achieving equilibrium. This is believed to be related to the initial state, in particular, to the initial state of the Universe, to which we now turn our attention.

6.3.4. *Approach to equilibrium in classical statistical mechanics 3: The initial sate, macroscopic states, Boltzmann versus Gibbs entropy. Examples: Reversible mixing systems and the evolution of densities*

In Sec. 6.3.2 we mentioned the important point that our Universe is a (by definition) closed system which is not in equilibrium. Indeed, the matter content of the Universe today consists of ordinary matter of which the stars are made, largely hydrogen atoms, the photons of the background radiation, a background of neutrinos, and a considerable amount of *dark matter* (ca. one third of the matter content of the Universe) of unknown origin. Since the present energy of the photons in the background radiation (≈ 2.9 K) is much below the ionization energy of hydrogen (about 3000 K), there is at most interaction of the hydrogen atoms with the tail of the black-body background radiation; the photons of the background radiation are therefore *not* in equilibrium with the matter in the Universe, i.e., the Universe is not in equilibrium, and the "temperature of the Universe" is only the *formal* temperature of the background radiation. The *hot big bang model of the Universe* (see [173, 16.4], for an elementary but very carefully presented account) asserts that the aforementioned equilibrium must have occurred when the Universe was about 1000 times smaller than today, i.e., a time before which hydrogen atoms would not have been able to exist.

According to the hot big bang model, the entropy per unit volume S_γ of the photons in the background radiation is (see [173])

$$S_\gamma = 3.6 n_\gamma, \tag{6.47a}$$

where n_γ is the photon density, and

$$S_\gamma R^3 = \text{const.}, \tag{6.47b}$$

where $R = R(t)$ (the "radius of the Universe") is the *scaling factor* (see [173, 16.3] et seq.), and t is the "cosmic time". The matter density $\rho = \text{const.} \, R^{-3}$, and thus

$$S_\gamma = 3.6 n_\gamma = \text{const.} \, R^{-3} = \text{const.} \, \rho. \tag{6.47c}$$

The ratio S_γ / ρ, or n_γ / ρ (up to a fixed constant), is called the *specific entropy* of the Universe. Equation (6.47c) shows that it does not change during the expansion of the Universe. Very surprisingly, most of the entropy of the Universe lies in the background radiation! Therefore, the *initial* (shortly after the big bang) specific entropy *determines* the nature of the Universe, as we now explain.

As is well-known, in stars there occurs the fusion of hydrogen to helium and other heavier elements, and nuclear energy is transformed into other forms of energy. Nuclear energy is liberated at a temperature of millions of degrees, and irradiated to the outside at temperatures of the order of only a few thousands of degrees (see [258]). Since they receive energy at high temperature and liberate it at low temperature, they are "entropy generating machines": the energy radiated and the intensity of light in space is a measure of the entropy generated by them. Everything radiates, even black holes, and thus *entropy increases*. We must, therefore, understand the entropy increase and consequent irreversibility (due to the second law) in a closed system: this is our aim in the present section.

The entropy increase in the Universe today is, however, not very pronounced, due to the aforementioned fact that most of the specific entropy of the Universe lies in the background radiation, and that part is constant, by (6.47c). This means that the initial specific entropy determines the nature of the Universe, as mentioned above. Indeed, if it had been much smaller, almost the whole hydrogen available would have been converted to helium shortly after the big bang, and the stars would not shine for long periods. On the other hand, if it had been much larger, the Universe would have been too hot for galaxy formation, i.e., stars would probably not exist. In both cases, there would be no life in the Universe, and thus the initial state should play a major role in the presently observed irreversibility. There is still some controversy about the latter assertion, and we refer to [167, Sec. 5] for an entertaining discussion, but should like to end this section with a quotation from Jost's essay "Boltzmann und Planck" [136, p. 48]: "The laws of nature appear surprisingly indifferent with regard to past and future. Boltzmann was indeed right after all: irreversibility rests upon a singular state in the past."

The initial state is not, however, solely responsible for the observed irreversibility: we must restrict ourselves to a "coarse-grained" or reduced description of the system — of its macrostates — as indicated by the Ehrenfest model. There are several approaches to this problem, and we recommend [167] for a lucid exposition. Consider the physical interpretation of Ehrenfest's urn model, mentioned in Sec. 6.3.3, forgetting about the dynamics of the model. Specifically, assume that we have a real gas inside a box of volume V, divided into two halves L and R (A and B in the Ehrenfest model), with volumes V_L and V_R, with $V_L + V_R = V$, $N_L + N_R = N$, and separated by a wall which permits exchange of both energy E and number of particles N. Following [167], assume that the *microstate* $X \equiv (x_1, p_1, \ldots, x_N, p_N)$ of the system is a point in its phase-space $\Gamma = V^N \otimes \mathbb{R}^{3N}$. A *macrostate*, denoted by M, will be a reduced description of the system; as an example, divide V into K cells, where K is still large but $K \ll N$, and specify the number of particles, the momentum, and the amount of energy in each cell. It is clear that M is determined by X, and we shall write $M = M(X)$, but there is a continuum of states corresponding to the same M. Let Γ_M denote the region in Γ consisting of all microstates X corresponding to a given macrostate M. Let Γ_M be the region in Γ consisting of all microstates X corresponding to a given macrostate M, and

$$|\Gamma_M| = (N!\hbar^N)^{-1} \int_{\Gamma_M} \prod_{i=1}^N dx_i dp_i \qquad (6.48a)$$

its symmetrized $6N$-dimensional Liouville volume in units of $\hbar^{3N}$.

Starting with the initial state with all particles constrained to L (all particles in A in the Ehrenfest model), if this constraint is removed at time t_a, for $t > t_a$ the phase-space volume available to the system is fantastically enlarged, by a factor of 2^N. Let

$$s = s\left(\frac{N}{V}, \frac{E}{V}\right) \qquad (6.48b)$$

be the specific entropy of the gas with N particles and energy E, in a box of volume V. The entropy of the constrained system, i.e., of a gas as above but with a completely restrictive wall of separation, which does not allow exchange of energy or particles, is

$$S = V_L s\left(\frac{N_L}{V_L}, \frac{E_L}{V_L}\right) + V_R s\left(\frac{N_R}{V_R}, \frac{E_R}{V_R}\right). \qquad (6.49a)$$

When the wall is lifted, and V_L and V_R united to form V, S is maximized, by the second law, subject to the constraints $E_L + E_R = E$, $N_L + N_R = N$, yielding the equilibrium entropy

$$S_{eq}(N, V) = V_L s(1/2, 1/2) + V_R s(1/2, 1/2) = V s(1/2, 1/2). \qquad (6.49b)$$

As remarked by Lebowitz, it was Boltzmann great insight to provide a microscopic definition of the operationally measurable S_{eq}: he connected S_{eq} to $\log|\Gamma_{M_{eq}}|$, where the macroscopic state M_{eq} corresponding to equilibrium is characterized by the fact that it is the unique macrostate for which $|\Gamma_{eq}|/|\Sigma_{eq}| \sim 1$, where $|\Sigma_E|$ is the total phase-space volume available under the constraint $H(X) \in (E - \Delta E, E)$, and $\sim$ means equality in the limit $N \to \infty$. We are referring here to the "thickened microcanonical" version (6.37c). After reaching M_{eq}, one sees (mostly) only small fluctuations in N_L/N and $E_L(t)/N$ about the value $1/2$ in (6.49b), of the order of the square root of the number of particles involved: this has been indicated by (6.46g) in the case of the Ehrenfest model.

In addition, Boltzmann defined an entropy also for macroscopic systems *not* in equilibrium. Again following [167], associate to each microscopic state X of a macroscopic system a number S_B — the *Boltzmann entropy*, depending only on $M(X)$, given up to multiplicative and additive constants (which may depend on N) by

$$S_B(X) = S_B(M(X)) \tag{6.50a}$$

with

$$S_B(M) = k \log |\Gamma_M|. \tag{6.50b}$$

The above splitting into two equations emphasizes the logical independence of (6.50a) and (6.50b).

It is to be remarked that there exist microstates X_0 which evolve in time in such a way that $|\Gamma_{X_0(t)}| \ll |\Gamma_{\text{eq}}|$. One example are those with their velocities directed away from the barrier lifted at $t = t_a$. The fraction of microstates leading to decrease of the Boltzmann entropy goes to zero exponentially in the number of particles. Indeed, interpreting $\text{Prob}\{M\}$ as the ratio of the fractions of time spent by the system in Γ_M and $\Gamma_{M_{\text{eq}}}$, according to (6.36a) and (6.36c) expressed for nonequilibrium densities $\bar{\rho}$, and identifying $S_f(\bar{\rho})$ with S_B (which we shall expound in greater detail next), (6.37c) and Gibbs' theorem yield

$$\text{Prob}\{M\} \sim \exp[S_B(M) - S_{\text{eq}}] \tag{6.51}$$

and the assertion follows from the extensivity of both $S_B(M)$ and S_{eq} with the number of particles N.

We now come to specific examples. These are rare, but may be found in the theory of dynamical systems when viewed in terms of densities, see App. A, (A.53) et seq. We first provide a motivation along the lines of [161].

Assume that one has a transformation $S : [0,1] \to [0,1]$, and pick a large number of initial states $x_1^0, \ldots, x_N^0$. To each of these states, we apply the map S, thereby obtaining N new states denoted by

$$x_1^1 = S(x_1^0), \; x_2^1 = S(x_2^0), \ldots, x_N^1 = S(x_N^0) \tag{6.52a}$$

and let

$$\chi_\Delta(x) = \begin{cases} 1 & \text{if } x \in \Delta, \\ 0 & \text{otherwise.} \end{cases} \tag{6.52b}$$

Intuitively speaking, $\rho_0(x)$ is the density function for the initial states $x_1^0, \ldots, x_N^0$ if, for every "not too small" interval $\Delta_0 \subset [0,1]$,

$$\int_{\Delta_0} \rho_0(u)\,du \sim \frac{1}{N} \sum_{j=1}^{N} \chi_{\Delta_0}(x_j^0). \tag{6.52c}$$

Likewise, the density function $\rho_1(x)$ for the states $x_1^1, \ldots, x_N^1$ satisfies, for $\Delta \subset [0,1]$,

$$\int_\Delta \rho_1(u)du \sim \frac{1}{N} \sum_{j=1}^N \chi_\Delta(x_j^1). \tag{6.52d}$$

Defining, as usual, $S^{-1}(\Delta) \equiv \{x : S(x) \in \Delta\}$, it follows by definition that $x_j^1 \in \Delta$ iff $x_1^0 \in S^{-1}(\Delta)$, and thus

$$\chi_\Delta(S(x)) = \chi_{S^{-1}(\Delta)}(x). \tag{6.52e}$$

Putting (6.52c) and (6.52e) into (6.52d) we find

$$\int_\Delta \rho_1(u)du \sim \frac{1}{N} \sum_{j=1}^N \chi_{S^{-1}(\Delta)}(x_j^0)$$

$$\sim \int_{S^{-1}(\Delta)} \rho_0(u)du. \tag{6.52f}$$

Equation (6.52f) relates the densities for the initial states and for the evolved states. Defining the action

$$\rho_1(x) = (P\rho_0)(x),$$

one obtains a bounded linear operator — the Ruelle–Perron–Frobenius operator (A.53), which satisfies (A.54), viz.,

$$\int_A (Pf)(x)d\mu(x) = \int_{T^{-1}(A)} f(x)d\mu(x) \tag{6.53}$$

with $d\mu(x) = dx$, $f = \rho$ and $S = T$. It is a Markov operator, i.e., satisfies (A.57) and (A.59) of App. A.

The above notation emphasizes the correspondence between the above concepts and the framework of statistical mechanics (6.35c), although we are dealing with only one degree of freedom: as remarked in App. A, this close analogy depends on the fact that we are describing the evolution of a density, which is entirely different from the evolution of individual trajectories. For instance, two iterations for the quadratic map $S(x) = 4x(1-x)$, $0 \le x \le 1$, on the initial density $\rho(x) = 1$ for all $x \in [0,1]$ yield a $P^2\rho$ almost indistinguishable from the Ulam–von Neumann stationary density, which is given by $\rho_\star(x) = \frac{1}{\pi\sqrt{x(1-x)}}$!! (see [161, Fig. 1.2.2, p. 8]). In comparison, see [161, Fig. 1.1.1], for two qualitatively very different, both extremely erratic, trajectories corresponding to close initial points! Thus, on passing from points to, e.g., characteristic functions of intervals, one already observes "macroscopic behavior", since an interval needs a macroscopic (in the limit infinite) number of points to build a density according to (6.52c).

The Gibbs entropy for classical dynamical systems is defined precisely as (6.37a), with $k = 1$ and $c = 1$ to agree with the convention in [161]: we write

$$S_G(f) = \int_X \eta(f(x)) d\mu(x) \tag{6.54a}$$

which we call the Gibbs entropy of f as before, with $f = \rho$ being the density, and

$$u \to \eta(u) \equiv -u \log u, \quad \eta(0) = 0. \tag{6.54b}$$

We assume we are given a finite measure space $(\Omega, \mathcal{B}, \mu)$, with $\mu(X) < \infty$ and $P : L^1 \to L^1$ is a Markov operator, as defined in (A.57) and (A.59) of App. A: $f \geq 0 : f \in L^1$ is a density. Notice that the integral in (6.54a) over the positive part of $\eta(f(x))^+ = \max[0, \eta(f(x))]$ is always finite, and thus $S_G(f)$ is either finite or equal to $-\infty$. For the following concepts and results on convex functions, see [215].

Definition 6.3. A real-valued bounded function defined on an interval I of $\mathbb{R}$ is *convex* if, for all $x_1, x_2 \in I$, each point $M(x)$ of the graph Γ of f such that $x \in [x_1, x_2]$ is *below* the segment $M(x_1)M(x_2)$:

$$f(\alpha_1 x_1 + \alpha_2 x_2) \leq \alpha_1 f(x_1) + \alpha_2 f(x_2) \tag{6.55a}$$

for all $\alpha_1, \alpha_2 \geq 0 : \alpha_1 + \alpha_2 = 1$.

Now f is said to be *concave* if $(-f)$ is convex. We have the follwing proposition.

Proposition 6.2. *Let f be a real-valued bounded function in an open set I and D an at most countable subset of I. In order that f be convex it is necessary and sufficient that it be continuous, have a right derivative f'_d at each point of $I \backslash D$, and that f'_d be increasing in $I \backslash D$. Thus, if a, b, c are three points of I with $a < b < c$, and $K_1 = \sup_{x \leq b} f'_d(x)$, then*

$$f(b) - f(a) \leq K_1(b - a). \tag{6.55b}$$

Proposition 6.3. *If f has a second derivative at each point of the open set I, convexity of f is equivalent to the condition*

$$f''(x) \geq 0. \tag{6.55c}$$

Problem 6.6. *Show that* (6.55a) *is violated if* (6.55c) *is violated, by setting* $a = x, b = y, 1/2(x + y) = t, 1/2(x - y) = h$ *with* $x > y$, $h > 0$.

Hint: If $f''(t) < 0$, there exist δ and h both strictly positive such that $f'(t + u) - f'(t - u) < -\delta u$, $0 < u \leq h$.

The function η, given by (6.54b), is concave on $(0, \infty)$ by Proposition 6.2, because $\eta''(u) = -1/u$.

Problem 6.7. *Show that concavity of* η *and* (6.55b) *imply Gibbs' inequality*

$$u - u \log u \leq v - u \log v, \quad for\ u, v > 0. \tag{6.55d}$$

Equation (6.55d) has the consequence that, if f and g are two densities s.t. $\eta(f(x))$ and $f(x) \log g(x)$ are integrable,

$$-\int_X f(x) \log f(x) d\mu(x) \leq -\int_X f(x) \log g(x) d\mu(x) \tag{6.55e}$$

with equality holding only for $f = g$. This yields immediately the following proposition.

Proposition 6.4 ([161, Proposition 9.1.1]). *Let* $\mu(X) < \infty$ *and consider all possible densities defined on* X. *Then, in the family of all such densities, the maximal entropy occurs for the constant density*

$$f_0(x) = \frac{1}{\mu(X)} \tag{6.55f}$$

and, for any other f, *the Gibbs entropy is strictly smaller.*

Problem 6.8. *Prove Proposition 6.3 using* (6.55e).

Proposition 6.5. *Let* $(X, \mathcal{B}, \mu)$ *be a finite measure space and* $S : X \to X$ *a reversible (invertible) measure-preserving transformation. If* P *is the Ruelle–Perron–Frobenius operator corresponding to* S, *then*

$$S_G(P^n f) = S_G(f) \quad for\ all\ n \geq 1. \tag{6.56}$$

Proof. It is easy to check that P, defined by

$$(Pf)(x) = f(S^{-1}(x))J^{-1}(x), \tag{6.57a}$$

where

$$J^{-1}(x) \equiv \frac{dS^{-1}(x)}{dx}, \qquad (6.57b)$$

satisfies (6.53) (or (A.54)) and, since (A.54) defines P uniquely by the Radon–Nikodym theorem, P is indeed given by (6.57a) and (6.57b). By assumption, $J^{-1} \equiv 1$, and (6.56) follows from (6.54a), performing the change of variable $y = S(x)$.   □

Proposition 6.5 shows that the Gibbs entropy is time-invariant whenever the dynamics is reversible: this is (6.36b). We now come to an analogue of Boltzmann's entropy. This was already touched upon by Dorfman [72, Chap. 7.2], for the baker map, and we just formulate a natural generalization. We assume $(X = X_1 \times X_2, \mathcal{B}, \mu)$, where X_1, X_2 are one-dimensional compact sets, $d\mu(x,y) = d\mu_1(x)d\mu_2(y)$ with μ_i, $i = 1, 2$, probability measures on X_i, $i = 1, 2$, and P is a Markov operator (6.57a) and (6.57b), with constant stationary density $P1 = 1$.

From now on, we assumed we are given a

density $f = f(x,y)$ which is continuous in $(x,y) \in X_1 \times X_2$ $\qquad$ (6.58a)

i.e.,

$$f \geq 0 \qquad (6.58b)$$

and

$$\int d\mu_1 d\mu_2 f(x,y) = 1 \qquad (6.58c)$$

and that, for f satisfying (6.58a),

$$\sup_{(x,y)\in X_1 \times X_2} |(Pf)(x,y)| \leq 1. \qquad (6.58d)$$

We now define the *reduced density*

$$\tilde{f}(x) \equiv \int d\mu_2(y) f(x,y). \qquad (6.58e)$$

Since μ_2 is finite and X_2 is compact, there exists

$$\int d\mu_2(y)(P^n f)(x,y) \quad \forall n = 1, 2, \ldots. \qquad (6.58f)$$

The choice of μ_2 instead of μ_1 is arbitrary (but may matter in specific examples). It follows from (6.58a)–(6.58c) that

$$\tilde{f} \text{ is a continuous density in } x \in X_1 \tag{6.59a}$$

i.e.,

$$\tilde{f} \geq 0 \tag{6.59b}$$

and

$$\int d\mu_1 \tilde{f}(x) = 1. \tag{6.59c}$$

Definition 6.4. The *Boltzmann entropy* of f is defined by

$$S_B(f) = -\int d\mu_1(x)\eta(\tilde{f}(x)). \tag{6.60a}$$

By (6.59b),

$$-\infty < S_B(f) \leq 0. \tag{6.60b}$$

It is clear from the above definition that S_B is the Gibbs entropy of the reduced density $\tilde{f}$, which agrees with (6.50a) and (6.50b): the latter expresses the fact that the Boltzmann entropy is the Gibbs entropy of the macroscopic state $M(X)$ in the case of the thickened microcanonical ensemble. In our case, the macroscopic state is identified with the reduced density.

Proposition 6.6. *Assume* (6.58a)–(6.58d) *and, in addition, that*

$$\lim_{n \to \infty} \int d\mu_2(y)(P^n f)(x, y) = 1 \quad \forall x \in X_1. \tag{6.60c}$$

Then

$$S_B(f) \leq 0 \quad \forall n \geq 1 \tag{6.60d}$$

and

$$\lim_{n \to \infty} S_B(P^n f) = 0. \tag{6.60e}$$

Proof. Equation (6.60d) follows from Proposition 6.3 because, by definitions (6.54a) and (6.60a), $S_B(f)$ is the Gibbs entropy for $\tilde{f}$, which is also a density by (6.59a). The function η is continuous and uniformly bounded in compact subsets of $\mathbb{R}_+$, which yields (6.60e) by (6.60c), (6.58d) and dominated convergence. $\qquad\square$

We now show that, under assumption (6.58a) (which includes (6.58b) and (6.58b), (6.58d) and (6.60c) hold for two prototypes of reversible, mixing dynamical systems, the baker's transformation (A.48) and the Arnold cat map (A.62), the later itself a paradigm of an Anosov system.

Equation (6.58d) follows from (6.57a) and (6.57b), and (A.48) (or (A.55)) and (A.62), for the baker transformation and the cat map, respectively.

For the baker's transformation, X is the unit square and $\mu_1 = \mu_2 =$ Lebesgue measure on $[0, 1]$. Define, for a density satisfying (6.58a),

$$W_0(x) \equiv \int_0^1 dy f(x, y). \tag{6.60f}$$

Then

$$W_n(x) \equiv \int_0^1 dy (P^n f)(x, y) \tag{6.60g}$$

satisfies the recursion relation

$$W_n(x) = \frac{1}{2}\left[W_{n-1}\left(\frac{1}{2}\right) + W_{n-1}\left(\frac{x+1}{2}\right)\right]. \tag{6.60h}$$

Problem 6.9. *Prove* (6.60h) *from* (A.55).

The solution of (6.60h) may be easily verified to be

$$W_n(x) = \frac{1}{2^n} \sum_{k=0}^{2^n-1} W_0\left(\frac{x+k}{2^n}\right). \tag{6.60i}$$

By the continuity hypothesis in (6.58a), (6.58c) and (6.60i), $\lim_{n\to\infty} W_n(x) = 1$, for all $x \in [0, 1]$, proving (6.60c) for the baker transformation.

For the Arnold cat map (A.62), with the same definitions (6.60f) and (6.60h),

$$W_n(x) = \int_0^1 dy f(G_{2n}x - G_{2n-1}y, G_{2n-2}y - G_{2n-1}x), \tag{6.60j}$$

where G_n are the Fibonacci numbers (3.112). We shall restrict ourselves to a subclass of continuous functions, viz., to those continuous periodic densities f of the form

$$f(x, y) = \sum_{(k,l)\in\mathbb{Z}^2} c_{k,l} \exp[2\pi i(kx + ly)] \tag{6.60k}$$

with

$$c_{0,0} = 1 \tag{6.60l}$$

coming from (6.58c), and such

$$\sum_{(k,l)\in\mathbb{Z}^2} |c_{k,l}| < \infty. \tag{6.60m}$$

Putting (6.60k) into (6.60j), we find that each of the integrals in y appearing in (6.60j) is of the form

$$\int_0^1 dy \exp[2\pi i(-kG_{2n-1} + lG_{2n-2})y] \tag{6.60n}$$

but, since $G_k = \alpha^k + (-1/\alpha)^k$, with $\alpha = \frac{1+\sqrt{5}}{2}$, see (3.112), we find

$$\lim_{n\to\infty} \frac{G_{2n-2}}{G_{2n-1}} = \frac{2}{1+\sqrt{5}} \tag{6.60o}$$

and

$$\lim_{n\to\infty} G_n = \infty. \tag{6.60p}$$

It follows from (6.60o) and (6.60p) that, for n sufficiently large, (6.60k) is zero unless $k = l = 0$, and thus, by (6.60j),

$$\exists\, n_0 \in \mathbb{N} : W_n(x) = 1 \quad \text{if } n \geq n_0 \tag{6.60q}$$

and thus,

$$\lim_{n\to\infty} W_n(x) = 1 \quad \forall\, x \in [0,1]$$

proving (6.60c) for the cat map.

For the cat map it is immaterial whether the x-coordinate or the y-coordinate is chosen in the definition (6.59a)–(6.60c) of the Boltzmann entropy, but for the baker's transformation it is not. Indeed, the x-coordinate of the baker's transformation (A.48) is transformed by the one-sided shift T_2, (A.43) with $r = 2$, which is noninvertible and leads to an increase of the Gibbs entropy (see [161, Theorem 9.3.2]). Thus it is reasonable that the loss of information of the y-coordinate leads to irreversibility, and not vice versa: the y-coordinate is an "irrelevant" variable, analogous to one whose characteristic variation time is a duration of a collision in the Bogoliubov theory (see the discussion in [72, p. 94]).

Mathematically speaking, the variable y is a "slow" variable in the theory of averaging, see [59, 264].

Some final remarks are in order. A first remark is that for dynamical systems the entropy is negative and its maximum value is zero. This has to do with the fact that the volumes become infinitesimally small instead of infinitely large, see (6.52a) et seq.. In order to bring the analogy with macroscopic systems in statistical mechanics closer, one may regard a system of N "particles" in a unit box, each of them subject to an "effective" map which preserves some of the properties it would have in a more realistic system, e.g., the aforementioned frictionless billiard ball moving on a table with circular obstacles of a given radius; the property is exponential sensitivity to initial conditions. The Boltzmann entropy then becomes

$$S_B^N(f) = -N \int d\mu_1(x)\tilde{f}(x)\log\tilde{f}(x) \tag{6.60r}$$

and (6.51) becomes

$$\text{Prob}\{\tilde{f}\} \sim \exp[S_B^N(f) - S_B^N(1)] = \exp[S_B^N(f)], \tag{6.60s}$$

where the probability is assumed to correspond to the "coarse graining" (6.36c). In this respect, densities distinct from the equilibrium density are exponentially suppressed, e.g., for $\tilde{f} = \frac{\chi_M}{|M|}$, where $M \subset [0, 1]$, (6.60s) yields $\text{Prob}\{\tilde{f}\} = |M|^N$. Again assuming that the "coarse graining" is given as in (6.36c), we see that not *every* microscopic state (point in the unit square) leads to the macroscopic equilibrium state (density). An example is explicitly given by the baker map, for which the dyadic rationals are excluded: they build the exception to Theorem A.5. The fact that they comprise a set of zero Lebesgue measure is the dynamical-system counterpart to the set of "bad microscopic states" being "small".

A second remark concerns the interpretation given above to the system of independent "particles". The same result (Proposition 6.6) is obtained for different classes of dynamical systems — the baker transformation is not (everywhere) a diffeomorphism, and is a Bernoulli system, the Arnold map is a diffeomorphism and is a K-system, but both possess the fundamental properties of being reversible and exhibiting exponential sensitivity to initial conditions. This suggests that a result such as Proposition 6.5 does not depend on the class of system, provided it has the fundamental properties, and also that the definition of reduced density or macrostate may be subject to variations: it is, however, important to keep track of the "slow" and "fast" variables, such as in the passage from "phase-space functions" to "single-particle distribution" functions [72].

6.3.5. *Approach to equilibrium in quantum systems: Analogies, differences, and open problems*

In quantum mechanics the entropy is given by the von Neumann entropy (6.38): as Lebowitz remarks [167], the definition of a macrostate M in quantum mechanics also goes back to von Neumann; one specifies, within some tolerance, the eigenvalues $\{M_\alpha\}$ of a set $\hat{M}$ of "rounded-off" commuting macroscopic observables, i.e., operators representing particle number, energy, etc. in each of the cells into which the box containing the system is divided: this yields a decomposition of the Hilbert space $\mathcal{H}$ describing the system as

$$\mathcal{H} = \bigoplus_\alpha \Gamma_\alpha = E_\alpha \mathcal{H} \qquad (6.60t)$$

into orthogonal components Γ_α, of dimensions $|\Gamma_\alpha|$; the E_α are the projectors onto Γ_α. Let, now,

$$p_\alpha(\rho) \equiv \mathrm{Tr}(E_\alpha \rho). \qquad (6.60u)$$

Then the macroscopic entropy S_{mac} of a system with density matrix ρ is defined by

$$S_{\mathrm{mac}}(\rho) = -k\mathrm{Tr}(\tilde{\rho}\log\tilde{\rho}), \qquad (6.60v)$$

where

$$\tilde{\rho} \equiv \sum_\alpha \left(\frac{p_\alpha}{|\Gamma_\alpha|}\right) E_\alpha. \qquad (6.60w)$$

Equation (6.60u) is the analogue of the "coarse-grained" or reduced density (6.57a) with (6.60v) the analogue of (6.57b). Taking the decomposition (6.60t) as the analogue of the partition of classical phase space $\Gamma = \bigcup_M \Gamma_M$, one is tempted to define the quantum analogue of $S_B(M)$ (6.50b) by

$$S_B^Q(M_\alpha) = k\log|\Gamma_{M_\alpha}|). \qquad (6.60x)$$

As Lebowitz remarks [167], one expects that $S_B^Q(M)$ will increase or stay constant with time after a constraint is lifted in a macroscopic system, until the system reaches the macroscopic equilibrium state M_{eq} corresponding to the microcanonical density matrix.

In spite of these similarities, there are also differences, as Lebowitz aptly points out in his review. Indeed, the analogue of (6.50a) does *not* hold in quantum theory: even when the system is in a microscopic state ρ corresponding to a definite macrostate at time $t = t_0$, a quantum system will not be in this unique macrostate for all times t. For instance, if $\rho = E_\Psi$, the projector onto a pure state Ψ, at $t = t_0$, may evolve to a linear combination of wave-functions corresponding to different macrostates. Entanglement (quantum correlations between separated systems) invalidates the classical concept of isolated states in quantum theory.

In spite of the above-mentioned difficulties, very great progress has been made in this issue, in connection to a generalization of von Neumann's ergodic theorem in quantum mechanics [96, 97, 255]. The treated examples are still very artificial (see [255]), but one also hopes for much further progress along these lines.

One should also mention the work [69] on the quantized version of the Kac ring model. We feel, however, that chaotic dynamics, which plays a major role in the classical theory of approach to equilibrium in classical mechanics, is also expected to play a major role in quantum mechanics, not only semiclassically, but in some intrinsic quantum mechanical fashion, viz. our discussion of quantum Anosov systems. It may, therefore, also be of interest to look at the quantum counterparts of maps such as those considered in Sec. 6.3.4, e.g., the quantum baker map (which is far from trivial, see [60]), or the Weigert model (6.63), which even contains a kinetic energy and has a physical interpretation, related to the cat map, in the hope that their treatment suggest possible Ansätze for realistic quantum systems.

6.4. Interlude: Systems with an Infinite Number of Degrees of Freedom

In this interlude, we restrict ourselves to explain the main concepts of the theory of systems with infinite number of degrees of freedom. We must be of necessity brief: fortunately, excellent textbook surveys exist, to which the reader may refer for further details (see [122, 237]).

6.4.1. *The Haag–Kastler framework*

In 1964, Haag and Kastler [106] proposed a theory of local algebras of observables which built on the following general ideas:

(i) it should be based on the algebraic structure of the observables;

(ii) it should work with local observables, i.e., observables describing events in *bounded* regions of space–time.

Idea (i) may be regarded as a variation on a theme which appeared in the early days of quantum mechanics, where the canonical commutation relations of the coordinates and momenta generate an algebra which determines the structure of nonrelativistic quantum mechanics, and idea (ii) was influenced by relativistic quantum field theory (rqft) where the basic objects are operator-valued distributions. As an example, the electric field $\vec{E}(x)$ has to be smeared with a smooth test function f on space–time, before one obtains a well-defined operator $\vec{E}(f) = \int dx f(x) \vec{E}(x)$ in the Hilbert space of states $\mathcal{H}$. Physically, this is due to the behavior of fluctuations of the electric field in, e.g., the (Fock) vacuum, see the discussion in [224, p. 33] and the reference to Bohr–Rosenfeld. One may associate such an operator with a space–time region $\mathcal{O}$, if f has compact support contained in $\mathcal{O}$. One may think of the Haag–Kastler algebra associated to the region $\mathcal{O}$ as consisting of bounded functions of observables such as the components of $\vec{E}(f)$.

Haag and Kastler (HK) assumed that to each bounded space–time region $\mathcal{O}$ there is associated an algebra $\mathcal{A}(\mathcal{O})$, and $\mathcal{O}_1 \subset \mathcal{O}_2$ implies $\mathcal{A}(\mathcal{O}_1) \subset \mathcal{A}(\mathcal{O}_2)$ (*isotony*). By taking the union over all $\mathcal{O}$ one gets the local algebra $\cup_{\mathcal{O}} \mathcal{A}(\mathcal{O})$. In addition, HK assumed that their algebras $\mathcal{A}(\mathcal{O})$ were C^*-*algebras*, which are Banach $\star$-algebras (see [31]): vector spaces over the complex numbers, which are also associative algebras over $\mathbb{C}$, having an operation with the properties which make them into a $\star$-algebra. The norm satisfies the properties of the norm in a Banach space, and completeness in this norm is assumed: this makes the algebra into a Banach $\star$-algebra. Finally, the property

$$\|A^{\star} A\| = \|A\|^2$$

is assumed, which makes the Banach $\star$-algebra into a C^*-algebra.

The norm-closure of the local algebra is a $C*$-algebra $\mathcal{A}$ is called the *quasi-local algebra*, which is the basic element defining the infinite system as an approximation of bounded systems: it is referred to as a net of local algebras. Its most basic property in the HK framework is Einstein causality or *local commutativity of the observables*:

$$[\mathcal{A}(\mathcal{O}_1), \mathcal{A}(\mathcal{O}_2)] = 0 \quad \text{if } \mathcal{O}_2 \subset \mathcal{O}_1'. \tag{6.61}$$

Above, $\mathcal{O}'$ denotes the space-like complement of $\mathcal{O}$, i.e., the set of points y such that $(y - x)$ is space-like for all $x \in \mathcal{O}$.

In order to complete the formulation of HK, a fundamental concept is necessary: a *state* on a $C*$-algebra $\mathcal{A}$ is defined to be a positive linear form

ω on $\mathcal{A}$, normalized to one on the unit element $\mathbb{I}$ of $\mathcal{A}$ (we assume it has one), i.e.,

$$\omega(\alpha A + \beta B) = \alpha \omega(A) + \beta \omega(B),$$
$$\omega(A^\star A) \geq 0,$$
$$\omega(\mathbb{I}) = 1.$$

The set of states on $\mathcal{A}$ will be denoted by $E_{\mathcal{A}}$. If $\omega, \omega' \in E_{\mathcal{A}}$, define

$$\|\omega - \omega'\| \equiv \sup_{A \in \mathcal{A}} \frac{|\omega(A) - \omega'(A)|}{\|A\|}.$$

Norm convergence and weak* convergence in $E_{\mathcal{A}}$ are defined as

$$\text{norm} \lim_{\alpha} \omega_\alpha = \omega \iff \lim_{\alpha} \|\omega_\alpha - \omega\| = 0,$$
$$\text{weak}^* \lim_{\alpha} \omega_\alpha = \omega \iff \lim_{\alpha} [\omega_\alpha(A) - \omega(A)] = 0 \quad \forall A \in \mathcal{A}.$$

An *automorphism* τ of a $C*$-algebra $\mathcal{A}$ is a transformation τ of $\mathcal{A}$ which preserves its algebraic structure, i.e.,

$$\tau(\alpha A + \beta B) = \alpha \tau(A) + \beta \tau(B),$$
$$\tau(A^\star) = (\tau(A))^\star,$$
$$\|\tau(A)\| = \|A\|.$$

A very important theorem is the following.

Theorem 6.5. $E_{\mathcal{A}}$ *is compact in the weak*-topology.*

Theorem 6.5 follows from the Banach–Alaoglu theorem, see [207, Theorem IV-21].

A **-representation* π of $\mathcal{A}$ on a Hilbert space $\mathcal{H}$ is a mapping π of $\mathcal{A}$ into $\mathcal{B}(\mathcal{H})$ (the algebra of bounded operators on the Hilbert space $\mathcal{H}$) such that

$$\pi(\alpha A + \beta B) = \alpha \pi(A) + \beta \pi(B),$$
$$\pi(AB) = \pi(A)\pi(B),$$
$$\pi(A^\star) = (\pi(A))^\star.$$

The representation π is *cyclic* with cyclic vector Ω if the set of vectors $\{\pi(A)\Omega$ with $A \in \mathcal{A}\}$ is dense in $\mathcal{H}$. The Gelfand–Naimark–Segal construction (GNS) [32] states that, given a state ω on $\mathcal{A}$, there exists a cyclic representation π_ω on a Hilbert space $\mathcal{H}_\omega$ with cyclic vector Ω_ω, such that

$$\omega(A) = (\Omega_\omega, \pi_\omega(A)\Omega_\omega). \tag{6.62a}$$

Furthermore, let $G : g \to \tau_g$ be a representation of a group G by automorphisms of $\mathcal{A}$, and ω be invariant under G in the sense that

$$\omega(\tau_g(A)) = \omega(A) \quad \forall g \in G \quad \text{and} \quad \forall A \in \mathcal{A}. \tag{6.62b}$$

Let $E_{\mathcal{A}}^G$ denote the set of states invariant under G. Then there exists a unitary representation of G, $G : g \to U(g)$, such that

$$U(g)\pi_\omega(A)U(g)^{-1} = \pi_\omega(\tau_g(A)) \tag{6.62c}$$

and

$$U(g)\Omega_\omega = \Omega_\omega. \tag{6.62d}$$

A natural candidate for $U(g)$ is

$$U(g)\pi_\omega(A)\Omega_\omega \equiv \pi_\omega(\tau_g(A))\Omega_\omega. \tag{6.62e}$$

Clearly the above defines $U(g)$ on a dense set in $\mathcal{H}_\omega$, where it is isometric

$$\|U(g)\pi_\omega(A)\Omega_\omega\|^2 = \omega((\tau_g(A))^\star \tau_g(A))$$
$$= \omega(\tau_g(A^\star A)) = \omega(A^\star A) = \|\pi_\omega(A)\Omega_\omega\|^2.$$

The inverse of $U(g)$ is $U(g^{-1})$, so that $U(g)$ has a unitary extension, which we also call $U(g)$: we say $U(g)$ *implements* the automorphism group τ_g. If $G = \mathbb{R}$, with addition as group operation, τ_t is the dynamical evolution and ω is τ_t-invariant, then

$$\pi_\omega(\tau_t(A)) = U_t \pi_\omega U_t^{-1}. \tag{6.62f}$$

By Stone's theorem

$$U_t = \exp(itH_\omega), \tag{6.62g}$$

where the self-adjoint operator H_ω is the *physical Hamiltonian*. Equations (6.62d) and (6.62g) imply

$$H_\omega \Omega_\omega = 0. \tag{6.62h}$$

Now $(\mathcal{H}, \pi_\omega, \Omega_\omega)$ is the "Gelfand triple" associated to ω.

A *W^*-algebra or von Neumann algebra* $\mathcal{M}$ is a weakly closed $*$-subalgebra of $\mathcal{B}(\mathcal{H})$, for some Hilbert space $\mathcal{H}$. Since norm convergence is stronger than weak convergence (see Chap. 1), this implies that W^*-algebras are special $C*$-algebras.

The operator H_ω is an example of an *infinitesimal generator*: if $\{\tau_t\}_{t\in\mathbb{R}}$ with $\tau_{t_1}\tau_{t_2} = \tau_{t_1+t_2}$ is a one-parameter group of a $C*$- (respectively, $W*$-) algebra $\mathcal{A}$ such that $\tau_t(A)$ is norm continuous (respectively, weakly continuous) in t, then this group has an infinitesimal generator which is a *derivation* δ, a mapping δ from a norm-dense (respectively, weakly dense) $*$-subalgebra $D(\delta)$ of $\mathcal{A}$ into $\mathcal{A}$ such that

$$\delta(\alpha A + \beta B) = \alpha\delta(A) + \beta\delta(B),$$
$$\delta(A^\star) = (\delta(A))^\star,$$
$$\delta(AB) = A\delta(B) + \delta(A)B.$$

The generator δ of τ is related to H_ω in the case τ_t is norm continuous by the formula

$$\pi_\omega(\delta(A))\Psi = i[H_\omega, \pi_\omega(A)]\Psi \quad \forall\,\Psi \in D(H_\omega) \quad \text{and} \quad \forall\,A \in D(\delta).$$

The HK theory is valid both in nonrelativistic theories and rqft, with some important changes for nonrelativistic theories. As remarked by Dubin and Sewell [75], for continuous Boson systems, like the free Bose gas, time translations are not $C*$-algebra automorphisms because they are nonlocal: a treatment using $W*$-algebras becomes necessary. In rqft, time translations are indeed automorphisms due to the finite propagation speed: see the beautiful proof by Glimm and Jaffe in the case of self-interaction of scalar Bose fields in two-dimensional space–time [94].

In rqft, there are natural further requirements: the first one is that there is a representation of the relativity group $G : g \to \tau_g$ by automorphisms of the quasi-local algebra $\mathcal{A}$ such that

$$\tau_g(\mathcal{A}(\mathcal{O})) = \mathcal{A}(g(\mathcal{O})). \tag{6.63}$$

For G the Poincaré group $\mathcal{P}$, i.e., the group of space–time symmetries whose action on points x of Minkowski space–time $\mathcal{M}$ is given by $x \to \Lambda x + a$ where Λ is a restricted Lorentz transformation (with positive determinant and preserving the forward light cone, see [135]) and $a \in \mathcal{M}$, an invariant state ω_0 then yields via the GNS construction a cyclic vector Ω_0, and a unitary representation of the Poincaré group, $g \to U(g)$, such that

$$\pi(\tau_g(A)) = U(g)\pi(A)U(g)^{-1} \quad \forall\,g \in \mathcal{P} \tag{6.64}$$

and

$$U(g)\Omega_0 = \Omega_0 \qquad (6.65)$$

that is to say, Ω_0 (or ω_0) is what is usually called a *vacuum state*.

Which states give rise to physically acceptable representations? This is a wide subject (see [11, 111]), depending on whether one speaks of particle content, stability properties, etc. An important necessary condition for theories invariant under $\mathcal{P}$ is to require two conditions: for each state ω there is a representation of the Poincaré group $g \to U_\omega(g)$ such that

$$\pi_\omega(\tau_g(A)) = U_\omega(g)\pi_\omega(A)U_\omega(g)^{-1} \quad \forall A \in \mathcal{A} \qquad (6.66)$$

and the *spectral condition*:

the energy–momentum operator associated with U_ω has spectrum

in the forward light-cone $\{p : p \cdot p \geq 0 \text{ and } p^0 \geq 0\}$. $\qquad (6.67)$

Note that for Minkowski space–time, $p \cdot p \equiv (p^0)^2 - \sum_{i=1}^3 (p^i)^2$, for $p = (p^0, p^1, p^2, p^3)$.

Many very important aspects of the HK theory were not touched upon, such as the (beginning theory of) superselection sectors and Haag duality: the books by Haag [111] and Araki [11] should be consulted for a comprehensive survey.

The only known example of a rqft in four space–time dimensions is the free quantum field of mass m (positive or zero), see [208, Chap. X.7]. We assume the reader is familiar with the basic notions of Fock space, see also [182] for an introduction. The Segal field operator $\Phi_S(f)$ on Fock space (see [208, p. 209]) yields a unitary operator $W(f) = \exp[i\Phi_S(f)]$ satisfying the so-called Weyl form of the canonical commutation relations (CCR) (see [270] for an introduction) $W(f + g) = W(f)W(g)\exp[-i\Im m(f, g)/2]$. It may be shown that there exists a unique up to $*$-isomorphism C^*-algebra generated by the Weyl operators, but the $W(f)$ are not norm-continuous, and the corresponding representation is not separable (see [32, Theorem 5.2.8]). In spite of the non-separability of the Weyl algebra, one may, however, choose for the algebra of local observables a $C*$-algebra such that norm continuity of the automorphisms of time translation, space-translation and transformation by the Lorentz boosts holds (see [111, pp. 129–132; 124, p. 052703-4]). This is an important technical point, which will be used in Sec. 6.5.2.

6.4.2. Quantum spin systems

Those nonrelativistic systems which fit most naturally in the HK framework are *quantum spin systems*. The bounded regions will be denoted by Λ, which are subsets of $\mathbb{Z}^n$, where $n \geq 1$ is the space dimension. Isotony is defined in the same way, but local commutativity becomes the condition of "causality":

$$[\mathcal{A}(\Lambda_1), \mathcal{A}(\Lambda_2)] = 0 \quad \text{if } \Lambda_1 \cap \Lambda_2 = \emptyset. \tag{6.68}$$

An important consequence of the "causality" condition is *asymptotic abelianness with respect to space translations*:

$$\lim_{|x| \to \infty} \|[\sigma_x(A), B]\| = 0 \tag{6.69}$$

for all $A, B \in \mathcal{A}$, where σ_x is the space-translations group of automorphisms of $\mathcal{A}$.

We now turn to the dynamics, assuming at first that we are given a Hamiltonian H_Λ for any finite region Λ of a lattice system. It is useful to keep in mind the generalized Heisenberg Hamiltonian as a prototype:

$$H_\Lambda = -\sum_{x,y \in \Lambda} J(x,y) \vec{S}_x \cdot \vec{S}_y, \tag{6.70a}$$

where $J(x,y) = J(y,x)$, $J(x,x) = 0$, and

$$\sup_{x \in \mathbb{Z}^n} \sum_{y \in \mathbb{Z}^n} |J(x,y)| < \infty \tag{6.70b}$$

and $\vec{S}_x \equiv (S_x^1, S_x^2, S_x^3)$, where $S_x^i = \sigma_x^i/2$ and σ_x^i are the Pauli matrices at x, $i = 1, 2, 3$.

Above, H_Λ acts on the Hilbert space $\mathcal{H}_\Lambda = \bigotimes_{x \in \Lambda} \mathbb{C}_x^2$, and $\vec{S}_x$ is short for $\mathbb{I} \otimes \cdots \otimes \vec{S}_x \otimes \cdots \otimes \mathbb{I}$. Clearly $\mathcal{H}_\Lambda$ carries the tensor product of the spin one-half representation of SU_2 with itself $|\Lambda|$ times, where $|\Lambda|$ denotes the number of sites in Λ. Equation (6.70b) is a thermodynamic stability requirement [32]. Now, let $A \in \mathcal{A}_L$, where $\mathcal{A}_L = \bigcup_\Lambda \mathcal{A}_\lambda$ is the algebra of strictly local observables (no norm closure), and define

$$\tau_t^\Lambda(A) = \exp(itH_\Lambda) A \exp(-itH_\Lambda). \tag{6.71a}$$

Under assumption (6.70b) (for the Hamiltonian (6.70a), otherwise similarly) it may be proved [32, 122] that

$$\tau_t(A) \equiv \text{norm} \lim_{\Lambda \nearrow \mathbb{Z}^n} \tau_t^\Lambda(A) \tag{6.71b}$$

exists for sufficiently small $|t|$ and, since it is norm preserving and satisfies the group property, extends uniquely to the quasi-local algebra $\mathcal{A}$ for all real t: it is, thus, a strongly continuous group of automorphisms of $\mathcal{A}$ which represents the dynamics of the infinite system.

We say that $A \in \mathcal{A}$ is an *analytic element* if there exists an entire function $z \to \tau_z(A)$ with values in $\mathcal{A}$ such that the restriction of $\tau_z(A)$ to $\mathbb{R}$ equals $\tau_t(A)$; our notation, following [122], is $A \in \tilde{\mathcal{A}}$. Now $\tilde{\mathcal{A}}$ is a *-algebra (see [122, Lemma 4.6]), which is norm-dense in $\mathcal{A}$ (see [122, Theorem 4.8]). In fact, if $\hat{\mathcal{D}}$ denotes the set of functions f of a real variable such that their Fourier transform $\hat{f}$ is an infinitely differentiable function of compact support, then, for each $A \in \mathcal{A}$ and each $f \in \hat{\mathcal{D}}$,

$$A(f) = \int_{-\infty}^{\infty} dt \tau_t(A) f(t) \qquad (6.72a)$$

exists as a Bochner integral and is an element of $\tilde{\mathcal{A}}$. Further, given $\epsilon > 0$, and $A \in \mathcal{A}$, there exists a positive $f_\epsilon \in \hat{\mathcal{D}}$ such that

$$\int_{-\infty}^{\infty} dt f_\epsilon(t) = 1 \quad \text{and} \quad \|A(f_\epsilon) - A\| \leq \epsilon \qquad (6.72b)$$

(see [122, Theorem 4.8]).

Two types of states arise in quantum statistics: ground states ω_∞ ($\beta = 1/kT = \infty$) and temperature states ω_β ($0 < \beta < \infty$). Their properties are quite different. Temperature (or thermal equilibrium) states satisfy, by definition, the equilibrium (or *KMS* condition) (for Kubo, Martin and Schwinger) (see [32, 122]):

$$\omega_\beta(\tau_t(A)B) = \omega_\beta(B\tau_{t+i\beta}(A)) \quad \forall A \in \tilde{\mathcal{A}} \quad \text{and} \quad \forall B \in \mathcal{A}. \qquad (6.73a)$$

As a consequence of Hadamard's three-line lemma (see [208, p. 33]), it is possible to provide an equivalent definition of KMS states without the use of $\tilde{\mathcal{A}}$, see [32, Proposition 5.37, p. 81]: for $A, B \in \mathcal{A}$ the functions F_{AB}, G_{AB} on $\mathbb{R}$, defined by

$$F_{AB}(t) = \omega_\beta(\tau_t(A)B),$$
$$G_{AB}(t) = \omega_\beta(B\tau_t(A))$$

may be extended to the complex plane in such a way that G_{AB} (respectively, F_{AB}) is analytic in the strip $\Im m\, z \in (0, \beta)$ (respectively, $\Im m\, z \in (-\beta, 0)$), and continuous on its boundaries, with

$$F_{AB}(z) = G_{AB}(z + i\beta). \qquad (6.73b)$$

This condition provides a prescription how to calculate the equilibrium state of an infinite system without any limiting procedure. For applications to mean-field models, see [262]. The motivation is that for finite systems the Gibbs state is

$$\omega_{\beta,\Lambda}(A) = \mathrm{Tr}_{\mathcal{H}_\Lambda}(\rho_\Lambda A) \quad \text{with } \rho_\Lambda = \frac{\exp(-\beta H_\Lambda)}{\mathrm{Tr}_{\mathcal{H}_\Lambda} \exp(-\beta H_\Lambda)}, \tag{6.74a}$$

where $\mathcal{H}_\Lambda$ is the Hilbert space used to describe the system, and $A \in \mathcal{B}(\mathcal{H}_\Lambda)$, the algebra of bounded operators on this Hilbert space. It is easy to see that $\omega_{\beta,\Lambda}$ satisfies (6.73a), and, moreover, it is the only state satisfying (6.73a).

Problem 6.10. *Prove this assertion directly from* (6.73a).

Ground states ω_∞ satisfy, by definition, the following condition: there exists a function F, analytic in $\Im m\, z < 0$, uniformly bounded for $\Im m\, z \le 0$, and continuous on the real axis, such that for t real:

$$F(t) = \omega_\infty(\tau_t(A)B) \quad \forall\, A, B \in \mathcal{A} \tag{6.74b}$$

(see [122] or [32, Proposition 5.3.19]). The latter reference shows also an alternative characterization of ground states in terms of H_ω, given by (6.62g) and (6.62h):

$$H_\omega \ge 0 \tag{6.74c}$$

(*positivity of the Hamiltonian*). Equations (6.62h) and (6.74c) may be obtained from the limiting sequence

$$\omega_{\infty,\Lambda}(\cdot) = (\Omega_\Lambda, \cdot\, \Omega_\Lambda), \tag{6.74d}$$

where ω_Λ is a ground state of H_Λ:

$$H_\Lambda \Omega_\Lambda = E_\Lambda \Omega_\Lambda, \tag{6.74e}$$

where

$$E_\Lambda \equiv \inf \mathrm{spec}(H_\Lambda). \tag{6.74f}$$

By thermodynamic stability,

$$E_\Lambda \ge -c|\Lambda|, \tag{6.74g}$$

where $|\Lambda|$ is the volume (number of sites in) of Λ, where c is a positive constant. In general, E_Λ is of $O(-d|\Lambda|)$ for some positive d, and therefore

(6.62h) requires a *renormalization* (infinite in the thermodynamic limit)

$$H_\Lambda \to \tilde{H}_\Lambda \equiv H_\Lambda - E_\Lambda. \tag{6.74h}$$

Note that with (6.74h), both (6.74a) and (6.74d) yield *invariant* states, i.e., satisfying (6.62b) (with G the group of time translations) in the thermodynamic limit. Such limit points are, in general, only cluster points of the net (see [207, Chap. IV]) of states $\{\omega_\Lambda\}$. Since ω_Λ is defined on $\mathcal{A}_\Lambda \subset \mathcal{A}$, one may extend it in several ways to $\mathcal{A}$. Let $\tilde{\omega}_\Lambda$ be such an extension. We then have a sequence of states on $\mathcal{A}$ which is compact in the weak*-topology. A cluster point therefore exists by the Bolzano–Weierstrass theorem (see [207, Theorem IV.3]). Since $\mathcal{A}$ is separable in the quantum spin case, we may select from the sequence $\{\tilde{\omega}_\Lambda\}$ a subsequence $\tilde{\omega}_{\Lambda_n}$ such that

$$\text{weak-}\star \lim_n \tilde{\omega}_{\Lambda_n} = \omega \tag{6.75}$$

(see the remark in [207, p. 99, before Example 1]). Thus a sequence Λ_n such that the thermodynamic limit exists can always be found, but is not unique: such nonuniqueness is associated with the occurrence of phase transitions.

We have thus found an analogy between the *C^*-dynamical system* $(\mathcal{A}, \tau_t, \omega)$ and the classical dynamical system (Ω, ϕ_t, μ) (Definition A.2), with μ an invariant measure.

It is very convenient to extend the formalism to include *strong* limit points (as opposed to norm limits so far), which, in principle, do not belong to $\pi_\omega(\mathcal{A})$. Let $\bar{\pi}_\omega(\mathcal{A})$ denote the strong closure of $\pi_\omega(\mathcal{A})$. By von Neumann's commutant theorem [31], $\pi_\omega'' = (\pi_\omega')' = \bar{\pi}_\omega(\mathcal{A})$, where

$$\pi(\mathcal{A})' \equiv \{B \in \mathcal{B}(\mathcal{H}) : [A, B] = 0 \ \forall A \in \pi(\mathcal{A})\}.$$

As previously defined, π_ω'' is a von Neumann algebra or W^*-algebra, which we refer to as the von Neumann algebra generated by $\pi_\omega(\mathcal{A})$, $(\pi_\omega(\mathcal{A}))'$ being the *commutant*. Given $\pi_\omega(\mathcal{A})''$, the *center* $Z \equiv \pi_\omega(\mathcal{A}) \cap \pi_\omega(\mathcal{A})'$ is a commutative algebra, which plays an important role in Secs. 6.5.1 and 6.5.2.

Definition 6.5. A $W*$-algebra is said to be a *factor* iff its center Z consists of multiples of the identity: $Z = \{\lambda\mathbb{I}\}$. The corresponding representation is called factor or primary representation.

The set of invariant states under a group G, briefly, G-invariant states, $E_\mathcal{A}^G$, forms a convex, compact subset of the set of states $E_\mathcal{A}$ in the weak*-topology, by Theorem 6.5. The same properties are, of course, shared by the set $E_\mathcal{A}$. An *extremal invariant* or *ergodic* state is a state $\omega \in E_\mathcal{A}^G$

which cannot be written as a proper convex combination of two distinct states $\omega_1, \omega_2 \in E_{\mathcal{A}}^G$:

$$\omega \neq \lambda\omega_1 + (1 - \lambda)\omega_2 \quad \text{with } 0 < \lambda < 1 \text{ unless } \omega_1 = \omega_2 = \omega. \quad (6.76)$$

There is an alternative useful characterization. We say that a state ω_1 *majorizes* another state ω_2 if $(\omega_1 - \omega_2)$ is a positive linear functional on $\mathcal{A}$, i.e., $(\omega_1 - \omega_2)(A^*A) \geq 0 \ \forall A \in \mathcal{A}$. Clearly, if ω is a convex combination of two states ω_1 and ω_2, it majorizes both, and a state ω is defined to be *pure* if the only positive linear functionals majorized by ω are of the form $\lambda\omega$ with $0 \leq \lambda \leq 1$. Let $P_{\mathcal{A}}$ denote the set of pure states (respectively, pure invariant $P_{\mathcal{A}}^G$). It may be proved that $E_{\mathcal{A}} = P_{\mathcal{A}}$ (respectively, $E_{\mathcal{A}}^G = P_{\mathcal{A}}^G$). And that $E_{\mathcal{A}}$ (respectively, $E_{\mathcal{A}}^G$) is the weak*-closure of the convex envelope of $E_{\mathcal{A}}$ (respectively, $E_{\mathcal{A}}^G$) (see [31, Theorem 2.3.15]). We shall therefore use the terms pure and extremal interchangeably. When $G = \mathbb{R}$ with addition, corresponding to time translations, it is natural to identify extremal invariant states with *pure thermodynamic phases* [237]: indeed, if (6.76) does *not* hold, it is natural to regard ω as a *mixture* of two pure phases ω_1 and ω_2, with proportions λ and $1 - \lambda$, respectively. In particular, if ω is time-translation invariant and

$$\lim_{T \to \infty} \frac{1}{T} \int_0^T [\omega(A\tau_t(B)) - \omega(A)\omega(B)] = 0, \quad (6.77)$$

then ω is also extremal $\mathbb{R}$-invariant [31]. We now see that the above condition is a natural generalization of ergodicity as it appears in a classical abelian context.

If E_β denotes the set of extremal τ_t-invariant KMS states at inverse temperature β, $\omega \in E_\beta$ iff it is a factor, or primary, state (see [32, Theorem 5.3.30(3)]). The importance of this is that if ω is a factor state and asymptotic abelianness with respect to space translations (6.69) holds, then

$$\lim_{|x| \to \infty} [\omega(A\sigma_x(B) - \omega(A)\omega(\sigma_x(B)] = 0. \quad (6.78)$$

This cluster property follows from the Kadison transitivity theorem [70, p. 44] (see [31, p. 394]). Notice that no spatial translation invariance is assumed. Equation (6.78) is important in Sec. 6.5.2 and reflects the absence of long-range correlations, which implies absence of fluctuations of space averaged observables over large regions, a property characteristic of a pure phase, see [237, p. 97]. As for ground states, an extremal ground state is pure and the corresponding representation is irreducible [31, Theorem 2.3.19], i.e., the only invariant subspaces of $\mathcal{H}_\omega$ are $\{0\}$ and $\mathcal{H}_\omega$. Hence, $\pi_\omega(\mathcal{A})' = \{\lambda \mathbb{I}\}$ and $\pi_\omega(\mathcal{A})''$ equals $\mathcal{B}(\mathcal{H}_\omega)$. Such a state is, of course, automatically a factor state, and hence satisfies (6.78) whenever (6.69) holds.

6.5. Approach to Equilibrium and Related Problems in Quantum Systems with an Infinite Number of Degrees of Freedom

6.5.1. *Quantum mixing, dynamical stability, return to equilibrium and weak asymptotic abelianness*

An important class of nonuniformly hyperbolic maps T has been shown to admit a unique absolutely continuous invariant measure μ_0 (which we have seen to be important to prove approach to equilibrium (A.42)), which is ergodic (and so is an SRB measure for T, and exhibits exponential decay of correlations in the sense of (A.63) (see [265] and references given there, as well as App. A). Moreover, (T, μ_0) is asymptotically stable in a strong sense under certain random perturbations: this type of stability has replaced structural stability [114] as a general stability criterion, see [265, (1.4)]. What are the corresponding properties/criteria for quantum systems with infinite number of degrees of freedom?

It turns out that, for quantum systems with infinite number of degrees of freedom, similar notions play an important role. Their treatment is enriched, or complicated, by the fact that, unlike systems with finite number of degrees of freedom (von Neumann's uniqueness theorem, see, e.g., [270] for a proof and readable introduction), there exists an uncountable infinity of nonequivalent representations (of the CCR, CAR or (infinite) tensor product of SU_2 representations, depending on whether one is dealing with Bosons, Fermions or spin systems).

We shall consider, for example, a quantum lattice system (e.g., a quantum spin system), and let ω be a spatially translation-invariant state, and A an observable. Let

$$\eta_\Lambda(A) \equiv \frac{1}{|\Lambda|} \sum_{x \in \Lambda} \pi_\omega(\sigma_x(A)) \tag{6.79a}$$

denote the *space average* of A in the region Λ. By von Neumann's ergodic theorem — Theorem A.2 of App. A — we have

$$\text{s} \lim_{\Lambda \nearrow \mathbb{Z}^n} \frac{1}{|\Lambda|} \sum_{x \in \Lambda} U_x = E_0, \tag{6.79b}$$

where s lim denotes strong limit, and E_0 is the projector onto the subspace of all U_x-invariant vectors, i.e., vectors v such that $U_x v = v \ \forall x \in \mathbb{Z}^n$, and U_x is the unitary operator implementing the automorphism σ_x in the representation π_ω (see (6.64)). By (6.79a) and (6.79b),

$$\eta_{\Lambda,\omega}(A)\Omega_\omega = \frac{1}{|\Lambda|}\sum_{x\in\Lambda} U_x \pi_\omega(A)\Omega_\omega \longrightarrow E_0 \pi_\omega(A)\Omega_\omega \quad \text{as } \Lambda \nearrow \mathbb{Z}^n. \quad (6.79c)$$

Further, by (6.69),

$$\lim_{\Lambda\nearrow\mathbb{Z}^n} \|[\eta_{\Lambda,\omega}(A), \pi_\omega(B)]\| = 0 \qquad (6.79d)$$

for all B in the strictly local algebra. We have the following proposition.

Proposition 6.7 ([262]). *The equation*

$$\text{s}\lim_{\Lambda\nearrow\mathbb{Z}^n} \eta_{\Lambda,\omega}(A) \equiv \eta_\omega(A), \qquad (6.80a)$$

exists in $\mathcal{B}(\mathcal{H}_\omega)$, *is given by*

$$\eta_\omega(A)\pi_\omega(B)\Omega_\omega = \pi_\omega(B)(E_0\pi_\omega(A)E_0)\Omega_\omega \qquad (6.80b)$$

and belongs to the center

$$\eta_\omega(A) \in Z_\omega(\mathcal{A}) = \pi_\omega(\mathcal{A})' \cap \pi_\omega(\mathcal{A})''. \qquad (6.80c)$$

We call $\eta_\omega(A)$ *the* space average *of A.*

Problem 6.11. *Prove Proposition 6.7 using (6.79c) and (6.79d).*

The notion of ergodicity (6.76) may be, of course, transcribed to the group of space-translations: ω is *ergodic* with respect to $\{\sigma_x\}$ iff

$$\lim_{\Lambda\nearrow\mathbb{Z}^n} \frac{1}{|\Lambda|}\sum_{x\in\mathbb{Z}^n} \omega(\sigma_x(A)B) = \omega(A)\omega(B) \quad \forall A \in \mathcal{A}. \qquad (6.81)$$

As in the classical theory (App. A), ω is ergodic iff there is a unique vector (which is then, by necessity, Ω_ω) in $\mathcal{H}_\omega$ which is invariant under $\{U_x\}_{x\in\mathbb{Z}^n}$. Hence, if ω is ergodic for space translations, $E_0 = P_{\Omega_\omega} = |\Omega_\omega)(\Omega_\omega|$ and we have the following corollary to Proposition 6.7.

Corollary 6.2. *If ω is ergodic for space translations, the space average of A is a c-number:*

$$\eta_\omega(A) = \omega(A)\mathbb{I}. \qquad (6.82)$$

Definition 6.6. We shall say that the states ω_1 and ω_2 are *disjoint* if no subrepresentation of π_{ω_1} is unitarily equivalent to any subrepresentation of π_{ω_2}, i.e., if ω_1 and ω_2 are *not* disjoint, then either

$$\pi_{\omega_1} \leq \pi_{\omega_2} \quad \text{on } \mathcal{H}_{\omega_2} \tag{6.83a}$$

or

$$\pi_{\omega_2} \leq \pi_{\omega_1} \quad \text{on } \mathcal{H}_{\omega_1}. \tag{6.83b}$$

Above, the signs $\leq$ mean "is majorized by", as previously defined for states.

We have, then, a further corollary of Corollary 6.2.

Corollary 6.3. *Let ω_1 and ω_2 be two ergodic states with respect to space translations. If, for some $A \in \mathcal{A}$,*

$$\eta_{\omega_1}(A) = a_1 \tag{6.84a}$$

and

$$\eta_{\omega_2}(A) = a_2 \tag{6.84b}$$

and

$$a_1 \neq a_2, \tag{6.84c}$$

then ω_1 and ω_2 are disjoint.

Proof (see [112, Lemma 6]). By Definition 6.6, if ω_1 and ω_2 are not disjoint, then either (6.83a) or (6.83b) must take place. Assume the former. Then there exists a projector $E \in \pi_{\omega_2}(\mathcal{A})'$ such that $\pi_{\omega_1}(A) = \pi_{\omega_2}(A)E \,\forall A \in \mathcal{A}$. Hence, by (6.84a) and (6.84b), $a_2 E = a_1$, and thus $a_2 = a_1$, contradicting (6.84c). □

The space average (6.80a) corresponds to a macroscopic "pointer position", e.g., the mean magnetization in the z-direction $\sum_{x \in \Lambda} S_x^z / |\Lambda|$ in the Heisenberg model (6.70a), with $A = S^z$. If $\eta_{\omega_+}(S^z) = a_+ = 1$ and $\eta_{\omega_-}(S^z) = -1$, the states $\omega_\pm$ are macroscopically different, i.e., differ from one another by flipping an infinite number of spins.

Given a state ω_1, the set of states ω_2 "not disjoint from" ω_1 forms a *folium*.

Definition 6.7. A folium of states on $\mathcal{A}$ is a subset $\mathcal{F}$ of $E_{\mathcal{A}}$ such that:

(i) If $\omega_1, \omega_2 \in \mathcal{F}$, and $\lambda_1, \lambda_2 \in \mathbb{R}$, with $\lambda_1 + \lambda_2 = 1$, then $\lambda_1 \omega_1 + \lambda_2 \omega_2 \in \mathcal{F}$;

(ii) If $\omega \in \mathcal{F}$ and $A \in \mathcal{A}$, the state ω_A, defined by

$$\omega_A(B) \equiv \frac{\omega(A^\star B A)}{\omega(A^\star A)} \quad \text{with } \omega(A^\star A) \neq 0 \qquad (6.85)$$

also belongs to $\mathcal{F}$; and

(iii) $\mathcal{F}$ is norm-closed, i.e., contains its limit points in the norm topology.

Clearly, (6.85) has the interpretation of a local perturbation of the state ω. We refer to ω as an equilibrium state if it is either a ground state or a thermal (KMS) state: in this case, ω_A is a local perturbation of equilibrium, and we define the following.

Definition 6.8.

(i) ω has the property of *return to equilibrium* iff

$$\lim_{t \to \pm\infty} \omega_A(\tau_t(B)) = \omega(B) \quad \forall\, B \in \mathcal{A}; \qquad (6.86a)$$

(ii) ω has the *generalized mixing property* iff

$$\lim_{t \to \pm\infty} \omega(A\tau_t(B)C) = \omega(AC)\omega(B) \quad \forall\, A, B, C \in \mathcal{A}; \qquad (6.86b)$$

(iii) ω is *weakly asymptotically abelian* iff

$$\lim_{t \to \pm\infty} \omega(A[\tau_t(B), C]D) = 0 \quad \forall\, A, B, C, D \in \mathcal{A}. \qquad (6.86c)$$

Proposition 6.8 ([192, Lemma 3.5]). *Let ω be an invariant state with center $Z_\omega = \{z\mathbb{I}; z \in \mathbb{C}\}$. Then the following assertions are equivalent:*

(i) *ω has the property of return to equilibrium;*

(ii) *ω has the generalized mixing property;*

(iii) *the following relation holds:*

$$\text{weak} \lim \pi_\omega(\tau_t(B)) = \omega(B)\mathbb{I};$$

(iv) *$(\mathcal{A}, \tau, \omega)$ is weakly asymptotically abelian.*

Proof. (i) $\Rightarrow$ (ii) It follows by polarization; (ii) $\Rightarrow$ (i) is trivial; (ii) $\Rightarrow$ (iii) is trivial; (ii) $\Rightarrow$ (iv) $\omega(A[\tau_t(B), C]D)\omega(ACD)\omega(B) - \omega((ACD)\omega(B) = 0$.

(iv) $\Rightarrow$ (ii) Any weak limit point of $\pi_\omega(\tau_t(B)) : t \in \mathbb{R}$ has to belong to Z_ω by (iv), and is therefore a multiple of the identity. Now weak $\lim \pi_\omega$ $(\tau_{t_n}(B)) = c\mathbb{I}$ implies that $c = \omega(B)$ by τ-invariance of ω, thus all weak limit points are c, and $\pi_\omega(\tau_t(B))$ tends weakly to $\omega(B)$, implying (ii), respectively (iii). □

The interesting remark was made by Narnhofer and Thirring that the *opposite* requirements of the system being completely quantal ($Z_\omega = c\mathbb{I}$) and to be classical for large times (weak asymptotic abelianness (iv)) constrain it to the extent that all observables have to approach their equilibrium value, i.e., all states tend to equilibrium. As we have seen, this does not hold in the classical case! This universality shows the great relevance and interest of the mixing property for quantum systems with infinite number of degrees of freedom, a point of view which has been emphasized by Narnhofer and Thirring [191].

For (infinite) quantum systems, an analogue of structural stability (see [114, p. 312]) in classical mechanics is *dynamical stability*, a condition which deals with the existence of the state obtained from ω by a local perturbation P of the *dynamics*. We demand that in the folium of ω (Definition 6.7) there exist exactly one τ^P-invariant sate, where the suffix refers to the dynamics with generator $H + P$, then obtained as $\lim_{t \to \pm\infty} \omega(\tau_t^P(A)) \equiv \omega^P(A)$. First-order perturbation theory in P yields as condition for the equality of the limits $t \to \pm\infty$ above:

$$\int_{-\infty}^{\infty} \omega([P, \tau_t(A)])dt = 0 \quad \forall\, P, A \in \mathcal{A}. \tag{6.87}$$

It was shown by Haag, Kastler and Trych-Pohlmeyer [107] that, under certain clustering conditions ω is an extremal KMS state for some $\beta \in \mathbb{R} \cup \{\pm\infty\}$ iff (6.87) holds. As we shall see, one of these clustering conditions is the condition of mixing (6.88) below. For $\beta = \infty$, however, it was necessary a deep analysis by Bratelli, Kishimoto and Robinson [30] to complete the picture, as we shall now see.

Several striking differences between ground states and temperature states exist [273]. For KMS states, we have that the generalized mixing property (ii) follows from the *mixing property*:

$$\lim_{t \to \pm\infty} \omega(A\tau_t(B)) = \omega(A)\omega(B) \quad \forall\, A, B \in \mathcal{A}. \tag{6.88}$$

Problem 6.12. *Prove that* (6.88) *implies* (6.86b) *if ω is A KMS state.*

Hint: It is sufficient to show (6.86b) for A, B, C in a norm-dense subalgebra of $\mathcal{A}$. Use the KMS condition in the form (6.73a).

For a ground state, (6.86b) and (6.88) are not, however, equivalent. Another much more important difference concerns dynamical stability

(6.87): a mixing KMs state, i.e., satisfying (6.88), satisfies (6.87) (see [32, Theorem 5.4.12]), but it turns out that a mixing ground state satisfies (6.87) iff, in addition to the ground state condition (6.74c), the spectrum $\sigma(H_\omega)$ of H_ω has a gap:

$$\sigma(H_\omega) \subseteq \{0\} \cup [\epsilon, \infty) \quad \text{for some } \epsilon > 0. \tag{6.89}$$

This is due to the fact that a bounded perturbation may introduce an infinite number of infraparticles, each with "infinitesimally small energy". By Goldstone's theorem (see [271] and references given there), ground states are therefore *dynamically unstable* as long as spontaneous symmetry breakdown of a continuous symmetry takes place in a system with short-range forces.

Note that (6.87) is a *decay* condition on the quantity in the integrand, and that (i), (ii), (iii) express decay properties of several correlation functions.

For a thermal state $\omega = \omega_\beta$ it is convenient to consider the enveloping von Neumann algebra $\mathcal{R}_\omega = \pi_\omega(\mathcal{A})''$, with a normal invariant state (we now omit the index β):

$$\tilde{\omega}(A) = (\Omega_\omega, A\Omega_\omega) \quad \text{with } a \in \mathcal{R}_\omega \tag{6.90a}$$

$\tilde{\omega}$ is a faithful state (i.e., $\tilde{\omega}(A) = 0 \Rightarrow A = 0$), in contrast to the ground state (see, e.g., [273]). Since ω is τ-invariant, the W^*-dynamics $\tilde{\tau}$

$$\tilde{\tau}_t(A) = U_\omega^t A (U_\omega^t)^\dagger$$

is implemented by a one-parameter group $\{U_\omega^t; t \in \mathbb{R}\}$ of unitary operators

$$U_\omega^t = \exp(itL_\omega)$$

acting on $\mathcal{H}_\omega$. The self-adjoint operator is known as the *Liouvillean*, and has the property (corresponding to (6.62h) for H_ω):

$$L_\omega \Omega_\omega = 0. \tag{6.90b}$$

We have the following proposition.

Proposition 6.9. *The variable*

(i) ω (ω_β, *respectively*, ω_∞) *is mixing iff*

$$\text{weak} \lim_{t \to \pm\infty} \exp(itL_\omega) = \Omega_\omega(\Omega_\omega, \cdot)$$

(respectively, same formula with L_ω replaced by H_ω);

(ii) *If the spectrum of L_ω (respectively, H_ω) on $\Omega_\omega^\perp$ is purely absolutely continuous, then ω is mixing.*

Proof. (i) It follows from Proposition 6.8. The mixing condition (6.88) may be written as

$$(\Theta_{\omega,A} \exp(itL_\omega)\Theta_{\omega,B}) \to 0 \quad \text{where } \Theta_{\omega,A} \equiv ((\pi_\omega(A) - \omega(A)\mathbb{I})\Omega_\omega$$

which, by polarization, may be written as linear combination of quantities of the form $(\Psi, \exp(itL_\omega\Psi))$, with $\Psi \in \Omega_\omega^\perp$. The assertion follows then from the Riemann–Lebesgue lemma (Lemma 3.8) and the spectral theorem. $\square$

Clearly, (ii) is not necessary and sufficient because of the existence of (s.c.) Rajchman measures (Definition 3.8 et seq.) and we see the connection of (ii) both with the classical theory (Theorem A.4 of App. A) and Chap. 3.

In this section we showed that, similarly to classical dynamical systems, the mixing condition is, for infinite quantum systems, rich in mathematical and physical content. In the next section we shall turn to examples where mixing can actually be proved.

6.5.2. *Examples of mixing and weak asymptotic abelianness: The vacuum and thermal states in rqft*

In a beautiful paper, Maison [177] proved that, in a unitary representation of the Poincaré group, the infinitesimal space–time translations have a spectral measure without singular continuous part:

Theorem 6.6 ([177, Satz 2]). *If U_g is a unitary representation of the Poincaré group with generator P on a separable Hilbert space, then, for an arbitrary unit vector e in E_4, the spectral measure of $e \cdot P$ restricted to the orthogonal subspace of the translation invariant vectors in the GNS Hilbert space $\mathcal{H}$ is purely absolutely continuous. Thus (by Proposition 6.9) the mixing property (6.88) for the vacuum holds. If the vacuum is the unique invariant state, and χ_i, for $i = 1, 2$, represent localized one-particle states of the form*

$$\chi_i = \int f_i(\vec{k})|\vec{k}, m_i) \frac{d^3\vec{k}}{\sqrt{\vec{k}^2 + m_i^2}}$$

with $f_i(\vec{k}) \in L^2(\mathbb{R}^3)$, then

$$(\chi_1, U_t\chi_2) \to 0 \quad \text{as } t \to \infty,$$

where $U_t = \exp(itH_\omega)$.

The last formula shows the *spreading* of the wave packets, an interesting complement to Chap. 2.

The ground state is invariant under the Poincaré group, and thus associated to a unitary representation by (6.62f). Thermal states are, however, not invariant under the Lorentz boosts, because the KMS condition distinguishes a rest frame.

In [124] Theorem 6.6 was generalized to a class which includes thermal quantum fields. The method was entirely different from Maison's, which was based on the structure of the irreducible representations of the Poincaré group: it consisted of an extension of the arguments by Narnhofer and Thirring (see [192] and references given there) and applied there to certain Galilean invariant theories. Their basic idea was to exploit that: (a) the boost relates space translations and time translations; (b) for space translations, the large distance behavior is under control for the class of models considered. In [124], two new elements were added: (a1) the introduction of a time-dependent scale in the boosts, and (b1) the explicit use of local commutativity.

We denote by x^μ, with $\mu = 0, 1, 2, \ldots, \nu$, the points of Minkowski space–time $\mathbb{R}^{1+\nu}$. Thus, ν is the space dimension and $x^0 = t$ ($c = 1$) denotes the time variable. The transformations $T(a)$ and $L(v)$ of Minkowski space–time corresponding to space–time translations by $a \in \mathbb{R}^{1+\nu}$ and velocity boosts by $u \in (-1, 1)$ along the x^1-axis are defined, respectively, by

$$T(a)x = x + a \tag{6.91a}$$

and

$$L(v)x = \begin{pmatrix} x^0 \cosh v - x^1 \sinh v \\ x^1 \cosh v - x^0 \sinh v \\ x^2 \\ \vdots \\ x^\nu \end{pmatrix}, \tag{6.91b}$$

where $u = \tanh v$ and $\cosh v = (1 - u^2)^{-1/2}$. The corresponding automorphisms of $\mathcal{A}$, denoted by $\chi(a) \equiv (\tau_{a^0}, \sigma_{\vec{a}})$ and λ_v, satisfy the relations

$$\chi_{(t, \vec{x})} = \tau_t \circ \sigma_{\vec{x}} = \sigma_{\vec{x}} \circ \tau_t \quad \forall t \in \mathbb{R} \text{ and } \forall \vec{x} \in \mathbb{R}^\nu \tag{6.91c}$$

and

$$\lambda_v \circ \chi_a \circ \lambda_{-v} = \chi_{L(v)a} \quad \forall a \in \mathbb{R}^{1+\nu} \text{ and } \forall v \in \mathbb{R}. \tag{6.91d}$$

We shall assume that we are given a rqft described in terms of a quasi-local algebra $\mathcal{A}$ satisfying: (A1) the Haag–Kastler axioms; (A2) norm continuity of the time and space translation automorphisms and the Lorentz boosts, together with a state ω defined on $\mathcal{A}$, which is: (A3) either a pure state or a factor state and (A4) invariant under the automorphism group of space–time translations, and extremal space–translation invariant.

In the relativistic case, space translations are asymptotically abelian in norm

$$\lim_{|\vec{x}|\to\infty} \|[A, \sigma_{\vec{x}}(B)]\| = 0 \qquad (6.91e)$$

due to local commutativity (6.61). As remarked at the end of Sec. 6.4.1, it is possible to choose an algebra such that (A2) holds: we have to refer to [124] for this (important but technical) point, as well as to references to thermal field theories in two-dimensional space–time to which our results apply.

The authors of [124] start from a family of states

$$\omega_f = \int dv f(v)\omega_v \qquad (6.92a)$$

with

$$\omega_v = \omega \circ \lambda_v \qquad (6.92b)$$

and f a C^∞ function of compact support, satisfying

$$\int_{-\infty}^{\infty} dx f(x) = 1, \qquad (6.92c)$$

$$f(x) \geq 0, \qquad (6.92d)$$

$$\operatorname{supp} f \in [-a + \delta, a - \delta] \quad \text{with } 0 < \delta < a. \qquad (6.92e)$$

Exploiting (a), (b), (a1), (b1), the following theorem was proved in [124].

Theorem 6.7. *Let ω be a state satisfying the assumptions (A1)–(A4). Then, for any ω_f of the form (6.92a)–(6.92e) (hence space- and time-translation invariant by (6.91a)–(6.91d)), among which there are some arbitrarily close to ω in the weak*-topology, the time evolution is weakly asymptotically abelian, i.e., (6.86c) holds. Moreover, if ω is Lorentz invariant, the mixing property (6.88) holds.*

The last assertion of Theorem 6.7 corresponds to Theorem 6.6 (Maison's result).

Unfortunately, (6.92a) corresponds to forming a convex combination of "boosted" states, leading to a nontrivial center, and thus the equivalences in Proposition 6.8 are no longer true. In spite of this, we have the following problem.

Problem 6.13. *Show that* (A3) *and* (A4) *imply that* ω_f *is an extremal space-translation invariant state.*

In order to prove Theorem 6.7, it is sufficient to prove, analogously to (iii) of Proposition 6.8, that there exists an element B_f in the center such that

$$\lim_{t\to\infty} \omega_f(A\tau_t(B - B_f)) = 0. \qquad (6.92f)$$

Given some $B \in \mathcal{A}$, we may weakly define a nontrivial element B_f which lies at the center by

$$\omega_f(AB_fC) \equiv \int dv f(v)\omega_v(AC)\omega_v(B) \quad \forall A, B, C \in \mathcal{A}. \qquad (6.92g)$$

Indeed, by (6.92g), $\omega_f(AB_fC) = \omega_f(ACB_f)$ and $\omega_f(B_fC) = \omega_f(CB_f)$. We then set $\alpha_t \equiv t^{-1/2-\epsilon}$, with $0 < \epsilon < 1/2$, and define

$$g_t(v) \equiv \frac{g(v/\alpha_t)}{\alpha_t} \quad \forall v \in \mathbb{R} \qquad (6.92h)$$

is an approximation of the Dirac delta function, and let

$$h_t(v) \equiv \int dv_1 f(v - v_1)g_t(v_1) \qquad (6.92i)$$

with g_t as above; since the width of the support of g_t is proportional to $t^{-1/2-\epsilon}$,

$$|h_t(v) - f(v)| \leq c \sup \left|\frac{df}{dv}\right| t^{-1/2-\epsilon} \qquad (6.92j)$$

By (6.92h)–(6.92j), one may approximate $\omega_f(A\tau_t(B - B_f))$ by

$$\int dv f(v) \int dv_1 g_t(v_1)\omega_v(\lambda_{v_1}(A)(\lambda_{v_1} \circ \tau_t(B) - \omega_v(B)\mathbb{I})). \qquad (6.92k)$$

Problem 6.14. *Show that*

$$\lambda_{v_1} \circ \tau_t(B) = \tau_{t \cosh v_1} \circ \sigma_{t \sinh v_1} \circ \lambda_{v_1}(B). \qquad (6.92l)$$

We have

$$\sup_{v_1 \in \text{supp}\, g_t} \|\tau_t(A) - \tau_{t \cosh v_1}(A)\| \le c_A |t|^{2\epsilon} \tag{6.92m}$$

The above follows by expanding $\cosh v_1$ *in a power series and taking the support property of* g_t *into account: note that* (A2) *is used here. Finally, it may be shown from* (6.92l), (6.92m) *and* (6.78) *(where we use the fact, shown in Problem* 6.13, *that* ω_f *is extremal invariant) that* (6.92f) *holds. For details, see* [124].

It should be noted that Theorem 6.7 relies only on (6.78), which, itself, follows from (6.91e) and the fact that ω is a pure or factor state. It holds, thus, also for massless theories. If ω is a pure vacuum (ground) state with a gap (6.74d) — as follows in a rqft with minimal mass $m > 0$ — the rate of clustering in (6.78) is precisely $O(\exp(-m|x|))$, by the cluster theorem (see [87] and references given there).

It follows from Theorem 6.7 that, for those vectors in the family (6.92a) such that the corresponding GNS vector is separating — the *modular states*, the property (6.86a) of return to equilibrium holds. This fact may be regarded as a bonus from rqft: indeed, as we shall see in the next section, for quantum spin systems weak asymptotic abelianness is not generally valid.

In classical dynamical systems, mixing is related to sensitive dependence on initial conditions. We refer to [260] for applications of this idea in the quantum domain: in particular, under the assumption that the observables of a Poincaré invariant theory are a mixing system (in a stronger sense than defined here), the authors argue that the outcome of a phase transition might be influenced by the fact that the initial state contains elements which are not completely determined. The symmetry breaking of a highly symmetric theory after the big bang might fall into this pattern, since one is dealing with many particles (10^{88}!) where not everything can be measured. The assertion refers, however, to *local* observables and measurements, for which there are complementary regions in which they remain undetermined: global quantities such as (6.79a) are not affected.

The Unruh effect [236] may be formulated in the following form [61]: the state of a two-level system, uniformly accelerated with proper acceleration a, and coupled to a scalar Bose field initially in the Minkowski vacuum state, converges asymptotically in the detector's proper time to the Gibbs state at inverse temperature $\beta = 2\pi/a$, i.e., it registers a thermal black-body radiation at this inverse temperature. De Bièvre and Merkli formulated the problem as one of return to equilibrium, and showed that the corresponding approach to equilibrium is exponentially fast. They also remark that the fundamental work of Sewell [236] may be viewed as return to equilibrium in the van Hove (weak coupling) limit. They show (ii) of Proposition 6.8 for the Liouvillean of the interacting detector-field system using, in part, the beautiful theory of level-shift operators mentioned in [184, Chap. 5], and the spectral analysis developed by Bach, Fröhlich and Sigal in their seminal paper [17] on return to equilibrium in cutoff-nonrelativistic QED.

We hope to have convinced the reader of the importance of the mixing property for infinite systems. Very few results exist in this area and many open problems remain, e.g., for quantum spin systems, to which we now turn.

6.5.3. *Approach to equilibrium in quantum spin systems — the Emch–Radin model, rates of decay and stability*

For quantum spin systems, there are, as remarked at the end of last section, very few results concerning decay. This is due to the difficulties in the mathematical analysis of quantum time evolution. One notable and very interesting example is the study of the quantum XY chain, already given as an example in Robinson's seminal paper on return to equilibrium [216]. The model has been further extensively studied from this point of view by Araki: we recommend the very readable article by Araki and Barouch [12] and the references given there.

On the other hand, Emch [79] introduced a very interesting model, too, which was further studied by Radin in his Ph.D. thesis [206]. The model provides an explicit example of the fact that weak asymptotic abelianness does not generally hold for quantum spin systems (see [206, Proposition 3]). A recent extension to random systems [274] further enriched the range of the possibilities of rates of decay, providing examples of exponential vs. nonexponential decay.

The nonrandom model. Following Radin's review [206] of the work of Emch [79], consider the experiment of [171]: a CaF_2 crystal is placed in a magnetic field thus determining the z-direction, and allowed to reach thermal equilibrium. A rf pulse is then applied which turns the net nuclear magnetization to the x-direction. The magnetization in the x-direction is then measured as a function of time. For a review of more recent experiments on crossover phenomena in spin glasses which might be relevant for the forthcoming extension to random systems, see [223].

The above-mentioned experiment was modeled by an *Ising model* with long-range interactions. It is important conceptually as well as practically to view the Ising model as an *anisotropic limit* of quantum models, e.g., of the forthcoming Heisenberg model (6.70a) when the couplings associated to the x- and y-components of the Pauli spin operators tend to zero. Indeed this is so physical: the reason why the critical exponents of the (nearest-neighbor) Ising model in three dimensions (see, e.g., [278]) are so close to those measured in real magnetic systems is that most of the latter are highly anisotropic. From the conceptual side, the only natural (i.e. not imposed) dynamical evolution is the quantum evolution. The present model is a realization of the concepts exposed above: for possible experimental applications of the random version, see [223].

As in [79, 206], we assume an interaction of the form

$$H_\Lambda = 1/2 \sum_{j,k} \epsilon(|j-k|)\sigma_j^z \sigma_k^z - B \sum_j \sigma_j^z, \qquad (6.93)$$

where $j \in \Lambda$ and $k \in \Lambda$ in (6.93), where $\Lambda \subset \mathbb{Z}^\nu$ is a finite region. The dimension ν will not play an important role. Moreover, H_Λ is defined on the Hilbert space $\mathcal{H} = \bigotimes_{i \in \Lambda} \mathbb{C}^2$. The state representing the system after the application of the rf pulse will be assumed to be the product state

$$\rho = \rho(0) = \bigotimes_{j \in \Lambda} \phi_j, \qquad (6.94)$$

where

$$\phi_j(\cdot) = \mathrm{Tr}_j(\cdot \exp(-\gamma\sigma_j^x))/\mathrm{Tr}_j(\exp(-\gamma\sigma_j^x)), \qquad (6.95)$$

where Tr_j is the trace on $\mathbb{C}_j^2$. Other choices of the state are possible [206], but we shall adopt (6.94) for definiteness. Let

$$S^x = \frac{1}{N(\Lambda)} \sum_{j \in \Lambda} \sigma_j^x \qquad (6.96)$$

be the mean transverse magnetization, with $N(\Lambda)$ denoting the number of sites in Λ. The real number γ in (6.95) may be chosen as in [79] such as to maximize the microscopic entropy subject to the constraint $\mathrm{Tr}\, S^x \rho(0)$ equal to a constant, i.e., a given value of the mean transverse magnetization. Since the state (6.94) is a product state, γ is independent of V, and has the same value if S^x is replaced by $\sigma_{i_0}^x$, for any $i_0 \in V$. Let

$$\rho_t^V \equiv U_\Lambda^t \rho(0) U_\Lambda^{-t} \qquad (6.97)$$

and

$$\langle \sigma_{i_0}^x \rangle_\Lambda(t) \equiv \rho_t^\Lambda(\sigma_{i_0}^x), \tag{6.98}$$

where $U_\Lambda(t) = \exp(itH_\Lambda)$. We may write, by (6.97),

$$\langle \sigma_{i_0}^x \rangle_\Lambda(t) = \rho(0)(U_\Lambda(-t)\sigma_{i_0}^x U_\Lambda(t)). \tag{6.99}$$

It is natural to define

$$\langle \sigma_{i_0}^x \rangle(t) = \lim_{\Lambda \nearrow \mathbb{Z}^\nu} \langle \sigma_{i_0}^x \rangle_\Lambda(t) \tag{6.100}$$

provided the limit on the r.h.s. of (6.100) exists, as the expectation value of the local transverse spin in the (time-dependent) nonequilibrium state ρ_∞^t of the infinite system. A weak-star limit of the sequence of states $\rho_\Lambda^t(\cdot) = \mathrm{Tr}(\cdot U_\Lambda^t \rho(0) U_\Lambda^{-t})$ exists, by compactness, on the usual quasi-local algebra $\mathcal{A}$ of observables. The expectation value of $\sigma_{i_0}^x$ in the equilibrium state associated to (6.93) is zero by the symmetry of rotation by π around the z-axis. We may now pose the question whether the limit

$$f(t) \equiv \lim_{N(\Lambda) \to \infty} \rho_\infty^t \left(\frac{1}{N(\Lambda)} \sum_{i_0 \in \Lambda} \sigma_{i_0}^x \right), \tag{6.101}$$

where Λ denotes a finite subset of $\mathbb{Z}^\nu$, which is interpreted as the mean transverse magnetization, exists. The property of approach to equilibrium is expressed by

$$\lim_{t \to \infty} f(t) = 0. \tag{6.102}$$

Of particular interest is the rate of approach to equilibrium. In the present nonrandom case, the limit at the l.h.s. of (6.102) equals $f(t) = \rho_\infty^t(\sigma_{i_0}^x)$, for any i_0, by translation invariance of ρ_∞^t. This is not so for random systems, in which case additional arguments are necessary to show the convergence of the l.h.s. of (6.102) for almost all configurations of couplings.

At this point we simplify the analysis as in [79] by restricting ourselves to the one-dimensional case. Generalization of the results to any dimension ν is straightforward, as remarked in [79] (see also [206]). The Hamiltonian becomes

$$H_n = H_{\Lambda=[-n,n]} = 1/2 \sum_{j,k=-n}^n \epsilon(|j-k|)\sigma_j^z \sigma_k^z + B \sum_{j=-n}^n \sigma_j^z, \tag{6.103}$$

where the function $\epsilon(\cdot)$ is assumed to satisfy:

$$\epsilon(0) = 0 \tag{6.104}$$

and either:

$$\sum_{n=1}^{\infty} |\epsilon(n)| < \infty, \quad \text{i.e., } \epsilon \in l_1 \tag{6.105a}$$

or

$$\sum_{n=1}^{\infty} |\epsilon(n)|^2 < \infty, \quad \text{i.e., } \epsilon \in l_2. \tag{6.105b}$$

The terms containing the interaction $\epsilon(j)$ in the operator $U_n^{-t}\sigma_{i_0}^x U_n^t$ (with $[-n, n] = \Lambda$) which appear on the r.h.s. of (6.99) are of the form (see [79, p. 1200]):

$$[\cos(\epsilon(j)t) - i\sigma_{i_0}^z \sigma_{i_0-j}^z \sin(\epsilon(j)t)]$$
$$\times [\sigma_{i_0}^x \cos(2\epsilon(j)t) + \sigma_{i_0}^y \sigma_{i_0+j}^z \sin(2\epsilon(j)t)]$$
$$\times [\cos(\epsilon(j)t) + i\sigma_{i_0}^z \sigma_{i_0-j}^z \sin(\epsilon(j)t)]. \tag{6.106}$$

By (6.94), and employing a product basis of eigenstates

$$\Psi_\alpha = \bigotimes_i \Psi_{\alpha_i} \quad \text{with } \sigma_i^x \Psi_{\alpha_i} = \alpha_i \Psi_{\alpha_i}, \tag{6.107}$$

$\alpha_i = \pm 1$, in which $\rho(0)$ is diagonal, we see that the terms in (6.106) which contribute to the right-hand side of (6.99) are (using $\sigma^z\sigma^x = i\sigma^y$ et cicl.):

$$(\cos^2(\epsilon(j)t) - \sin^2(\epsilon(j)t))\rho(0)(\sigma_{i_0}^x)\cos(2\epsilon(j)t) = \cos^2(2\epsilon(j)t)\phi_{i_0}(\sigma_{i_0}^x). \tag{6.108}$$

The above formula is true, however, only if $i_0 - j \geq -n$, $i_0 + j \leq n$ in (6.106), due to the restriction in (6.103). If, e.g., $i_0 - j < -n$, with $i_0 + j < n$, the first and third terms in (6.106) will be replaced by the identity, and the resulting term, instead of (6.108), will be

$$\cos(2\epsilon(j)t)\phi_{i_0}(\sigma_{i_0}^x). \tag{6.109}$$

For this reason it is important to perform the limit on the r.h.s. of (7) before the one on the r.h.s. of (8): the asymmetry which yields a few terms of the form (17) in the case of a finite system for $i_0 \neq 0$ disappears upon taking the limit $n \to \infty$. For this reason it is also important to take the infinite volume limit in (6.100) in the sense of Fisher, which means roughly that the domains V have to increase at about the same rate in all directions (see [261] for comments and references): in one dimension, this condition is satisfied by the symmetric choice in (6.103).

With the above remarks, we are allowed to restrict ourselves to $i_0 = 0$ in (6.100) and obtain, with (6.103), (6.95):

$$\langle \sigma_0^x \rangle_n(t) = \phi_0(\sigma_0^x) \prod_{j=1}^{n} \cos^2(2\epsilon(j)t) \times \cos(2Bt). \tag{6.110}$$

We now have the following proposition.

Proposition 6.10. *Let*

$$\phi_0(\sigma_0^x) \neq 0. \tag{6.111}$$

Then, if $\epsilon \in l_2$, i.e., (6.105b) holds,

$$\exists \lim_{n \to \infty} \langle \sigma_0^x \rangle_n(t) \equiv \langle \sigma_0^x \rangle(t). \tag{6.112}$$

If $\epsilon \in l_1$, i.e., if (6.105a) holds, for no constants $0 < \alpha < \infty, 0 < C < \infty$, the inequality

$$|\langle \sigma_0^x \rangle(t)| \leq C \exp(-|\alpha|t) \quad if \, t \geq 0 \tag{6.113}$$

is satisfied.

Proof. By (6.105b), the product on the r.h.s. of (6.110) converges to the infinite product

$$\wp \equiv \prod_{j=1}^{\infty} \cos^2(2\epsilon(j)t). \tag{6.114a}$$

Above, $\wp$ exists by the following simple bound: by (6.105b), for any t there exists $M < \infty$ such that $2|\epsilon(j)|t < 1 \, \forall j \geq M$ and thus

$$\wp \leq \prod_{j=1}^{\infty} |\cos^2(2t\epsilon(j))| \leq d \prod_{j \geq M} (1 - ct^2\epsilon(j)^2) \leq d \exp\left(-ct^2 \sum_{j \geq M} \epsilon(j)^2\right). \tag{6.114b}$$

Above, c and d denote positive constants which may vary at each step and we used the elementary inequalities: $0 < \cos(x) < 1 - cx^2$, $1 - x <$

$\exp(-x)$, for some $c > 0$ if $0 < x < 1$. We have thus proved (6.112), with

$$\langle \sigma_0^x \rangle (t) = \phi_0(\sigma_0^x)\wp. \tag{6.114c}$$

The fact that inequality (6.113) may not hold under assumption $\epsilon \in l_1$, i.e., (6.105a), follows from [206, p. 2954, and Proposition 1], who proved that

$$\langle \sigma_0^x \rangle (t) = (\Pi_\rho(\sigma_0^x)\Phi_\rho, \exp(it\Pi_\rho(2\tilde{H}_0)\Phi_\rho), \tag{6.115a}$$

where

$$\tilde{H}_0 = \sigma_0^z \sum_{k=-\infty}^{\infty} \epsilon(|k|)\sigma_k^z. \tag{6.115b}$$

Above, $\tilde{H}_0$ is the norm limit, as $n \to \infty$; in the quasi-local algebra $\mathcal{A}$ of the sequence of elements $\sigma_0^z \sum_{k=-n}^{n} \epsilon(|k|)\sigma_k^z$, which is the generator of the group of time-translation automorphisms of $\mathcal{A}$, and $\rho = \rho_\infty$ the state on $\mathcal{A}$, $\rho = \bigotimes_{j\in\mathbb{Z}} \phi_j$ corresponding to (6.94), with Π_ρ the GNS representation associated to ρ with cyclic vector Φ_ρ.

By (6.105a) and (6.115b), $\Pi_\rho(2\tilde{H}_0)$ in (6.115a) is a bounded self-adjoint operator, and, by Theorem 6.1 the impossibility of (6.113) is proved.   □

Remark 6.1. The above proof of nonexponential decay relies on the form (6.115a) and (6.115b), which is meaningless when only (6.105b) (but not (6.105a)) is valid, and, indeed, we see from (6.114b), by choosing $M \geq t^{1/\alpha}$ that, for potentials of class l_2 of the form $\epsilon(j) = j^{-\alpha}$ with $1/2 < \alpha \leq 1$ and $k \geq 1$ (which will occur in the sequel) exponential decay (in the sense of inequality (6.113)) is possible when only (6.105b) is required. The latter suffices to prove (6.112), as we saw, but it is clear from (6.93) or (6.103) that it precludes thermodynamic stability for the present model. We shall see in the next section that this is remedied by the introduction of randomness.

The random model. We now introduce the Hamiltonian of a disordered system corresponding to (6.103) (with $B = 0$ for simplicity):

$$\tilde{H}_n = 1/2 \sum_{j,k=-n}^{n} J_{(j,k)}\epsilon(|j - k|)\sigma_j^z\sigma_k^z, \tag{6.116}$$

where $J_{(j,k)}$ are, again for simplicity, independent identically distributed random variables (i.i.d. r.v.). We shall use $\mathrm{Av}(\cdot)$ to denote averaging with respect to the random configuration J. The J are assumed to satisfy [261]:

$$\text{Av}(J_{(j,k)}) = 0, \tag{6.117a}$$

$$|\text{Av}(J_{(j,k)}^n)| \leq n! c^n \quad \forall n = 2, 3, 4, \ldots . \tag{6.117b}$$

Models of type (6.116) have been used to describe *spin glasses*, dilute solutions of atoms of large magnetic moment (e.g. Fe) in a paramagnetic substrate (e.g. Cu). The case in which $\epsilon(|j|) = 0$ if $|j| > 1$, i.e., of nearest neighbors in a cubic lattice of unit spacing is known as the *Edwards–Anderson* spin glass, which exhibits the phenomenon of *frustration* [275]. Let the free energy per site f_n be defined by

$$f_n(J) \equiv \frac{-kT}{2n+1} \log Z_n(J), \tag{6.118a}$$

where

$$Z_n(J) = \text{Tr}(\exp(-\beta \tilde{H}_n)) \tag{6.118b}$$

is the partition function and the trace is over the Hilbert space $\mathcal{H}$.

Theorem 6.8 ([261]). *Under assumptions* (6.105b), (6.117a) *and* (6.117b), *the thermodynamic limit of the free energy per site*

$$f(J) = \lim_{n \to \infty} f_n(J) \tag{6.119a}$$

exists and equals its average:

$$f(J) = \text{Av}(f(J)) = \lim_{n \to \infty} \text{Av}(f_n(J)) \tag{6.119b}$$

for almost all configurations J (a.e. J).

In [261] general ν-dimensional systems were considered (as remarked, we could equally have done so here), with the limit of finite regions taken in Fisher's sense. The reason why (6.105b) suffices for the existence of the thermodynamic limit is that, in order to obtain a uniform lower bound for the average free energy per site, the cumulant expansion (see, e.g., [240, (12.14), p. 129], for the definition of Av_c, there called Ursell functions): $\text{Av}(\exp(tJ_{(i,j)})) = \exp(\sum_{n=2}^{\infty} \text{Av}_c(J_{(i,j)}^n)t^n/n!)$ was used [261], which, by (6.117a), starts with the second cumulant $\text{Av}_c(J_{(i,j)}^2) = \text{Av}(J_{(i,j)}^2)$, which is the variance of $J_{(i,j)}$. Condition (6.117b) was used to control the sum in the exponent above.

We now investigate whether model (6.116), which is thermodynamically stable in the sense that the thermodynamic limit for the free energy exists under assumption (6.105b) for the lattice potential by Theorem 6.8, also yields a well-defined thermodynamic limit for the dynamics, i.e., of $\langle \sigma_{i_0}^x \rangle(t)$. Instead of (6.106), a typical term containing the interaction $\epsilon(j)$ in the operator $U_n^{-t} \sigma_{i_0}^x U_n^t$ appearing on the r.h.s. of (6.99) is, now,

$$(\cos(tJ_{i_0-j,i_0}\epsilon(j))) - i\sigma_{i_0}^z \sigma_{i_0-j}^z \sin(J_{i_0-j,i}\epsilon(j)t))$$
$$\times (\sigma_{i_0}^x \cos(2J_{i_0,i_0+j}\epsilon(j)t) + \sigma_{i_0}^y \sigma_{i_0+1}^z \sin(2J_{i_0,i_0+j}\epsilon(j)t))$$
$$\times (\cos(J_{i_0-j,i_0}\epsilon(j)t) + i\sigma_{i_0}^z \sigma_{i_0-j}^z \sin(J_{i_0-j,i_0}\epsilon(j)t)). \tag{6.120}$$

We ignore the differences commented after (6.107), which are analogous in the present case but, again, will disappear upon taking the limit $\lim_{n\to\infty}$ as discussed there. For any fixed configuration J, an estimate of the same type (6.114b) in Proposition 6.9 yields a formula analogous to (6.114c) in the limit $n \to \infty$:

$$\langle \sigma_i^x \rangle(t) = \rho_\infty^t(\sigma_i^x) = \phi_i(\sigma_i^x)\wp_i \tag{6.121}$$

where

$$\wp_i \equiv \prod_{k=1}^{\infty} \cos(2tJ_{i,i+k}\epsilon(k)) \cos(2tJ_{i-k,i}\epsilon(k)) \tag{6.122}$$

and we use the same notation as in (6.101). We now obtain for the finite version of the r.h.s. of (6.101), with $\Lambda = [-m, m]$ and $N(\Lambda) = 2m + 1$:

$$\rho_\infty^t(m) \equiv \rho_\infty^t \left(\frac{1}{2m+1} \sum_{i=-m}^{m} \sigma_i^x \right) = \delta \frac{1}{2m+1} \sum_{i=-m}^{m} \wp_i, \tag{6.123}$$

where

$$\delta = \phi_i(\sigma_i^x) \tag{6.124}$$

is a nonzero number independent of i. We have now the following theorem.

Theorem 6.9. *Let the r.v. J be Bernoulli or Gaussian. Then there exists a subsequence $(m_r)_{r=1}^{\infty}$ such that*

$$\lim_{r\to\infty} \rho_\infty^t(m_r) = \tilde{\wp}_{i_0} \equiv Av(\wp_{i_0}) \tag{6.125}$$

a.e. J.

Proof. Consider the $(2m+1) \times (2m+1)$ matrix with the r.v. $\wp_i$ as entries. The random variables $\wp_i$ and $\wp_j$, for $i \neq j$, are not independent. Let $j > i$.

Then,

$$\wp_i \equiv \prod_{k=1}^{\infty} \cos(2tJ_{i,i+k}\epsilon(k)) \cos(2tJ_{i-k,i}\epsilon(k)) \tag{6.126a}$$

and

$$\wp_j \equiv \prod_{k=1}^{\infty} \cos(2tJ_{j,j+k}\epsilon(k)) \cos(2tJ_{j-k,j}\epsilon(k)). \tag{6.126b}$$

If $i+k = j$, $j-k = i$, and there will be one common factor $\cos(2tJ_{i,j}\epsilon(k))$ in $\wp_i$ and $\wp_j$. We now consider the r.v. on the r.h.s. of (6.123):

$$X_m \equiv \frac{1}{2m+1} \sum_{i=-m}^{m} \wp_i. \tag{6.127}$$

Since the $J_{i,j}$ are identically distributed, $\tilde{\wp}_i$ is independent of i. Consider

$$\tilde{X}_m \equiv \frac{1}{2m+1} \sum_{i=-m}^{m} (\wp_i - \tilde{\wp}_i). \tag{6.128}$$

We have

$$\lim_{|i-j|\to\infty} \mathrm{Av}((\wp_i - \tilde{\wp}_i)(\wp_j - \tilde{\wp}_j)) = 0. \tag{6.129}$$

To prove (6.129), let $j > i$ for definiteness. By (6.126a) and (6.126b), with $k = j - i$,

$$|Av((\wp_j - \tilde{\wp}_j)(\wp_k - \tilde{\wp}_k))|$$
$$\leq \mathrm{const.}|\mathrm{Av}(\cos^2(2J_{i,i+k}\epsilon(k)t))$$
$$- (\mathrm{Av}(\cos(2J_{i,i+k}\epsilon(k)t)))^2|. \tag{6.130}$$

The r.h.s. of (6.130) is identically zero for Bernoulli r.v.. For the Gaussian,

$$\mathrm{Av}(\cos^2(2J_{i,i+k}\epsilon(k)t)) = \frac{1}{\sqrt{\pi}} \int_{-\infty}^{\infty} e^{-x^2} \cos^2(2x\epsilon(k)t)dx$$
$$= \frac{1 + \exp(-4t^2\epsilon(k)^2)}{2} \tag{6.131a}$$

while

$$(\mathrm{Av}(\cos(2J_{i,i+k}\epsilon(k)t)))^2 = \exp(-2\epsilon(k)^2t^2). \tag{6.131b}$$

Putting (6.131a) and (6.131b) into (6.130) we find that the r.h.s. of (6.130) is $O((\epsilon(k))^4)$ for large k, which tends to zero by (6.105b), thus proving

(6.129). By (6.128) and (6.129),

$$\lim_{m\to\infty} \text{Av}((\tilde{X}_m)^2) = 0. \tag{6.132}$$

Thus, by Chebyshev's inequality (see, e.g., [42, p. 48]), $\tilde{X}_m \to 0$ in probability, which implies by the Borel–Cantelli lemma that a subsequence converges to zero a.e. J (see, e.g., [42, Theorem 4.2.3]). This proves (6.125). □

The convergence a.e. J is the well-known property of self-averaging, which we conjecture holds even without the necessity of restricting to a subsequence. As discussed in [261] it is essential for the reproducibility of the outcomes of experiments, in this case the measurement of the transverse magnetization.

By (6.101), (6.125) and (6.131b), we have the results:

$$f_B(t) = \prod_{k=1}^{\infty} \cos^2(2t\epsilon(k)) \tag{6.133a}$$

for the Bernouilli distribution and

$$f_G(t) = \exp(-2t\,(\epsilon(k))^2) \tag{6.133b}$$

for Gaussian r.v..

Remark 6.2. By (6.133a), (6.114a) and (6.114b), the rate of approach to equilibrium agrees with the nonrandom rate in the Bernouilli case. This leads immediately to the interpretation of (6.114a) and (6.114b) for lattice potentials of class l_2, i.e., satisfying (6.105b), in view of Theorem 6.8. See also Remark 6.1.

The rates of approach to equilibrium. By (6.133b) f_G is explicit in the Gaussian case:

$$f_G(t) = \exp\left(-2t^2 \sum_{k=1}^{\infty}(\epsilon(k))^2\right) = O(\exp(-t^2)). \tag{6.134}$$

From (6.134) it is exponential in the sense of inequality (6.113) whatever the range of the interaction. We now have the following proposition.

Proposition 6.11. *The approach to equilibrium for Gaussian r.v. is faster than for the Bernouilli distribution in the case of potentials of class l_2, i.e.,*

$$\lim_{t\to\infty} \frac{f_B(t)}{f_G(t)} = \infty \tag{6.135a}$$

if

$$\epsilon(k) = k^{-\alpha} \quad with \; 1/2 < \alpha \le 1 \quad and \quad k \ge 1. \qquad (6.135b)$$

The same holds in the prototypical example of infinite, but extreme short range case

$$\epsilon(k) = 2^{-k-1} \quad \text{with } k \ge 1. \qquad (6.135c)$$

Proof of Proposition 6.11. We first consider (6.135b) and use the elementary inequalities:

$$\cos x > 1 - x^2/2 \quad \text{if } 0 < x < 1 \qquad (6.136a)$$

and

$$\exp(-x) < 1 - x/2 \quad \text{if } 0 < x < 1.5936 \qquad (6.136b)$$

to find a lower bound to $f_B(t)$, given by (6.133a). We assume that the time values are sampled at points differing from the zeros of the cosine function in both cases (6.136a) and (6.136b). Then, by (6.136a) and (6.136b),

$$|f_B(t)| = \prod_{k=1}^{\infty} (\cos^2(2tk^{-\alpha}))^2 \ge (2t)^{1/\alpha} \prod_{2tk^{-\alpha} < 1} (1 - 1/2(2tk^{-\alpha})^2)^2$$

$$\ge \text{const.}(2t)^{1/\alpha} \exp\left(\sum_{2tk^{-\alpha} < 1} 8t^2 k^{-2\alpha} \right)$$

$$\ge \text{const.}(2t)^{1/\alpha} \exp\left(-8t^2 \int_{(2t)^{1/\alpha}}^{\infty} dx \, x^{-2\alpha} \right)$$

$$= \text{const.} \exp(-dt^{1/\alpha} + c \log(t)), \qquad (6.137)$$

where d, c are positive constants. Equation (6.135a) follows from (6.134), (6.135b) and (6.137). We now turn to (6.135c), for which an exact result exists for $f_B(t)$:

$$f_B(t) = \prod_{k=1}^{\infty} (\cos(2t/2^{k+1}))^2 = \prod_{k=1}^{\infty} (\cos(t/2^k))^2 = (\sin(t)/t)^2 \qquad (6.138)$$

by Vieta's formula (A.1) of App. A, and (6.135a) is obvious from (6.134) and (6.138). □

The above model has some points in common with general aspects brought about in Sec. 6.3: initial state not invariant, and evolution of a macroscopic quantity (the mean magnetization along the x-axis): the equilibrium value (i.e., corresponding to the dynamics (6.116)) is zero, and hence there is approach to this equilibrium value. Equation (6.138) shows, in addition, a very important feature: the *non-Markovian approach to equilibrium*, pointed out by Emch and Radin in their pioneering work.

In [227] the phase diagram of the mean-field random Ising model was studied, and it was proved that a tricritical point exists there for the Gaussian but not in case of a Bernouilli distribution. Thus, equilibrium properties are, in general, also quite sensitive to the nature of the probability distribution. It was suggested [227] that the observed differences might be due to the fact that a discrete distribution of probabilities samples just a few values of the couplings and therefore introduces some short-ranged elements into the problem. From this point of view, it may be conjectured that discrete distributions have a closer connection with real materials and, indeed, in the nonrandom case — which is a caricature of a real old experiment [171] — a decay of type (6.138) provides a better qualitative description than (6.134). In contrast, Gaussian distributions sample many values of the couplings, and "reinforce" in a sense the long-range nature of the interactions, when they exist: they may be more adequate for long-range spin glasses, and Bernoulli distributions for short-range spin glasses. Assuming this conjecture, Proposition 6.11 predicts for the former a faster approach to equilibrium.

We finally come to the interesting conceptual question of exponential versus nonexponential decay in the sense of inequality (6.113). By (6.134), exponential decay occurs in the Gaussian case for interactions of both short and long range, and in the Bernoulli case for very long-range interactions of class l_2 by (6.114b). In both cases the Hamiltonian is not bounded from below, in agreement with Theorem 6.1. It is to be remarked that even in the well-known model of the Edwards–Anderson spin glass, in the classical or quantum version (see e.g. [275]), the Hamiltonian is not bounded below in the Gaussian case: this would also lead to exponential decay in the present model. On the contrary, for nonrandom systems exponential decay is excluded by Proposition 6.10.

A final point is that, in spite of the simplicity of the present model, it is a genuine quantum many-body problem, i.e., of infinite number of degrees of freedom, about which almost no results on time evolution with good ergodic properties exist: one exception is [192]. There it is pointed out that the condition for existence of a state-dependent time evolution in the Heisenberg picture is thermodynamic stability in the sense of the N-particle (in our case N-site) Hamiltonian having a lower bound of the form $-\text{const}.\,N$. This is called *stability of the second kind* in [170], and would require, in our case, lattice potentials of class l_1. Accordingly, the Heisenberg time evolution is well-known to exist as an automorphism of the quasi-local algebra in the l_1 case (see [206]), i.e., in a state-independent sense, and it is at best state-dependent in the l_2 case: an example is the present construction, which is valid for the same class of states considered in [79, 206].

Appendix A

A Survey of Classical Ergodic Theory

In this appendix we discuss some of the elements of classical ergodic theory, together with some rudiments of number theory, which are complementary to the discussion in Chaps. 3 and 6. For complete and readable expositions, see [48, 178, 266], and for a discussion in the spirit of this appendix, which emphasizes the relationship with number theory and probability, see [138].

The ergodic problems of classical mechanics have been covered in various classical monographs, notably [15, 247]. A lucid treatment of statistical mechanics aspects may be found in [247] and the still pedagogically valuable [168].

To François Viète (1540–1603), who is by some considered as the "father of algebra" (see the fascinating history in the article by J. J. O'Connor and E. F. Robertson in Wikipedia) is attributed the formula (Vieta's formula):

$$\frac{\sin x}{x} = \prod_{k=1}^{\infty} \cos(2^{-k}x). \tag{A.1}$$

This formula is used twice in this book: in the beginning of Chap. 4 as the basis of the most elementary example that the convolution of two s.c. measures may be a.c., and in the approach to equilibrium of the spin model in Chap. 6. It is therefore quite adequate to use it, following [138], in order to provide a first illustration of the connection between number theory, analysis and probability.

Given a real number $x \in [0, 1)$, and an integer $q \geq 2$, there exists one and only one representation of x in the basis q given by the expansion

$$x = \sum_{k=1}^{\infty} \frac{\epsilon_k}{q^k} = \frac{\epsilon_1}{q} + \frac{\epsilon_2}{q^2} + \cdots + \frac{\epsilon_k}{q^k} + \cdots, \tag{A.2}$$

where ϵ_k are integers such that

$$0 \leq \epsilon_k < q \quad \text{for } k = 1, 2, \ldots \tag{A.3}$$

and

$$\epsilon_k < q - 1 \quad \text{for any infinity of } k\text{'s.} \tag{A.4}$$

If $q = 10$, then (A.2) coincides with the decimal representation

$$x = 0.\epsilon_1 \epsilon_2 \cdots \epsilon_k \cdots . \tag{A.5}$$

We shall use the same notation to represent x in basis q, with $q \neq 10$. Except when explicitly mentioned, we shall use $q = 2$. In this case, (A.2) and (A.3) establishes a one-to-one correspondence between a real number $x \in [0, 1]$ and a binary sequence

$$\omega = (\epsilon_j)_{j \geq 1} \quad \text{with } \epsilon_j \in \{0, 1\} \tag{A.6}$$

with an infinity of digits ϵ_j is equal to zero: note that the latter ((A.4) in general) chooses one among the two possible representations, e.g.,

$$3/4 = 1/2 + 1/2^2 + 0/2^3 + 0/2^4 + \cdots = 1/2 + 0/2^2 + 1/2^3 + 1/2^4 + \cdots .$$

Note also that in (A.6), $\{\epsilon_j\}_{j \geq 1}$ varies with x, and we denote this by

$$x \Leftrightarrow \{\epsilon_j(x)\}_{j \geq 1}. \tag{A.7}$$

For example,

$$\epsilon_1(x) = \begin{cases} 0 & \text{if } x \in [0, 1/2), \\ 1 & \text{if } x \in [1/2, 1), \end{cases} \tag{A.8}$$

$$\epsilon_2(x) = \begin{cases} 0 & \text{if } x \in [0, 1/4) \cup [1/2, 3/4), \\ 1 & \text{if } x \in [1/4, 1/2) \cup [3/4, 1), \end{cases} \tag{A.9}$$

etc. We have the following definition.

Definition A.1. For each integer k, the kth Rademacher function is given by

$$r_k(x) = 1 - 2\epsilon_k(x). \tag{A.10}$$

For example, from (A.8)

$$r_1(x) = \begin{cases} 1 & \text{if } x \in [0, 1/2), \\ -1 & \text{if } x \in [1/2, 1), \end{cases} \tag{A.11}$$

$$r_2(x) = \begin{cases} 1 & \text{if } x \in [0, 1/4) \cup [1/2, 3/4), \\ -1 & \text{if } x \in [1/4, 1/2) \cup [3/4, 1), \end{cases} \tag{A.12}$$

etc. Whatever n, $r_n(x)$ is constant, and is equal to ± 1 in the intervals

$$I_{n,j} \equiv [j/2^n, (j+1)/2^n) \tag{A.13}$$

with $j = 0, 1, \ldots, 2^n - 1$, alternating sign when x varies from one interval to the next, with $r_n(0) = 1$.

Equation (A.10) may thus be expressed as

$$r_j(x) = \text{sign}(\sin 2^j \pi x) \tag{A.14}$$

where

$$\text{sign}(y) \equiv \begin{cases} 1 & \text{if } y > 0, \\ 0 & \text{if } y = 0, \\ -1 & \text{if } y < 0. \end{cases} \tag{A.15}$$

From the definition or (A.14), $r_j(x)$ changes sign an odd number of times in each interval $I_{i,k} = [k/2^i, (k+1)/2^i)$ of continuity of $r_i(x)$ if $i < j$ (look at (A.11)). It follows that

$$\int_0^1 r_i(x) r_j(x) dx = \delta_{ij}. \tag{A.16}$$

The Rademacher functions do not form a complete orthonormal basis of $L^2(0, 1)$, because they may be expressed as a particular linear combination of the so-called Haar functions, which do form a complete orthonormal basis, see [139].

We now use (A.2) with $q = 2$ and (A.10) to write

$$1 - 2x = 1 - 2 \sum_{k=1}^{\infty} \epsilon_k / 2^k = \sum_{k=1}^{\infty} \frac{1 - 2\epsilon_k}{2^k} = \sum_{k=1}^{\infty} \frac{r_k(x)}{2^k}.$$

On the other hand,

$$\int_0^1 \exp\left(i\xi(1 - 2x)\right) dx = \frac{\exp(i\xi)(1 - \exp(-2i\xi))}{2i\xi} = \frac{\sin \xi}{\xi}$$

and therefore

$$\frac{\sin \xi}{\xi} = \int_0^1 dx \prod_{k=1}^{\infty} \exp(i\xi 2^{-j} r_j(x)). \tag{A.17}$$

Problem A.1. *Use the identity* $\sin \alpha = 2 \sin(\alpha/2) \cos(\alpha/2)$ *successively to obtain*

$$\sin \xi = 2^n \sin(\xi/2^n) \cos(\xi/2^n) \cos(\xi/2^{n-1}) \cdots \cos(\xi/2),$$

and then use the above to obtain Vieta's formula (A.1).

Problem A.2. *Prove that*

$$\int_0^1 \exp[i\xi 2^{-k} r_k(x)] dx = \cos(x/2^k). \tag{A.18}$$

Equations (A.1), (A.17) *and* (A.18) *suggest the validity of the following.*

Proposition A.1.

$$\int_0^1 \prod_{k=1}^{\infty} \exp[i\xi c_k r_k(x)] dx = \prod_{k=1}^{\infty} \int_0^1 \exp[i\xi c_k r_k(x)] dx \ . \tag{A.19}$$

Proposition A.1 with $c_k = 2^{-k}$ is Vieta's formula (A.1) written in terms of the orthogonal (see (A.16)) Rademacher functions.

Probabilistic interpretation. We shall provide two proofs of Proposition A.1. Initially, let us remark that the linear combination $\Psi(x) = \sum_{k=1}^{n} c_k r_k(x)$ is a constant function in each interval $I_{n,j} = [j/2^n, (j+1)/2^n)$, for all $j = 0, 1, \ldots, 2^n - 1$, of length $|I_{n,j}| = 2^{-n}$. Observe that the intervals in which $r_k : k < n$ assume constant values ± 1 are compatible with the intervals $I_{n,j}$ and contain them. Using (A.2) and (A.10), we may establish a one-to-one relation between each value of j which labels the interval $I_{n,j}$ and the vector $\vec{\sigma} = (\sigma_1, \ldots, \sigma_n)$ of components $\sigma_m \in \{-1, 1\}$:

$$j/2^n = \frac{1 - \sigma_1}{2} \frac{1}{2} + \frac{1 - \sigma_2}{2} \frac{1}{2^2} + \cdots + \frac{1 - \sigma_n}{2} \frac{1}{2^n}. \tag{A.20}$$

Problem A.3. *Derive the dependence of j in $\vec{\sigma} = \vec{\sigma}(j)$ given by* (A.20) *explicitly.*

If $x \in I_{n,j}$, then $\Psi(x)$ is a constant equal to

$$\Psi(x) = \Psi(j) = \sum_{k=1}^{n} c_k \sigma_k = \vec{c} \cdot \vec{\sigma}$$

and therefore, for $F : \mathbb{R} \to \mathbb{R}$,

$$\int_0^1 F(\Psi(x))dx = \sum_{j=0}^{2^n-1} |I_{n,j}| F(\Psi(j))$$

$$= \frac{1}{2^n} \sum_{\vec{\sigma} \in (\{-1,1\})^n} F(\vec{c} \cdot \vec{\sigma}),$$

where in the second line we used the relation $j \Leftrightarrow \sigma$ shown in Problem A.3 and (A.20). Substituting $F(\Psi) = \exp(i\xi\Psi)$, we obtain

$$\int_0^1 \exp\left(i\xi \sum_{k=1}^n c_k r_k(x)\right) dx = \frac{1}{2^n} \sum_{\vec{\sigma} \in (\{-1,1\})^n} \exp(i\xi\vec{c} \cdot \vec{\sigma})$$

$$= \prod_{k=1}^n \left[\frac{1}{2} \sum_{\sigma \in \{-1,1\}} \exp(i\xi c_k \sigma) \right]$$

$$= \prod_{k=1}^n \cos(\xi c_k)$$

$$= \prod_{k=1}^n \int_0^1 \exp\left(i\xi c_k r_k(x)\right) dx \qquad \text{(A.21)}$$

by (A.18). We conclude the proof of proposition A.1 upon setting $c_k = 2^{-k}$ and then taking the limit $n \to \infty$. $\qquad \square$

We now introduce the probability space $(\Omega, \mathcal{B}, \mu)$ with $\Omega = [0,1), \mathcal{B}$ the Borel algebra generated by the subintervals of $[0,1)$, and μ Lebesgue measure on $[0,1)$. A random variable (r.v.) $f : \Omega \to \mathbb{R}$ is a measurable function, i.e., s.t. $f^{-1}(A) \in \mathcal{B}$ for any Borel set A. We see that $r_1(x), \ldots, r_n(x)$ are r.v. defined on the space $([0,1), \mathcal{B}, \mu)$. The probability that a r.v. f assumes some value in the interval $I = [a,b) \subset \mathbb{R}$ is the Lebesgue measure of the set

$$f^{-1}(I) = \{x \in [0,1) : a \le f(x) < b\}$$

$$= \mathbb{P}(a \le f < b)$$
$$= \mu(\{x \in [0,1) : a \le f(x) < b) = |f^{-1}(I)|.$$

Since the Rademacher functions $r_k(x)$ assume only the values ± 1,

$$\mathbb{P}(r_k = \pm 1) = \mu(\{x \in [0,1) : r_k(x) = \pm 1) = \int_0^1 \frac{1 \mp r_k(x)}{2} dx = 1/2.$$

Vieta's formula (A.1) is satisfied due to the following property of the Rademacher functions. Consider the event E_k that $r_k(x)$ assumes the value $\sigma_k \in \{-1,1\}$ for each $k = 1, \dots, n$ individually:

$$E_k = \{x \in [0,1) : r_k(x) = \sigma_k\}$$

and the event $E^{(n)}$ that $r_k(x)$ assumes the value σ_k for $k = 1, \dots, n$ jointly:

$$E^{(n)} = \{x \in [0,1) : r_k(x) = \sigma_k \text{ for all } k = 1, \dots, n\} = \bigcap_{k=1}^{n} E_k.$$

It follows from the definition of the r_k that the probability of these events fulfills the relation

$$\mathbb{P}(E^{(n)}) = \mu(\{x \in [0,1) : r_k(x) = \sigma_k, k = 1, \dots, n\}$$
$$= \prod_{k=1}^{n} \mu(x \in [0,1) : r_k(x) = \sigma_k) = \prod_{k=1}^{n} \mathbb{P}(E_k) = \frac{1}{2^n}. \qquad \text{(A.22)}$$

As an example, in the case $n = 3$, we may verify by inspection that the Lebesgue measure of the set of x s.t. $r_1(x) = 1, r_2(x) = -1, r_3(x) = -1$ is the length of the interval $I_{3,3} = [3/8, 1/2)$ (see (3.78) et ff.):

$$1/8 = |I_{3,3}| = |I_{1,0}| \times |I_{2,1} \cup I_{2,3}| \times |I_{3,1} \cup I_{3,3} \cup I_{3,5} \cup I_{3,7}|$$
$$= 1/2 \times 1/2 \times 1/2.$$

In Sec. 3.3 we have actually shown (A.22) in general.

Equation (A.22) shows that $r_1, \dots, r_n$ are (by definition, see, e.g., [42]) *independent* r.v., and therefore the integrand on the r.h.s. of (A.21) may be written as the *expectation* $\mathbb{E}$ (in the language of probability, see again [42]):

$$\mathbb{E} \exp\left(i\xi \sum_{k=1}^{n} c_k r_k \right) = \int_0^1 \exp\left(i\xi \sum_{k=1}^{n} c_k r_k(x) \right) dx$$
$$= \sum_{\vec{\sigma} \in (\{-1,1\})^n} \exp\left(i\xi \sum_{k=1}^{n} c_k r_k \right)$$

$$\times \mu(\{x : r_k(x) = \sigma_k, \ k = 1, \ldots, n\})$$

$$= \sum_{\vec{\sigma} \in (\{-1,1\})^n} \exp(i\xi\vec{c} \cdot \vec{\sigma}) \prod_{k=1}^{n} \mu(x : r_k(x) = \sigma_k)$$

$$= \prod_{k=1}^{n} \mathbb{E} \exp(i\xi c_k r_k),$$

which equals the r.h.s. of (A.21), concluding the second proof of Proposition A.1. □

By (3.78), Definition A.1 and (A.22) provide an alternative wording (in terms of Rademacher functions) of (3.70e).

Proposition A.1 expresses a connection between probability theory and elementary number theory. We now come to the interplay between ergodic theory and dynamical systems; some connection between ergodic theory and number theory appears in Theorem A.5, and the final paragraphs of the appendix are reserved to some further elementary notions of number theory which were used in Chap. 3.

Given a probability space $(\Omega, \mathcal{B}, \mu)$, and a group $\{\Phi_t; t \in \mathbb{R}\}$ of transformations with $\Phi_{t+s} = \Phi_t \Phi_s$, $\Phi_0 = \mathbb{I}$ — a *flow* (respectively, *semiflow*), which preserves the measure μ:

$$\mu(\Phi_t(A)) = \mu(A) \quad \text{for all } t \in \mathbb{R} \text{ and } A \in \mathcal{B}$$

(respectively, for all $t \in \mathbb{R}_+$), we say that $A \subset \mathcal{B}$ is *invariant* w.r.t Φ_t iff $\Phi_t(A) = A$ for all $t \in \mathbb{R}$ (respectively, $t \in \mathbb{R}_+$). In case of discrete times, we define $T = \Phi_1$, $T^n = \Phi_1^n = \Phi_1 \circ \cdots \circ \Phi_1 \equiv \Phi_n$ where the $\circ$ indicates composition, and the $\cdots$ refer to composition n times. In this case, A is invariant if $T(A) = A = T^{-1}(A)$ if T is an *automorphism*, i.e., a bijection $T : \mathcal{B} \to \mathcal{B}$ such that $\mu(A) = \mu(T(A)) = \mu(T^{-1}(A))$. If T is an *endomorphism*, i.e., a surjection s.t. for all $A \in \mathcal{B}$, $T^{-1}(A) \in \mathcal{B}$ and $\mu(A) = \mu(T^{-1}(A))$, A is said to be invariant if $T^{-1}(A) = A$.

Definition A.2. A *dynamical system* is a group or semigroup of transformations which preserves the measure μ of a probability space $(\Omega, \mathcal{B}, \mu)$. It is said to be *ergodic* if for all A invariant, either $\mu(A) = 0$ or $\mu(A) = 1$.

Sometimes ergodic dynamical systems are also referred to as *quasi-ergodic*. This terminology originates from the following basic theorem.

Theorem A.1 (Birkhoff's ergodic theorem). *If $f \in L^1(\Omega, d\mu)$, the "time average"*

$$\lim_{T \to \infty} \frac{1}{T} \int_0^T dt f(\Phi_t(x)) \equiv \hat{f}(x) \tag{A.23}$$

exists for almost every x with respect to μ. Moreover, if the system is ergodic, $\hat{f}$ is a.e. a constant and thus equal to its "space average"

$$\hat{f}(x) = \int_\Omega f d\mu \tag{A.24}$$

a.e. with respect to μ.

For a relatively simple proof, see [266, p. 33]. If the system is ergodic, each integral of motion (i.e., a function g s.t. $g(\Phi_t(x)) = g(x)$) is constant a.e.. Indeed, let $C_a = \{x : g(x) < a\}$, then C_a is invariant for all $a \in \mathbb{R}$. Hence $g(x) < a$ a.e. or $g(x) \geq a$ a.e. Since this is true for all a, $g(x) = $ const. a.e. This explains the last assertion of Birkhoff's theorem. In the discrete case, the time averages are

$$\lim_{N \to \infty} \frac{1}{N} \sum_{i=0}^{N-1} f(T^i x). \tag{A.25}$$

Putting $f = \chi_A$, the characteristic function of $A \subset \mathcal{B}$, the time average, i.e., the mean time spent by the system in region A, equals $\mu(A)$.

It may also happen that (A.25) exists for x in some manifold M, and $f : M \to \mathbb{R}$ a continuous function, and is independent of the point x in some set $B \subset M$: call it $E_x(f)$. Then $f \to E_x(f)$ defines for all $x \in B$ a non-negative linear operator on the space $C^0(M, \mathbb{R})$ of real continuous functions from M to $\mathbb{R}$, which, by Theorem 1.2 may be written in the form $E_x(f) = \int f d\mu$ for a Borel measure μ on M, for all $f \in C^0(M, \mathbb{R})$, and for all $B \subset M$, which we assume is of positive Lebesgue measure (on M). Here B is the called the ergodic *basin* of μ, and μ a *SRB measure* (Sinai–Ruelle–Bowen measures, see [29, 246]). Note that such a measure may be "physically observed" by computing time-averages of continuous functions for randomly chosen points $x \in M$ (positive probability of getting $x \in B$). SRB measures are expected to exist in great generality and are of foremost importance in nonequilibrium statistical mechanics (see [72, Chap. 13] for a concise but lucid exposition and further references). In the following exposition, the measure μ may be thought to stand for a SRB measure.

Definition A.3. An automorphism (or flow) is *mixing* iff, for all $f, g \in L^2(\Omega, d\mu)$,

$$\lim_{n \to \pm\infty} \int_\Omega f(T^n x) g(x) d\mu = \int_\Omega f d\mu \int_\Omega g d\mu \qquad \text{(A.26)}$$

in the case of an automorphism, or

$$\lim_{t \to \pm\infty} \int_\Omega f(\Phi_t(x)) g(x) d\mu = \int_\Omega f d\mu \int_\Omega g d\mu \qquad \text{(A.27)}$$

in the case of a flow. Setting $f = \chi_A$ and $g = \chi_B$, A, B both in $\mathcal{B}$, we obtain

$$\lim_{n \to \pm\infty} \mu(T^n(A) \cap B) = \mu(A)\mu(B), \qquad \text{(A.28)}$$

which is sometimes used as definition of mixing. In the case of an endomorphism, the definition is

$$\lim_{n \to \pm\infty} \mu(T^{-n}(A) \cap B) = \mu(A)\mu(B). \qquad \text{(A.29)}$$

Conversely, one may pass from (A.28) to (A.26) using the fact that finite linear combinations of characteristic functions are dense in $L^2(\Omega, d\mu)$.

The best "physical" interpretation of the mixing property, e.g., (A.28), has been given in the famous example of Arnold and Avez [15], a glass containing initially 20% of rum and 80% of coca-cola. Initially, the rum is in region A, which is disjoint of a region B, fully occupied by coca-cola. After n "mixings" the percentage of rum in B will be $\mu(T^n(A) \cap B)/\mu(B)$. In this situation, one expects that for $n \to \infty$ all parts of the glass will contain approximately 20% of the rum, which is the content of (A.28).

It is very useful to express the mixing property in other, occasionally more manageable forms. For this purpose, we define the following.

Definition A.4. The *Koopman operator* U_t is defined by

$$(U_t f)(x) = f(\Phi_t(x)) \qquad \text{(A.30)}$$

for flows, and similarly in the other cases.

Clearly, for each t, U_t is isometric, i.e., preserves the norm in $L^2(\Omega, d\mu)$:

$$\|U_t f\|^2 = \int_\Omega |f(\Phi_t(x))|^2 d\mu(x) = \int_\Omega |f(\Phi_t(x))|^2 d\mu(\Phi_t(x))$$

$$= \int_\Omega |f(x)|^2 d\mu(x) = \|f\|^2$$

and it is a bijection: each $g \in L^2$ may be written as some Uf: $f(x) = g(\Phi_t^{-1}(x))$. Therefore, the Koopman operator is unitary. Now (Φ, μ) is ergodic iff 1 is a *simple* eigenvector of U, because f is invariant under Φ_t if $U_t f = f$: Φ is ergodic iff the invariant functions are constant almost everywhere. Since the latter are scalar multiples of one another, Φ is ergodic iff the subspace of solutions of $Uf = f$ has dimension one. Related to this, and basic to ergodic theory, is the following mean ergodic theorem of von Neumann.

Theorem A.2. *Let U be a unitary operator on a Hilbert space $\mathcal{H}$, and P be the orthogonal projection onto $\{\Psi : \Psi \in \mathcal{H} \text{ and } U\Psi = \Psi\}$. Then, for any $f \in \mathcal{H}$,*

$$\lim_{n\to\infty} \frac{1}{n} \sum_{i=0}^{n-1} U^i f = Pf. \tag{A.31}$$

Problem A.4. *Prove Theorem A.2, using the decomposition*

$$\mathcal{H} = \operatorname{Ran}(\mathbb{I} - U) \oplus \operatorname{Ker}(\mathbb{I} - U^\dagger),$$
$$\operatorname{Ker}(\mathbb{I} - U^\dagger) = \operatorname{Ker}(\mathbb{I} - U).$$

Alternatively, see [207, Theorem II.11].

Together with our remarks preceding Theorem A.2, we have shown that T is ergodic iff, $\forall f, g \in L^2(\Omega, d\mu)$,

$$\lim_{n\to\infty} \frac{1}{n} \sum_{i=0}^{n-1} (U^i f, g) = (f, 1)(1, g). \tag{A.32}$$

The assertion in the form corresponding to (A.26) is

$$\lim_{n\to\infty} \frac{1}{n} \sum_{i=0}^{n-1} \int f(T^i x) g(x) d\mu = \int f d\mu \int g d\mu \tag{A.33}$$

or, in the continuum case,

$$\lim_{T\to\infty} \frac{1}{T} \int_0^T \int f(\phi_t(x)) g(x) d\mu(x) = \int f d\mu \int g d\mu. \tag{A.34}$$

We now come to the important problem of relating the *spectrum* of U on $\{1\}^{\perp}$ to ergodic properties. We need a final definition.

Definition A.5. We say that an endomorphism T is *weak mixing* iff

$$\lim_{n \to \infty} \frac{1}{n} \sum_{i=0}^{n-1} |\mu(T^{-i}(A) \cap B) - \mu(A)\mu(B)| = 0$$

with analogous definitions in the other cases (automorphism, flow). As usual, equivalent definitions similar to (A.26) and (A.27) follow, and therefore T is weakly mixing if, $\forall f, g \in L^2(\Omega, d\mu)$,

$$\lim_{n \to \infty} \frac{1}{n} \sum_{i=0}^{n-1} |(U^i f, g) - (f, 1)(1, g)| = 0.$$

We now have the following theorem.

Theorem A.3. *If T is an automorphism and U the corresponding Koopman operator, T is weak mixing iff 1 is the only eigenvalue of U, and on $\{1\}^{\perp}$ the spectrum of U is continuous.*

Problem A.5. *Prove Theorem A.3 using the spectral theorem for U and Fubini's theorem, in a very similar fashion to the proof of Theorem 3.1.*

For a mixing automorphism, (A.26) may be written in terms of the Koopman operator U as

$$\lim_{t \to \pm\infty} (U_t f, g) = (f, 1)(1, g) \quad \text{for all } f, g \in L^2(\Omega, d\mu). \tag{A.35}$$

As a consequence of (A.35), Lemma 3.6 and the spectral theorem we have (see [207, Theorem VII.15, p. 241]) the following.

Theorem A.4. *Let T be an automorphism and U the associated Koopman operator. If U has purely a.c. spectrum on $\{1\}^{\perp}$, then T is mixing.*

We have mentioned previously the "physical" significance of the mixing property. We now inquire into its relation with the ergodic property and the related significance in a real physical framework, that of statistical mechanics. It is clear from (A.28) or (A.29) that mixing implies ergodicity, for, take A invariant, $T^{-1}(A) = A$, and let $B = A$ in (A.29), for instance, then the latter yields $\mu(A) = \mu(A)^2$ and therefore $\mu(A) = 0$ or $\mu(A) = 1$, i.e., the system is ergodic. The converse does not hold, however, as we now show. Let M be the circle $\{z \in \mathbb{C} : |z| = 1\}$, μ Lebesgue measure over M, Φ the translation $\Phi(z) = \theta z$ with $\theta = \exp(2\pi i\omega)$ where $\omega \in \mathbb{R}$. Consider the orthonormal basis of $L^2(M, d\mu)$ given by $\{z^p : p \in \mathbb{Z}\}$; we have

$$(U^n z^p, z^p) = \theta^{pn}, \tag{A.36}$$

Φ is ergodic iff 1 is a simple eigenvalue of U, i.e., iff $p\omega$ is *not* in $\mathbb{Z}$ if $p \neq 0$, i.e., if ω is irrational. However, Φ is not mixing under the latter condition, because, taking $f = g = z^p$ in (A.35), we obtain, for $p \neq 0$, $\lim_{n\to\pm\infty}(U^n z^p, z^p) = 0$ as the condition for mixing, but, by (A.36), it is not satisfied because $\lim_{n\to\pm\infty} \theta^{pn}$ does not exist. These results extend immediately to higher-dimensional tori and show that ergodicity does not imply "approach to equilibrium": the translations of the torus do not deform a region A, they are such that the intersections of $T^n(A)$ with B are alternately empty or of positive measure. We now show that mixing does imply approach to equilibrium in the framework of classical statistical mechanics.

Consider the dynamical system generated by the motion of N matter points (gas molecules) in a fixed volume V, and let $\Gamma = \{x \equiv (q_i, p_i), i = 1, \ldots, 3N\}$ be the corresponding phase space, with q_i denoting the generalized coordinates, and p_i the momenta. If the system is isolated (fixed total energy E), Γ will be compact and there exists [254] an invariant measure μ which may be interpreted as the distribution in thermodynamic equilibrium. Starting from initial conditions in a certain volume V_1, the set of admissible coordinates and momenta of the molecules forms a subset $A \subset \Gamma$. The formula

$$\mu_0(B) = \frac{\mu(A \cap B)}{\mu(A)} \tag{A.37}$$

defines a *initial* distribution (which is not the equilibrium distribution), interpreted as *conditional* distribution relatively to the system's known initial condition. We may now relate μ_0 to the knowledge of the state of the system at time t by the measure μ_t defined by

$$\mu_t(B) = \mu_0(T_t(B)), \tag{A.38}$$

where T is the flow which leaves μ invariant. In the given example,

$$\mu(F) = \mu_E(F) = \int_F \delta(H(x) - E)dx \tag{A.39}$$

the *microcanonical Gibbs measure* where $H(x)$ is the Hamiltonian describing the system and (A.38) is *Liouville's theorem*. Suppose, now, that μ_0 is a.c. with respect to μ, i.e., there exists $\rho_0 \in L^1(M = \Gamma, d\mu)$ such that

$$d\mu_0 = \rho_0(x)d\mu. \tag{A.40}$$

Then,

$$\mu_t(B) = \int_M \chi_B(T_t x)d\mu_0 = \int_M \chi_B(T_t x)\rho_0(x)d\mu. \tag{A.41}$$

If the system is mixing, we obtain from (A.35), as $t \to \infty$,

$$\lim_{t \to \infty} \mu_t(B) = \int_M \chi_B d\mu \int_M \rho_0 d\mu = \mu(B)1 = \mu(B), \tag{A.42}$$

which means that, whatever the initial distribution μ_0, normalized and a.c. w.r.t. μ, the time-translates μ_t of μ_0 under T_t converge, for $t \to \infty$, to the equilibrium distribution. Equation (A.42) may be taken as the definition of the *approach to equilibrium*. It means that mixing systems are "memoryless", i.e., they possess a stochastic character which justifies equilibrium statistical mechanics. The microscopic mechanism of this process of loss of memory is the sensitive (exponential) dependence on initial conditions (later on precisely defined) produced by "defocalizing shocks" between the gas molecules, first pointed out by Krylov, see [247] and references given there. It may also be felt in the *exponential rate* at which the mixing condition takes place, a topic to which we also return later.

We may picture the gas molecules as a system of hard spheres enclosed in a cube with perfectly reflecting walls or periodic boundary conditions. This is supposed to be a K-system (see [266, Definition 4.7, p. 101]). Rising still one step in this so-called ergodic hierarchy (see [168]), we come to *Bernoulli systems*, such as the one we presently introduce along the lines of [15].

Define T_r by $T_r : X \to Y$, where $X = [0, 1]$ and $Y = [0, 1]$, by

$$T_r x = \mathrm{fr}(rx) \equiv rx \bmod 1. \tag{A.43}$$

Note that this mapping is not one-to-one, it is an endomorphism. In fact, if $r = 2$, we see that the inverse image of a point x is $x/2$ or $(x + 1)/2$. Now T_r is called the r-adic transformation and leaves Lebesgue measure μ invariant. Indeed, if

$$A = \left[\frac{m}{r^n}, \frac{m + 1}{r^n}\right] \quad \text{for } m \in [0, r^n - 1]$$

we have

$$T_r^{-1}(A) = \bigcup_{s=0}^{r-1} \left[\frac{m+s}{r^{n+1}}, \frac{m+s+1}{r^{n+1}} \right]$$

and

$$\mu(A) = r^{-n} = \mu(T_r^{-1}(A)) = rr^{-(n+1)}.$$

Consider the case $r = 2$. Then T_2 has a simple alternative description: writing $x = .\epsilon_1\epsilon_2 \cdots$, i.e., $x = \sum_{n=1}^{\infty} \frac{\epsilon_n}{2^n}$ with $\epsilon_n \in \{0, 1\}$ as in (A.5), then $T_2 x = .\epsilon_2\epsilon_3 \cdots$, $T_2^n x = .\epsilon_{n+1}\epsilon_{n+2} \cdots$ (check!) and therefore T_2 is also known as the one-sided shift. Let $M = [0, 1]^{\mathbb{Z}}$ be the probability space. The sigma-algebra of subsets of M generated by the cylinders $C = \{(x_n)_{n\in\mathbb{Z}} : x_i = \epsilon_{i_s}, s = 1, \ldots, k\}$, where i_s is an increasing finite sequence of integers, and $\epsilon_{i_s} \in \{0, 1\}$ coincides with the Borel sigma-algebra of M, equipped with the product topology. If μ_0 is a probability measure s.t. $\mu_0(\{0\}) = p_1$ and $\mu_0(\{1\}) = p_2$, with $p_1 + p_2 = 1$, $p_1 > 0$, $p_2 > 0$, we take as μ the product measure; in particular, $\mu(C) = \prod_{s=1}^{k} p_{i_s}$. The mixing property (A.29) need only be verified for A, B cylinders. But, for n sufficiently large, $T_2^{-n}(A)$ and B are disjoint and, since $\mu(T_2^{-n}(A)) = \mu(A)$, we have that $\mu(T_2^{-n}(A) \cap B) = \mu(A)\mu(B)$ for n sufficiently large. Thus we have proven that T_2 is mixing.

Take, now, as our dynamical system the set of autonomous n first-order difference equations

$$\vec{x}_{n+1} = T(\vec{x}_n), \tag{A.44}$$

where $\vec{x} \equiv (x_1, \ldots, x_n)$, $T(\vec{x}) \equiv (T_1(\vec{x}), \ldots, T_n(\vec{x}))$. The mapping $T : M \to \mathbb{R}^n$, where M is a manifold (which we often identify with an open set in $\mathbb{R}^n$), is a *vector field*: we assume it is an automorphism. Let $J(\vec{x})$ denote the Jacobian matrix $J_{i,j}(\vec{x}) = \frac{\partial T_i}{\partial x_j}(\vec{x})$ associated with (A.44). Defining $T^n(\vec{x}) \equiv T(T \cdots T)(\vec{x})$ as before and using the chain rule we have

$$J(T^n)(\vec{x}) = J(T^{n-1})(\vec{x}) \cdots J(T)(\vec{x})J(\vec{x}). \tag{A.45}$$

Let μ denote the measure invariant under the automorphism T, which completes the definition of the dynamical system. We shall assume it has compact support and that it is ergodic under the automorphism T.

Then, Oseledec's multiplicative ergodic theorem (see [78, 196]) implies the existence, for almost all $\vec{x}$ with respect to μ, of the limit

$$\lim_{n\to\infty} [J(T^n)^\dagger(\vec{x})J(T^n)(\vec{x})] \equiv \Lambda_{\vec{x}}. \tag{A.46}$$

The positive matrix $\Lambda_{\vec{x}}$ has eigenvalues $\lambda_1 > \lambda_2 > \cdots$.

Definition A.6. λ_i, with multiplicity m_i, is called a *Lyapunov exponent*.

By ergodicity of μ it follows that these exponents are constants for almost all values of $\vec{x}$ (for SRB measures this will hold for almost all $\vec{x}$ in the ergodic basin of μ). If $E_{\vec{x}}^i$ is the subspace of $\mathbb{R}^n$ corresponding to the eigenvalues smaller than $\exp(\lambda_i)$, $\mathbb{R}^n = E_{\vec{x}}^1 \supset E_{\vec{x}}^2 \supset \cdots$, one can show that for almost all $\vec{x}$ with respect to μ,

$$\lim_{n\to\infty} \frac{\log \|J(T^n(\vec{x}))u\|}{n} = \lambda_i \quad \text{for } i = 1, 2, \ldots, \tag{A.47}$$

if $u \in E_{\vec{x}}^i \backslash E_{\vec{x}}^{i+1}$. In (A.47) the norm is the Euclidean norm in $\mathbb{R}^n$. In particular, for all $u \in \mathbb{R}^n \backslash E_{\vec{x}}^2$, the limit equals the *largest Lyapunov exponent* λ_1. We shall see explicit examples shortly. But note that existence of a positive largest Lyapunov exponent is a precise definition of what is meant by sensitive or exponential dependence on initial conditions, which is an important element of chaotic behavior: see [78] for a detailed discussion. It is however easy to illustrate the phenomenon on the basis of the example T_2: consider two x_1, x_2 both in $[0, 1]$, close in the sense that the first n digits are identical. By the interpretation of T_2 as one-side shift, it follows immediately that $T_2^n x_1$ and $T_2^n x_2$ differ already in the first digit: an initially exponentially small difference 2^{-n} is magnified by the evolution to one of order $O(1)$. This is the best way to approach exponential sensitivity to initial conditions for quantum systems: see Chap. 6.

It turns out, however, that T_2 has more than sensitive dependence on initial conditions: the sequence of iterations $T_2^n x_0$ has, for almost every x_0, the same random character as the sequence of successive tossings of a coin; this is what characterizes Bernoulli systems and places them at the top of the ergodic hierarchy.

Theorem A.5 (Borel's theorem on normal numbers). *For a.e. $x_0 \in [0, 1]$, the nth digit in its binary expansion has relative frequency $1/2$, i.e., almost all x_0 is normal to base 2.*

Proof. Let $f : [0,1] \to [0,1]$ be defined by

$$f(x) = \begin{cases} 0 & \text{if } x < 1/2, \\ 1 & \text{if } x \geq 1/2. \end{cases}$$

Since T_2 is ergodic (being mixing),

$$\lim_{n \to \infty} \frac{1}{n} \sum_{i=0}^{n-1} f(T_2^i x) = \int_0^1 f(x)dx = 1/2$$

by Theorem A.1, for a.e $x \in [0,1]$, but

$$f(T_2^i x) = \begin{cases} 0 & \text{if } \epsilon_i = 0, \\ 1 & \text{if } \epsilon_i = 1, \end{cases}$$

which proves the theorem.   □

Problem A.6. *Which numbers comprise the set of zero Lebesgue measure in Theorem A.5?*

The two-sided shift (i.e., in both directions) is a.e. an automorphism of $[0,1] \to [0,1]$: the baker's map [161]. It is a caricature of the Smale horseshoe, which describes the local behavior of Hamiltonian systems near homoclinic points (which we shall define shortly): see [15, 59]. We now introduce it because it plays an important illustrative role in Chap. 6.

Let X be the unit square $X = [0,1] \times [0,1]$. The Borel sigma-algebra $\mathcal{B}$ is now generated by all possible rectangles of the form $[0,a] \times [0,b]$, and the Borel measure μ is the unique measure on $\mathcal{B}$ such that $\mu([0,a] \times [0,b]) = ab$. We define a transformation $S : X \to X$ — the *baker's transformation* by

$$S(x,y) = \begin{cases} (2x, y/2) & \text{if } 0 \leq x < 1/2 \text{ and } 0 \leq y \leq 1, \\ (2x-1, y/2+1/2) & \text{if } 1/2 \leq x \leq 1 \text{ and } 0 \leq y \leq 1, \end{cases} \quad \text{(A.48)}$$

The reader might wish to look at [168, Fig. 5] or verify for himself the effect of S on some figure symmetrically disposed w.r.t. the line $x = 1/2$: as in the kneading of a piece of dough, one first squashes the unit square to half its original height and twice its original width, and then cuts the resulting in half and moves the right half of the rectangle above the left. Repeating this operation several times, the result is to very quickly scramble, or mix, various parts of the original figure: this provides a very colorful illustration of the mixing property!

The Koopman operator, defined by (A.30), is unitary from $L^2(\Omega, d\mu)$ to itself, and leads (for automorphisms) to a characterization of the mixing property by (A.35). It is interesting, from the point of view of approach to equilibrium, to consider an alternative description. Consider, to fix ideas, the case of endomorphisms, and define the Koopman operator (we shall keep the same name for simplicity) now as an operator from $L^\infty(\Omega, d\mu)$ to $L^\infty(\Omega, d\mu)$ by (see [161, p. 42]):

$$(Uf)(x) = f(Tx). \tag{A.49}$$

It is well-defined because $f_1(x) = f_2(x)$ a.e. implies $f_1(Tx) = f_2(Tx)$ a.e. by the definition of endomorphism. It has the property

$$\|Uf\|_\infty \le \|f\|_\infty. \tag{A.50}$$

If $f \in L^1$, the functional

$$g \in L^\infty \to (f, Ug) \tag{A.51}$$

defines a continuous linear functional on L^∞, with

$$|(f, Ug)| \le \|f\|_1 \|g\|_\infty. \tag{A.52}$$

We may thus define a bounded linear operator — the *Ruelle–Perron–Frobenius operator* (see, e.g., [161, 222]) $P : L^1 \to L^1$ by

$$(Pf, g) = (f, Ug). \tag{A.53}$$

From (A.53), for $g = \chi_A$, where A is any Borel subset of Ω, we obtain

$$\int_A (Pf)(x)d\mu(x) = \int_{T^{-1}(A)} f(x)d\mu(x). \tag{A.54}$$

Conversely, (A.54) defines P uniquely by the Radon–Nikodym theorem, see [161, p. 37]. In [161], P is mentioned as being the adjoint of the Koopman operator U (defined by (A.49)), but the Banach space adjoint (see, e.g., [207, p. 185]) is an operator defined from the dual space (the space of continuous linear functionals on the given space) to the dual space: in the present case, the dual of L^∞ is a huge space, which contains L^1 properly. In fact, by [207, Exercise 8(b), p. 86], there exists a bounded linear functional λ on $L^\infty(\mathbb{R})$ such that $\lambda(f) = f(0) \, \forall f \in C(\mathbb{R})$, i.e., the dual space of $L^\infty(\mathbb{R})$ contains the Dirac measure at the origin. But, as we shall see, it is absolutely crucial for the developments here and in Chap. 6 that P be considered as an operator mapping densities (in contrast to individual orbits, formally characterized by the evolution of delta measures at some point) to densities!

The baker transformation is a.e. invertible (it is not invertible on the line $y = 1/2$): it is only a.e. a diffeomorphism. Taking $A = [0, x] \times [0, y]$ we have for $0 \le x \le 1$ and $0 \le y < 1/2$, $T^{-1}(A) = [0, x/2] \times [0, 2y]$ and thus, by (A.54),

$$(Pf)(x,y) = \frac{\partial^2}{\partial x \partial y} \int_0^{x/2} ds \int_0^{2y} dt f(s,t) = f(x/2, 2y), \quad \text{if } 0 \le y < 1/2$$

$$(A.55)$$

and for $1/2 \le y \le 1$ and $0 \le x < 1$

$$T^{-1}(A) = [0, x/2] \times [0, 1] \cup [1/2, 1/2 + x/2] \times [0, 2y - 1]$$

and hence, again by (A.54),

$$(Pf)(x,y) = \frac{\partial^2}{\partial x \partial y} \left[\int_0^{x/2} ds \int_0^1 dt f(s,t) + \int_{1/2}^{1/2+x/2} ds \int_0^{2y-1} dt f(s,t) \right]$$

$$= f(1/2 + x/2, 2y - 1), \quad \text{if } 1/2 \le y \le 1. \tag{A.56}$$

Note that, by (A.54), the Ruelle–Perron–Frobenius operator maps *densities* to *densities*: a density (function) f any $f : f \in L^1$ and $f \ge 0$ a.e. We denote the class of density functions by D. The positivity-preserving character of P, i.e.,

$$Pf \ge 0 \text{ a.e.} \quad \text{if } f \ge 0 \text{ a.e.} \tag{A.57}$$

is immediate from the Radon–Nikodym theorem (see, again, [161, p. 37]). It leads to the inequality

$$\|Pf\|_1 \le \|f\|_1. \tag{A.58}$$

Let $f^+(x) \equiv \max\{0, f(x)\}$ and $f^-(x) \equiv \min\{0, f(x)\}$. We have that $(Pf)^+ = (Pf^+ - Pf^-)^+ \le (Pf)^+$ and $(Pf)^- \le Pf^-$. From these inequalities it follows that $|Pf| \le P|f|$, and it is now simple to show.

Problem A.7. *Prove* (A.58).

Analogous to Problem A.7, it may also be proved that equality in (A.58), i.e., $\|Pf\|_1 = \|f\|_1$, occurs iff Pf^+ and Pf^- have disjoint supports; in particular,

$$\|Pf\|_1 = \|f\|_1 \quad \text{if } f \ge 0 \text{ a.e.} \tag{A.59}$$

Properties (A.57) and (A.59) define a *Markov operator*. If P is a Markov operator and for some $f \in L^1$, $Pf = f$, then f is called a *fixed point* of P. If $Pf = f$, then $Pf^+ = f^+$ and $Pf^- = f^-$ (see [161, Proposition 3.1.3]). If $f \in D$, and $Pf = f$, f is called a *stationary density* of P. We have the following important theorem (see [161, Theorem 4.4.1]).

Theorem A.6. *T is mixing iff* $\{P^n f\}$ *is weakly convergent to 1 for all* $f \in D$, *i.e.,*

$$\lim_{n \to \infty} (P^n f, g) = (f, 1)(1, g) \quad \forall f \in D \quad and \quad \forall g \in L^\infty. \tag{A.60}$$

If P is a Markov operator with stationary density 1, i.e.,

$$P1 = 1, \tag{A.61}$$

then $\{P^n f\}$ converges weakly to the stationary density: the weak limit is a fixed point of P and, being unique, it is the function $f = 1$. This is an approach to equilibrium in the formerly defined sense.

How does all this apply to the baker's transformation? Property (A.57) and hence (A.59) follow from (A.55). We now consider the mixing property. The latter is simplest to see by noting that, if $(x_0, y_0) \in [0, 1] \times [0, 1]$, s.t. $x_0 = .a_1 a_2 a_3 a_4 \cdots$ and $y_0 = .b_1 b_2 b_3 b_4 \ldots$, by juxtaposing the representations in the following way: $\cdots b_4 b_3 b_2 b_1 a_1 a_2 a_3 a_4 \ldots$, the baker transformation may be seen to correspond to the shift of the decimal point to the right — the so-called two-sided shift, see [266, p. 18] (check!). Then mixing follows by the same proof as for the one-sided shift. Finally, (A.61) also follows from (A.55). Thus, by Theorem A.6, there is approach to equilibrium for the baker's map. Note that $f \in L^1$ excludes a delta measure at some point in the square: this is expected because the baker map is (a.e.) invertible, and individual orbits are therefore reversible. We see hereby that approach to equilibrium in the theory of dynamical systems in the sense defined above depends on two key features: (Q1) one looks only at the evolution of densities, and not individual orbits: this is a reduced description of the system, similar to Boltzmann's approach (see Chap. 6); (Q2) the initial state (density) is not the (invariant, stationary) density, which is the uniform distribution on the unit square. We return to these points in Sec. 6.4.

We finish this appendix with two subjects: Anosov systems and rate of mixing (with some remarks on nonuniformly hyperbolic systems) and a few basic results of number theory used in the main text.

The baker transformation is a prototype of an important class of transformations, the *Anosov systems* [10]. Another special example of Anosov system is the Arnold cat map [15]:

$$T_A(x, y) = (x + y, x + 2y) \mod 1. \tag{A.62}$$

The fixed points F of T_A are given by $(x + y, x + 2y) = (x, y)$ hence $F = (0, 0)$. This is an example of a *hyperbolic point*. In the more general case (A.44), but now with T a mapping from a complex manifold M to $\mathbb{C}^n$, let $\vec{x_0}$ be an invariant or equilibrium point of the vector field T:

Definition A.7. A linear vector field on $\mathbb{C}^n$ given by $\vec{x} \to L\vec{x}$, where L is a $n \times n$ matrix with complex entries, is said to be *hyperbolic* iff the spectrum $\sigma(L)$ of L has empty intersection with the imaginary axis. An equilibrium point $\vec{x_0}$ of the vector field T is said to be hyperbolic iff the Jacobian matrix $J(\vec{x_0})$ is a hyperbolic vector field. By the Jordan decomposition (see [114]), it may be shown that $\sigma(\exp(L)) = \exp(\sigma(L))$: hence, $\vec{x_0}$ is hyperbolic iff $J(\vec{x_0})$ does not have any eigenvalue of unit modulus.

Due to the Hartman–Grobman theorem (see [104]) the phase portrait of a vector field near a hyperbolic point has the same topological structure as its linearization at this point: therefore a hyperbolic fixed point of a vector field is unstable. Such is not the case for *elliptic* fixed points, for which $J(\vec{x_0})$ has modulus one: their stability depends on the character of the nonlinear terms [104]. The Hartman–Grobman theorem also guarantees the existence of the so-called local (stable and unstable) manifolds at any hyperbolic point (see also [247, Lecture 17]). We now return to the cat map. The Jacobian matrix

$$\begin{pmatrix} 1 & 1 \\ 1 & 2 \end{pmatrix}$$

has eigenvectors $v^\pm$ given, respectively, by:

$$\begin{pmatrix} 1 \\ \dfrac{1 \pm \sqrt{5}}{2} \end{pmatrix}$$

corresponding to eigenvalues $r_\pm = (3 \pm \sqrt{5})/2$. The corresponding invariant (local) manifolds are $W^\pm(\vec{0}) = E_{\vec{0}}^\pm$ with $E^\pm$ generated by $v^\pm$, with angular coefficients s and $-s^{-1}$, with $s = (1 + \sqrt{5})/2$. Since the latter are irrational numbers, these manifolds cover the torus densely. Because $r_+ > 1$, $(0,0)$ is a hyperbolic point.

Problem A.8. *Identify the sets E_0^1 and E_0^2 in the definition of Lyapunov exponents and compute them for the cat map.*

The sum of the Lyapunov exponents for an area-preserving map, such as the cat map above, is zero: this follows in a straightforward way from their definition. In Hamiltonian and other measure-preserving maps chaos has its origins in *homoclinic points* [15] defined as points of intersection of the local (stable and unstable) manifolds associated to a given hyperbolic point, other than the point itself (see also [59]). The usual proofs of the existence of the "homoclinic tangle" rely on the (often verified, such as in the cat map) hypothesis that the manifolds intersect *transversally*, i.e., are not tangential at the homoclinic point (Melnikov's method, see [128] for a pedagogic discussion). See also the monograph of Palis and Takens for the discussion of homoclinic tangencies [197]. In the case of the cat map, considering the image of a region B after a large number n of iterations of the map T_A, it will consist of the product of r_+^n expansions in the unstable direction v^+, and r_-^n contractions in the stable direction v_-, i.e., a highly "stretched" band in direction v^+ (mod 1): for n large $T_A^n B$ covers the torus and the mixing property holds (see [161, Example 4.4.3, p. 71] for the proof). The two invariant manifolds intersect in homoclinic points; the image of a homoclinic point is, by definition, also a homoclinic point, and, thus, the cat map possesses an infinity of homoclinic points; it is also immediate (check!) that the manifolds intersect transversally at these points.

Problem A.9. *Which points of the unit square yield periodic orbits of the cat map? What can one say about the asymptotic distribution of their periods?*

Hint for the second part: See the discussion in [59] on the Ozorio–Hannay uniformity principle.

Anosov systems such as the cat map (A.62) are "uniformly hyperbolic". For such systems there is *exponential decay of correlations*

$$\left| \int_\Omega f(T^n x)g(x)d\mu - \int_\Omega f d\mu \int_\Omega g d\mu \right| \le Cr^n \quad \text{for some } r < 1 \qquad (A.63)$$

(see [265] for a nice review). Much of the recent work in dynamical systems has been, however, devoted to systems displaying only a weak form of hyperbolicity. One of them is that T is expanding in its critical orbit, see [265, Theorem 5.1, p. 125] or the original early article of Baladi and Viana [19] for a readable account for an important class of models. Together with the other assumptions, it is proved that T admits a unique absolutely continuous invariant measure which is ergodic (and so is an SRB measure for T) and exhibits exponential decay of correlations in the sense of (A.63).

Physically more "realistic" systems remain, however, very difficult to analyze. A prototype of the latter is the famous *standard or Chirikov–Taylor map* (see [247, p. 138]), which is the classical analogue of the kicked rotor (3.137a), (3.137b): it is a map of the cylinder $\mathbb{R} \times \mathbb{T}$ to itself, defined by:

$$z_{n+1} = z_n + k \sin(2\pi\phi_n),$$
$$\phi_{n+1} = \phi_n + z_{n+1} \bmod 1. \tag{A.64}$$

The importance of this map is reviewed by Sinai in [247, p. 138]. There exist very few rigorous results for this map: the beautiful Aubry–Mather theory shows that for $k > 1$ the standard map has no invariant curves which may be represented by a continuous function $z = f(\phi)$ (for a simple proof of this, see [247, p. 142]): the structures occurring for $k > 1$ are fractal objects denominated *Cantori*. Another result, much more difficult, due to Duarte [73, 74] proves, very roughly speaking, the existence of an abundance of of elliptic islands even deep inside the region $k > 1$ of "hard chaos" in the model (actually this region is conjectured to be $k > k_{\rm cr}$ with $k_{\rm cr}$ given by Greene's conjecture, see the discussion in [247]). Duarte's remarkable result throws considerable doubt on the validity of exponential decay of correlations (A.63) in the chaotic region: it may be algebraic instead, see Chap. 6 for further remarks. Estimates on the rate of decay of correlations are among the most challenging open problems, both in the theory of dynamical systems and in the theory of "quantum chaos".

We conclude this appendix with a brief review of some of the most fundamental aspects of number theory used in the main text, in particular Secs. 3.1 and 3.2. The reader will find a complete and very readable account of these features in Khinchin's book [150].

Irrational numbers may be approximated by rationals with arbitrary precision. For example, π is approximated by the sequence

$$s/r = 3/1, 31/10, 314/100, 3142/1000, 31416/10000, \ldots,$$

which yields the better approximants the greater the value of r. For such decimal approximations, we have for an irrational number μ: $|\mu - \frac{s_n}{r_n}|$

$< 1/r_n$, with $r_n = 10^n$ and a similar inequality is valid for any base. These approximants are not, however, the best possible, because they are contaminated by the intrinsic properties of the base. We are thus led to search a sequence of approximants $\{\mu_n\}_{n \geq 1}$ which is independent of the base: such a sequence is proportionated by the *continued fractions*:

$$\mu_n = a_0 + \cfrac{1}{a_1 + \cfrac{1}{a_2 + \cdots}}. \qquad (A.65)$$

The approximants in (A.65):

$$\mu_n = s_n/r_n \qquad (A.66)$$

obey the recursion relations

$$\begin{aligned} s_n &= a_n s_{n-1} + s_{n-2} \quad \text{for } n \geq 2, \\ r_n &= a_n r_{n-1} + r_{n-2} \quad \text{for } n \geq 2. \end{aligned} \qquad (A.67)$$

In order to prove (A.67) by induction, suppose it valid up to order $(n-1)$. Defining

$$\frac{s'_{n-1}}{r'_{n-1}} = a_1 + \cfrac{1}{a_2 + \cfrac{1}{a_3 + \cfrac{1}{\cdots + \frac{1}{a_n}}}}$$

we have:

$$\begin{aligned} s_n &= a_0 s'_{n-1} + r'_{n-1}, \\ r_n &= s'_{n-1}, \end{aligned} \qquad (A.68)$$

and by the induction hypothesis

$$\begin{aligned} s'_{n-1} &= a_n s'_{n-2} + s'_{n-3}, \\ r'_{n-1} &= a_n r'_{n-2} + r'_{n-3}. \end{aligned} \qquad (A.69)$$

Introducing (A.69) into (A.68)

$$\begin{aligned} s_n &= a_0(a_n s'_{n-2} + s'_{n-3}) + a_n r'_{n-2} + r'_{n-3}, \\ r_n &= a_n s'_{n-2} + s'_{n-3}. \end{aligned}$$

or

$$\begin{aligned} s_n &= a_n(a_0 s'_{n-2} + r'_{n-2}) + a_0 s'_{n-3} + r'_{n-3} = a_n s'_{n-1} + s'_{n-2}, \\ r_n &= a_n r'_{n-1} + r'_{n-2}. \end{aligned}$$

Clearly (A.67) is valid for $n = 2$ and thus the result is demonstrated.

The difference between two approximants μ_n and μ_{n-1} is

$$\frac{s_{n-1}}{r_{n-1}} - \frac{s_n}{r_n} = \frac{s_{n-1}r_n - s_n r_{n-1}}{r_n r_{n-1}} = \frac{(-1)^n}{r_n r_{n-1}}. \qquad (A.70)$$

In fact, multiplying the first formula in (A.67) by r_{n-1}, the second one by s_{n-1} and subsequently subtracting the resulting first one from the resulting second one, we obtain

$$s_{n-1}r_n - r_{n-1}s_n = -\left(s_{n-2}r_{n-1} - r_{n-2}s_{n-1}\right)$$

and, since $(s_{-1} = 1,\ r_{-1} = 0)$, $r_0 s_{-1} - s_0 r_{-1} = 1$, and therefore it follows that

$$s_{n-1}r_n - r_{n-1}s_n = (-1)^n$$

from which (A.70) follows.

Problem A.10. *Show that for all $n \geq 2$,*

$$\frac{s_{n-2}}{r_{n-2}} - \frac{s_n}{r_n} = \frac{(-1)^{n-1}a_n}{r_n r_{n-2}}.$$

The two results above show that the convergents of even order form an increasing sequence (we assume $a_n > 0$ for all $n > 1$), and those of odd order form a decreasing sequence. It follows therefore that both sequences have the same limit. Indeed, since, by (A.67), each convergent of odd order is greater than the immediately subsequent convergent of even order, it follows that *each* convergent of odd order is greater than *each* convergent of even order, and that the two sequences of convergents (of odd and even order) tend to the same limit.

The above result culminates (see [150]) in the following.

Theorem A.7 (Hurwitz's theorem). *For any μ, there exist rational approximants s_n/r_n such that*

$$\left|\mu - \frac{s_n}{r_n}\right| \leq \frac{1}{\sqrt{5}r_n^2}. \qquad (A.71)$$

Clearly the approximants converge the quicker the a_n diverge (for a rational number, some $a_n = \infty$ at some $n = n_0$ finite). As an illustration, we suggest that the reader show the explicit results for π (precise up to the

k_nth digit):

$$s_0/r_0 = 3,$$
$$s_1/r_1 = 22/7 \simeq 3.1429 \quad (k_1 = 2),$$
$$s_2/r_2 = 333/106 \simeq 3.1415 \quad (k_2 = 3),$$
$$s_3/r_3 = 355/113 \simeq 3.1515929 \quad (k_3 = 6)$$

(the last result, according to Berry, was known to Lao–Tse (604–531 B.C.)). The slowest convergence, corresponding to the irrational number most poorly approximated by rationals, corresponds to the *golden mean*: $a_0 = a_1 = \cdots = 1$,

$$1 + \cfrac{1}{1 + \cfrac{1}{1 + \cdots}} = \frac{\sqrt{5} - 1}{2}.$$

This number is important for several reasons: it appears in Problem 3.8 in connection with the P.V. numbers, and is also conjectured to be the rotation number corresponding to the critical coupling k_{cr} in the standard map, which signals the disappearance of the last "KAM torus", i.e., in the standard map the last continuous invariant curve. It saturates the inequality (A.71) provided by Hurwitz's theorem.

This last remark brings us to our last topic: *diophantine numbers*.

Definition A.8. An irrational number μ is said to be *diophantine of type* σ iff there exists a $\gamma > 0$ such that

$$\left| \mu - \frac{s}{r} \right| \geq \frac{\gamma}{r^\sigma} \quad \text{for all } s/r \in \mathbb{Q}. \tag{A.72}$$

In passing, we note that μ is said to be a Liouville number if it is neither rational nor diophantine. Alternatively, μ is Liouville iff there exists a sequence $\{s_n/r_n\} \in \mathbb{Q}$ such that $|\mu - \frac{s_n}{r_n}| < r_n^{-n}$ for all $n \geq 1$. By a theorem due to Liouville, every algebraic number (zero of a polynomial with rational coefficients) is diophantine. On the other hand, the number $\mu = \sum_{n=1}^\infty 2^{-n!}$ is by definition Liouville. Thus there exist nonalgebraic numbers: they are the transcendental numbers. For much more about this, see [186].

The set of diophantine numbers of type σ has full Lebesgue measure if $\sigma > 2$: for almost all $\mu \in [0, 1]$ and for all $\tau > 0$, there exists a number $K = K(\mu, \sigma = \tau + 2)$ such that

$$\left| \mu - \frac{s}{r} \right| \geq \frac{K}{r^{2+\tau}} \tag{A.73}$$

for all $s/r \in \mathbb{Q}$. In order to show (A.73), fix r and s. Consider the set of numbers $\mu \in [0, 1]$ such that (A.73) is violated. They form a set of length $\leq \frac{2K}{r^{2+\tau}}$, and, since $s \leq r$, the union of all such intervals with the same r has measure $\leq \frac{2K}{r^{1+\tau}}$. Summing over all the r we obtain a set with Lebesgue measure $\leq CK$, where $C = 2 \sum_{r=1}^{\infty} r^{-1-\tau} < \infty$ since $\tau > 0$. Choosing K arbitrarily small, we may render the measure of the set of numbers violating (A.73) as small as wished. Thus, the set of numbers violating (A.73) *for all* K has zero Lebesgue measure.

Appendix B

Transfer Matrix, Prüfer Variables and Spectral Analysis of Sparse Models

Weyl–Titchmarsh m-function of a Jacobi matrix. We may think of a *Jacobi matrix*

$$J = \begin{pmatrix} v_1 & p_1 & 0 & 0 & \cdots \\ p_1 & v_2 & p_2 & 0 & \cdots \\ 0 & p_2 & v_3 & p_3 & \cdots \\ 0 & 0 & p_3 & v_4 & \cdots \\ \vdots & \vdots & \vdots & \vdots & \ddots \end{pmatrix}, \tag{B.1a}$$

as an operator acting on the space of complex-valued square-summable sequences $u = (u_n)_{n \geq 0}$, denoted by $l_2(\mathbb{Z}_+)$, and the Schrödinger equation associated with J reads

$$((J - z\mathbb{I})u)_n = p_n u_{n+1} + p_{n-1} u_{n-1} + (v_n - z)u_n = 0 \tag{B.1b}$$

for $n \geq 1$ with $p_0 \equiv 1$ and $z \in \mathbb{C}$. A Jacobi matrix J^ϕ is said to satisfy a ϕ-boundary at 0 if

$$u_0 \cos\phi - u_1 \sin\phi = 0, \tag{B.1c}$$

for some $\phi \in [0, \pi)$; (B.1a), as an operator, satisfies Dirichlet 0-boundary phase condition $u_0 = 0$ and we write $J = J^0$. The phase boundary plays an important role on the characterization of singular part of the spectrum, as we shall see in the following. We call J an admissible Jacobi matrix if it is of the form (B.1a) and satisfies $0 < p_n \leq 1$ with $\sum_{n \geq 0} p_n^{-1} = \infty$ and $|v_n| \leq M < \infty$. Sooner, we shall restrict the admissible class to two sparse models dealt in this monography (see (4.1a) et seq.): (i) $p_n = 1 \; \forall n$ and $v_n = 0$ except for a lacunary subsequence $\mathcal{A} = (a_j)_{j \geq 1}$; (ii) $v_n = 0 \; \forall n$ and $p_n = 1$ except for a lacunary subsequence $\mathcal{A} = (a_j)_{j \geq 1}$.

The set of all solutions $u^\phi(z)$ of (B.1b) and (B.1c) forms a two-dimensional vector space whose base: $y(z) = u^\alpha(z)$ and $w(z) = u^{\alpha+\pi/2}(z)$ may be chosen by fixing the initial conditions

$$
\begin{aligned}
y_0 &= \sin\alpha, \quad y_1 = \cos\alpha, \\
w_0 &= \cos\alpha, \quad w_1 = -\sin\alpha
\end{aligned}
\tag{B.2}
$$

for some $\alpha \in [0, \pi)$. For $\Im z \neq 0$, there are two alternatives: either we have the *limit-circle* (in which case both $y(z)$, $w(z) \in l_2(\mathbb{Z}_+)$) or the *limit-point* case, for which there exist exactly one linear independent $l_2(\mathbb{Z}_+)$-solution. We have the following proposition.

Proposition B.1. *If J is an admissible Jacobi matrix, then J is limit-point.*

Proof. Let $y(z)$ and $w(z)$ be a base for (B.1b). By Green's formula (see, e.g., [256, (1.20)])

$$
0 = (\bar{w}, Jy)_n - (J\bar{w}, y)_n = -W_n[y, w] + W_0[y, w],
\tag{B.3a}
$$

where $W_n[f, g]$ is the (weighted) Wronskian function:

$$
W_n[f, g] = p_n(f_n g_{n+1} - f_{n+1} g_n)
\tag{B.3b}
$$

of two sequences $f = (f_n)_{n \geq 0}$ and $g = (f_n)_{n \geq 0}$ at n and $(f, g)_N = \sum_{n=1}^N \bar{f}_n g_n$ is the inner product restricted to $\{1, \ldots, N\}$. Equation (B.3a) implies that the Wronskian of two linearly independent (L.I.) solutions of the Schrödinger equation is constant and does not vanish:

$$
W_n[y, w] = W_0[y, w] = p_0(y_0 w_1 - y_1 w_0) = -1
\tag{B.3c}
$$

by (B.2). Together with (B.3b) and the Schwartz inequality, we have

$$
\sum_{n=0}^\infty \frac{1}{p_n} = \sum_{n=0}^\infty \frac{-1}{p_n} W_n[y, w] = \left| \sum_{n=1}^\infty (y_n w_{n+1} - y_{n+1} w_n) \right| \leq 2\|y\|\|w\|
\tag{B.3d}
$$

where $\|\cdot\|$ is the $l_2(\mathbb{Z}_+)$-norm. If J is admissible, then the l.h.s. of (B.3d) diverges and necessarily one of the two L.I. solutions: y or w does not belong to $l_2(\mathbb{Z}_+)$, concluding the proof. $\square$

We set $\alpha = 0$ in the following propositions and refer to Chap. 2 and [256, App. B] for references and further directions.

Proposition B.2. *The $l_2(\mathbb{Z}_+)$-solution of (B.1b) with $\Im z > 0$ can be written as*

$$\chi(z) = w(z) - m(z)y(z), \tag{B.4}$$

where $m(z) = m_0(z)$, the Weyl–Titchmarsh m-function related to J^α with $\alpha = 0$, is defined by

$$m(z) := \left(e_1, \frac{1}{J^0 - z\mathbb{I}}e_1\right) \tag{B.5}$$

and $\{e_j, j \geq 0\}$ denotes the canonical base of $l_2(\mathbb{Z}_+)$: $(e_j)_i = \delta_{ij}$.

Proof. Let $\chi(z)$ be the $l_2(\mathbb{Z}_+)$-solution of (B.1b). Since $y(z)$ is not an eigenvector of J when $\Im z > 0$, it cannot be an $l_2(\mathbb{Z}_+)$-solution of (B.1b) so, $\chi(z)$ is a linear combination of $y(z)$ and $w(z)$ of the form (B.4). The Green's function associated with J

$$g_{ij}(z) = \left(e_i, \frac{1}{J - z\mathbb{I}}e_j\right)$$

can be written in terms of the two L.I. solutions of (B.1b) $y(z)$ and $\chi(z)$:

$$g_{ij}(z) = \begin{cases} \dfrac{y_i(z)\chi_j(z)}{-W_j[y,\chi]} & \text{if } i < j, \\[3mm] \dfrac{y_j(z)\chi_i(z)}{-W_j[y,\chi]} & \text{if } j < i. \end{cases} \tag{B.6}$$

Setting $i = j = 1$ in (B.6), together with $W_j[y,\chi] = W_0[y,w] = -1$ and $y_1(z)\chi_1(z) = m_0$, the proof is concluded. □

Remark B.1. Since J^ϕ is a rank-one perturbation of J^0: $(J^\phi u)_n = (J^0 u)_n + \delta_{0,n}u_0 \tan\phi$, the Weyl–Titchmarsh m-function $m_\alpha(z)$, defined by (B.5) with J^0 replaced by J^α, satisfies

$$m_\alpha(z) = \frac{m(z)}{1 - m(z)\tan\alpha}. \tag{B.7}$$

Definition (B.5) implies that $m(z)$ is holomorphic in $\mathbb{C}\backslash\sigma(J)$ and, since it maps the upper half-plane $\mathbb{H}$ into itself, it is, in addition, a *Herglotz function* (called also Pick or Nevanlinna–Pick function).

Problem B.1. *Prove the above statements.*

Hint: Use the first resolvent equation $(J - z\mathbb{I})^{-1} - (J - \zeta\mathbb{I})^{-1} = (z - \zeta)(J - z\mathbb{I})^{-1}(J - \zeta\mathbb{I})^{-1}$ for the former.

As a consequence, $m(z)$ admits a unique canonical integral representation ([137, Chap. 2, Theorem I]):

$$m(z) = az + b + \int_{-\infty}^{\infty} \left(\frac{1}{\lambda - z} - \frac{\lambda}{\lambda^2 + 1} \right) d\rho(\lambda), \quad \Im z > 0,$$

where $a \geq 0$, $b \in \mathbb{R}$ are constants and $d\rho(\lambda)$ is a Borel–Stieltjes measure such that

$$\int_{-\infty}^{\infty} \frac{d\rho(\lambda)}{\lambda^2 + 1} < \infty.$$

It follows from (B.5) that

$$m(z) = -z^{-1} + O(z^{-2})$$

which implies $a = 0$. Since the spectrum of a bounded operator is compact, the spectral measure μ of J is supported in a compact set Σ,

$$m(z) = \left(e_1, \frac{1}{J^0 - zI} e_1 \right) = b - \int_{\Sigma} \frac{\lambda}{\lambda^2 + 1} d\rho(\lambda) + \int_{\Sigma} \frac{1}{\lambda - z} d\rho(\lambda),$$

where

$$\int_{\Sigma} \frac{1}{\lambda - z} d\rho(\lambda) = -z^{-1} \int_{\Sigma} d\rho(\lambda) + O(z^{-2})$$

as $z \to \infty$, from which we conclude that $b - \int_{\Sigma} \frac{\lambda}{\lambda^2+1} d\rho(\lambda) = 0$, and

$$m(z) = \int_{-\infty}^{\infty} \frac{1}{\lambda - z} d\rho(\lambda) \tag{B.8}$$

is a *Borel transform* of the spectral measure $\mu = d\rho$.

The Weyl m-function can be approached, as originally has been done by Weyl, as a sequence limit: $m(z) = \lim_{N \to \infty} m_N(z)$ of holomorphic functions in $\mathbb{C}\backslash\sigma(J^{0,\beta})$. Here, $J^{0,\beta}$ is a finite Jacobi matrix satisfying 0-Dirichlet boundary condition at 0 and β-boundary condition at N:

$$u_N \cos \beta - u_{N+1} \sin \beta = 0. \tag{B.9}$$

A solution $\chi(z) = \chi(z; N)$ of Schrödinger equation (B.1b),

$$\chi(z) := w(z) - m_N(z)y(z),$$

satisfies the phase β-boundary condition (B.9) at N iff

$$m_N(z) = \frac{w_N(z) - \zeta w_{N+1}(z)}{y_N(z) - \zeta y_{N+1}(z)}, \quad \zeta = \tan\beta. \tag{B.10}$$

Since the r.h.s. of (B.10) is a linear fractional map $L : \mathbb{C} \longrightarrow \mathbb{C}$, $L = (a\zeta + b)/(c\zeta + d)$ with $ad - bc \neq 0$, as ζ varies over $\mathbb{R}$, $m_N(z)$, for fixed $N \in \mathbb{N}$ and $z \in \mathbb{C}$ with $\Im z \neq 0$, varies over a circle K_N in $\mathbb{C}$, called *Weyl circle*. The fact that $\chi(z)$ satisfies (B.9) may also be expressed as $\zeta = \chi_N(z)/\chi_{N+1}(z)$.

Proposition B.3. *The inequality*

$$\|\chi(z)\|_N^2 \leq \frac{\Im m_N(z)}{\Im z} \tag{B.11}$$

is precisely the condition for m_N to lie inside a Weyl circle K_N:

$$|c_N - m|^2 = r_N^2 \tag{B.12}$$

of center $c_N = W_N[w, \bar{y}]/W_N[y, \bar{y}]$ and radius $r_N = |W_N[y, \bar{y}]|^{-1}$.

Proof. Let us show that (B.12) is the Weyl circle. We write the circle equation $\Im\zeta = \Im(\chi_N/\chi_{N+1}) = 0$ as

$$0 = W_N[\chi, \bar{\chi}] = W_N[v, \bar{v}] - m_N W_N[y, \bar{v}]$$
$$- \bar{m}_N W_N[v, \bar{y}] + |m_N|^2 W_N[y, \bar{y}]. \tag{B.13a}$$

Together with (B.3d), $W_N[g, \bar{h}] = -W_N[\bar{h}, g] = -\overline{W_N[h, \bar{g}]}$ and

$$|W_N[g, h]|^2 = W_N[g, \bar{h}]W_N[\bar{g}, h] + W_N[g, \bar{g}]W_N[h, \bar{h}] \tag{B.13b}$$

with $g = y$ and $h = w$, (B.13a) multiplied by $1/W_N[y, \bar{y}]$ ($W_N[y, \bar{y}] \neq 0$, by (B.13e) below) reads

$$-\frac{1}{|W_N[y, \bar{y}]|^2} + \frac{W_N[y, \bar{w}]}{W_N[y, \bar{y}]}\frac{\overline{W_N[y, \bar{w}]}}{\overline{W_N[y, \bar{y}]}}$$
$$- m_N \frac{W_N[y, \bar{w}]}{W_N[y, \bar{y}]} - \bar{m}_N \frac{\overline{W_N[y, \bar{w}]}}{\overline{W_N[y, \bar{y}]}} + |m_N|^2 = 0$$

which is exactly $-r_N^2 + |c_N - m_N|^2 = 0$. Hence, by (B.13a), m_N lies in K_N iff

$$\frac{W_N[\chi, \bar{\chi}]}{W_N[y, \bar{y}]} = 0. \tag{B.13c}$$

Using again the fact that m_N is a linear fractional map, either m_N satisfies $|c_N - m_N| \leq r_N$ or it is outside the Weyl circle K_N. Replacing m_N in the circle equation (B.13c) by its center c_N, the l.h.s. of (B.13c) is the negative number $-r_N^2$. So, m_N is inside or at K_N iff

$$\frac{W_N[\chi, \bar{\chi}]}{W_N[y, \bar{y}]} \leq 0. \tag{B.13d}$$

Applying Green's identity (B.3a) with w replaced by $\bar{y}$, together with $W_0[y, \bar{y}] = 0$, yield

$$W_N[y, \bar{y}] = -\Im z \, \|y\|_N^2. \tag{B.13e}$$

Similarly, with $W_0[\chi, \bar{\chi}] = \Im m_N$, we have

$$W_N[\chi, \bar{\chi}] = \Im m_N - \Im z \|\chi(z)\|_N^2. \tag{B.13f}$$

Replacing (B.13e) and (B.13f) into (B.13d), we conclude that m_N is inside or at K_N iff

$$\frac{\Im z \|\chi(z)\|_N^2 - \Im m_N}{\Im z \|y\|_N^2} \leq 0$$

which is equivalent to (B.11). $\qquad\qquad\qquad\qquad\qquad\qquad\qquad\qquad$ $\square$

Problem B.2. *Show* (B.13b).

The radius $r_N = r_N(z) = (\Im z \|y(z)\|_N^2)^{-1}$ is, by (B.13e) and definition (B.12), a monotone decreasing function of N and the circles $K_N(z)$, $N \geq 1$, contract to a limit point by Proposition B.1. It thus follows that $m_N(z)$ converges, as $N \to \infty$, to the Weyl–Titchmarsh m-function $m(z)$, uniformly in compact sets of $\mathbb{H}$, and $w(z) - m(z)y(z)$, $\Im z > 0$, is the (only one) $l_2(\mathbb{Z}_+)$-solution of (B.1b).

Spectral decomposition in terms of the values of $m(\lambda + i0)$. The spectral measure $\mu = d\rho$ may be decomposed into absolutely continuous, singular, singular continuous and pure point components: $\mu = \mu_{ac} + \mu_s$, $\mu_s =$

$\mu_{\text{sc}} + \mu_{\text{pp}}$, according to the boundary value $\Im m^+(\lambda) = \lim_{\varepsilon \downarrow 0} \Im m(\lambda + i\varepsilon)$ for which the Stieltjes inversion formula (see [137, Chap. 2, Lemma 1])

$$\rho(\lambda_+) - \rho(\lambda_-) = \lim_{\varepsilon \downarrow 0} \frac{1}{\pi} \int_{\lambda_-}^{\lambda_+} \Im m(\lambda + i\varepsilon) d\lambda$$

plays a fundamental role.

A function $f(z)$ is said to possess a *normal limit* at point $\lambda \in \mathbb{R}$ if $f(z)$ converges to a (finite or infinite) limit as $z \downarrow \lambda$ along perpendicular to real axis direction. Since $m(z)$ is Herglotz, we have (see [92]) the following lemma.

Lemma B.1. *If the Radon–Nikodym derivative $(d\mu/d\lambda)(\lambda)$ of the spectral measure μ at λ w.r.t. Lebesgue measure $d\lambda$ exists, finite or infinitely, then $\Im m^+(\lambda)$ exists and $(d\mu/d\lambda)(\lambda) = (1/\pi)\Im m^+(\lambda)$.*

Definition B.1. A set $\Sigma \subset \mathbb{R}$ is called *minimal* (or *essential*) *support* of a measure ν in $\mathbb{R}$ if

(i) $\nu(\mathbb{R}/\Sigma) = 0$ (i.e. Σ is the support of ν);
(ii) any subset $\Sigma_0 \subset \Sigma$ which does not support Σ has ν and Lebesgue measure 0: $\nu(\Sigma_0) = \mathcal{L}(\Sigma_0) = 0$.

It follows from de la Vallée–Poisson (see [224, Chap. IV, Theorem 9.6]), Lebesgue–Radon–Nikodym theorem (see [220, Theorem 6.9]) and Lemma B.1 that (see [92, Proposition 1]):

Proposition B.4. *The minimal supports Σ, Σ_{ac}, Σ_{s}, Σ_{sc} and Σ_{pp} of μ, μ_{ac}, μ_{s}, μ_{sc} and μ_{pp}, the spectral measure μ of a Jacobi matrix J, and the absolutely continuous (a.c.), singular (s), singular continuous (s.c.) and pure point (p.p.) parts, are, respectively, given by*

(i) $\Sigma = \{\lambda \in \mathcal{E} : 0 < \Im m^+(\lambda) \leq \infty\}$,
(ii) $\Sigma_{\text{ac}} = \{\lambda \in \mathcal{E} : 0 < \Im m^+(\lambda) < \infty\}$,
(iii) $\Sigma_{\text{s}} = \{\lambda \in \mathcal{E} : 0 < \Im m^+(\lambda) = \infty\}$,
(iv) $\Sigma_{\text{sc}} = \{\lambda \in \mathcal{E} : 0 < \Im m^+(\lambda) = \infty, \ \mathcal{L}(\lambda) = 0\}$,
(v) $\Sigma_{\text{pp}} = \{\lambda \in \mathcal{E} : 0 < \Im m^+(\lambda) = \infty, \ \mathcal{L}(\lambda) > 0\}$,

where $\mathcal{E} = \{\lambda \in \mathbb{R} : \Im m^+(\lambda) \text{ exists}\}$.

Now, we introduce a key notion in the Gilbert–Pearson theory relating solutions of the Schrödinger equation (B.1b) and the decomposition of the spectral measure μ of J.

Definition B.2. A solution $u = (u_n)_{n \geq 0}$ of $(J^\phi - \lambda \mathbb{I})u = 0$ (regardless the phase boundary ϕ) is said to be *subordinate* iff

$$\lim_{L \to \infty} \frac{\|u\|_L}{\|w\|_L} = 0 \tag{B.14}$$

holds for any linearly independent solution $w = (w_n)_{n \geq 0}$ of the equation, where $\|u\|_L^2 = \sum_{n=1}^{L} |u_n|^2$ denotes the norm over an interval of length L.

Remark B.2. Gilbert–Pearson's theory deals with boundary values of the Weyl–Titchmarsh m-function. In [92, Theorems 1 and 2] (see [149] for the discrete case) prove that $m^+(\lambda) = \lim_{\varepsilon \downarrow 0} m(\lambda + i\varepsilon)$ exists if and only if a subordinate solution of $(J - \lambda \mathbb{I})u$ exists which (when it exists) is given by $\chi(\lambda) = w(\lambda) - m^+(\lambda)y(\lambda)$ if $|m^+(\lambda)| < \infty$ and by $y(\lambda)$ if $|m^+(\lambda)| = \infty$. By (B.7), if $m^+(\lambda)$ exists, then $m_\alpha^+(\lambda)$ exists for every $\alpha \in (0, \pi)$; if it exists and $|m^+(\lambda)| = \infty$, then $m_\alpha^+(\lambda)$ is finite for every $\alpha \in (0, \pi)$; if $m^+(\lambda)$ exists and is finite, then there is only one $\alpha \in (0, \pi)$ such that $|m_\alpha^+(\lambda)| = \infty$.

It turns out that the absolutely continuous part $d\rho_{ac}$ of spectral measure $d\rho$ is supported on a set Σ_{ac} of λ's for which no subordinate solution of $(J^\phi - \lambda \mathbb{I})u$ exists and the singular part $d\rho_s$ is supported on a set Σ_s of λ's for which u obey the phase boundary ϕ and the Radon–Nikodym derivative $d\rho/d\lambda = (1/\pi)\Im m_\phi^+(\lambda)$ diverges. More precisely, we have the following proposition.

Proposition B.5. *Let J^ϕ be an admissible operator in $l_2(\mathbb{Z}_+)$ satisfying ϕ-phase b.c. (B.1c) and whose Schödinger equation is (B.1b). Let $\mu = d\rho$ be its spectral measure (B.8). Then, the minimal supports Σ, Σ_{ac}, Σ_s, Σ_{sc} and Σ_{pp} of μ, μ_{ac}, μ_s, μ_{sc} and μ_{pp} are, respectively, given by*

(i) *$\Sigma = \mathbb{R}/\Sigma_0$ where $\Sigma_0 = \{\lambda \in \mathbb{R}$: there exists a subordinate solution but it does not satisfy (B.9)\},*

(ii) *$\Sigma_{ac} = \{\lambda \in \mathbb{R}$: no subordinate solution exists\},*

(iii) *$\Sigma_s = \{\lambda \in \mathbb{R}$: there exists a subordinate solution that satisfies ϕ-phase b.c.\},*

(iv) *$\Sigma_{sc} = \{\lambda \in \mathbb{R}$: there exists a subordinate solution satisfying ϕ-phase b.c. but does not belong to $l_2(\mathbb{Z}_+)$\},*

(v) *$\Sigma_{pp} = \{\lambda \in \mathbb{R}$: there exists an $l_2(\mathbb{Z}_+)$ subordinate solution that satisfies ϕ-phase b.c.\}.*

Transfer matrix. We restrict ourselves to the two sparse models we have dealt with. For our convenience, we consider off-diagonal (4.1a)–(4.1d) analogous version of the model $J^\omega = J_0 + V^\omega$ defined by (4.4a)–(4.4c) but the formulas derived for the latter can be translated to the former without difficulties.

Definition B.3. Given $\lambda \in \mathbb{R}$ and $n \in \mathbb{N}$, an (n-step) *transfer matrix* $T(n; \lambda)$ is a product

$$T(n; \lambda) := T(n, n-1; \lambda) T(n-1, n-2; \lambda) \cdots T(1, 0; \lambda) \qquad \text{(B.15)}$$

of the (one-step) 2×2 transfer matrix

$$T(k, k-1; \lambda) = \frac{1}{p_k} \begin{pmatrix} \lambda & -p_{k-1} \\ p_k & 0 \end{pmatrix} \qquad \text{(B.16)}$$

(replaced by $\begin{pmatrix} \lambda - v_k & -1 \\ 1 & 0 \end{pmatrix}$ for the diagonal model (4.4a)–(4.4c)).

Any solution $u = (u_n)_{n \geq 0}$ of the Schrödinger equation

$$((J^\phi - \lambda \mathbb{I})u)_n = p_n u_{n+1} + p_{n-1} u_{n-1} - \lambda u_n = 0 \qquad \text{(B.17)}$$

for $n \geq 1$, $p_0 \equiv 1$, with ϕ-phase boundary condition at 0 (B.1c) satisfies

$$\begin{pmatrix} u_{k+1} \\ u_k \end{pmatrix} = T(k, k-1; \lambda) \begin{pmatrix} u_k \\ u_{k-1} \end{pmatrix}. \qquad \text{(B.18)}$$

We denote by $u^\phi = (u_n^\phi)_{n \geq 0}$ the solution of (B.17) "normalized" by $u_0^2 + u_1^2 = 1$.

Proposition B.6. *Let $(p_n)_{n \geq 0}$ be a sparse sequence:*

$$p_n = \begin{cases} p & \text{if } n = a_j \in \mathcal{A}, \\ 1 & \text{otherwise}, \end{cases}$$

with $0 < p < 1$ and $\mathcal{A}$ defined by (4.1c). Then, there exists a real 2×2 matrix U so that the n-step transfer matrix conjugated by U reads

$$UT(n; \lambda)U^{-1} = R((n - a_j)\varphi)P_{+-}R(\beta_j\varphi) \cdots P_{+-}R(\beta_1\varphi), \qquad \text{(B.19)}$$

for $a_j < n < a_{j+1}$ and $\lambda = 2\cos\varphi$, $\varphi \in (0, \pi)$, where

$$R(\theta) = \begin{pmatrix} \cos\theta & \sin\theta \\ -\sin\theta & \cos\theta \end{pmatrix} \qquad \text{(B.20)}$$

and

$$P_{+-} = \begin{pmatrix} p & 0 \\ (1/p - p)\cot\varphi & 1/p \end{pmatrix}. \qquad \text{(B.21)}$$

Proof. Let $\lambda = 2\cos\varphi$ and let

$$U = \begin{pmatrix} 0 & \sin\varphi \\ 1 & -\cos\varphi \end{pmatrix} \tag{B.22}$$

be a 2×2 nonsingular matrix that conjugates the free transfer matrix $T_0(\lambda) = \begin{pmatrix} \lambda & -1 \\ 1 & 0 \end{pmatrix}$ into a clockwise rotation matrix.

Problem B.3. *Show that* $UT_0(\lambda)U^{-1} = R(\varphi)$.

Equation (B.19) follows by (B.15), (B.20) and $P_{+-} = R(-\varphi)UT(a_j + 1, a_j; \lambda)T(a_j, a_j - 1; \lambda)U^{-1}R(-\varphi)$ for each $a_j \in \mathcal{A}$. $\qquad\square$

Prüfer variables.

Definition B.4. Let u_n^ϕ and U be as in Definition B.3 and (B.22). The real-valued functions $R_j = R_j(\varphi, \phi)$ and $\theta_j = \theta_j(\varphi, \phi), j = 0, 1, 2, \ldots$, given by

$$R_j \begin{pmatrix} \cos\theta_j \\ \sin\theta_j \end{pmatrix} := U \begin{pmatrix} u_{a_j+1}^\phi \\ u_{a_j}^\phi \end{pmatrix}, \tag{B.23}$$

are called *Prüfer variables.*

Proposition B.7 ([180, pp. 770, 771 and 776]). *Let* $(p_n)_{n \geq 1}$ *be as in Proposition B.6. Then,*

$$\theta_j = g(\varphi, \theta_{j-1}) - (\beta_j + \omega_j - \omega_{j-1})\varphi, \tag{B.24a}$$

$$\frac{1}{R_j^2} = F(\varphi, \theta_j)\frac{1}{R_{j-1}^2} \tag{B.24b}$$

hold for any $j \in \mathbb{N}$ *with* $(R_0, \theta_0) \in \mathbb{R}_+ \times [0, \pi]$, β_j *given by (4.1d),* $\{\omega_j\}_{j \geq 1}$ *i.i.d. random variables,*

$$g(\varphi, \theta) = \tan^{-1}\left(\frac{1}{p^2}(\tan\theta + \cot\varphi) - \cot\varphi\right), \tag{B.24c}$$

$$F(\varphi, \theta) = \frac{p^2}{a + b\cos 2\theta + c\sin 2\theta}, \tag{B.24d}$$

and a, b *and* c *are functions of* p *and* φ:

$$2a = (1 - p^2)^2 \cot^2\varphi + 1 + p^4,$$
$$2b = (1 - p^2)^2 \cot^2\varphi - 1 + p^4,$$
$$c = (1 - p^2)\cot\varphi.$$

Proof. It follows directly from (B.19) and (B.23).   □

Problem B.4. *Prove the recursive relations* (B.24a) *and* (B.24b).

An explicit computation yields

$$a + b\cos 2\theta + c\sin 2\theta \geq \min_{\theta}(a + b\cos 2\theta + c\sin 2\theta) = a - \sqrt{a^2 - p^4} > 0,$$

(B.25)

and implies that $F(\varphi, \theta)$ is uniformly bounded in $[0, \pi] \times [0, \pi]$ and has unit mean:

$$\bar{F} := \frac{1}{\pi} \int_0^\pi F(\varphi, \theta)d\theta = \frac{p^2}{\sqrt{a^2 - b^2 - c^2}} = 1$$

(B.26)

by [200, Eqs. (26) and (29)] together with $b^2 + c^2 = a^2 - p^4$.

Write

$$\mathbf{v}_0 \equiv R_0 \begin{pmatrix} \cos\theta_0 \\ \sin\theta_0 \end{pmatrix} := U \begin{pmatrix} \cos\phi \\ \sin\phi \end{pmatrix},$$

(B.27)

i.e., $R_0 = \sqrt{1 - \sin 2\phi \cos \varphi}$ and $\tan\theta_0 = \cot\phi/\sin\varphi - \cos\varphi$ with $0 < \theta_0 < \pi$. For n and j such that $a_N < n < a_{N+1}$, the Euclidean norm

$$\|UT(n; \lambda)U^{-1}\mathbf{v}_0\|^2 = R_N^2,$$

(B.28)

can be written as

$$R_N^2 = \frac{R_N^2}{R_{N-1}^2} \cdots \frac{R_1^2}{R_0^2} R_0^2 = \exp\left(\sum_{k=1}^N f(\varphi, \theta_k)\right) R_0^2,$$

(B.29)

where

$$f(\varphi, \theta) = -\log F(\varphi, \theta),$$

(B.30)

by (B.24b), (B.24d) and (B.25), is uniformly bounded in $[a, b] \times [0, \pi]$, $0 < a < b < \pi$, and has mean

$$\bar{f} = \frac{1}{\pi} \int_0^\pi f(\varphi, \theta)d\theta = \log\left(\frac{a + \sqrt{a^2 - b^2 - c^2}}{2p^2}\right).$$

(B.31)

The quantity between parenthesis $r = r(p, \varphi) = \frac{a + \sqrt{a^2 - b^2 - c^2}}{2p^2}$ is explicitly given by

$$r = 1 + \frac{v^2}{4} \csc^2\varphi$$

with $v = (1 - p^2)/p$ or, using $\lambda = 2\cos\varphi$, by

$$r = 1 + \frac{v^2}{4 - \lambda^2},\tag{B.32}$$

and coincides with the analogous expression for the diagonal model (4.4a)–(4.4c). We shall employ it in the following for both diagonal and off-diagonal sparse models.

Spectral nature criteria. We gather a collection of results on spectral analysis useful for the sparse models in consideration. We begin with the following (see [163, Lemma 3.1]).

Proposition B.8. *Let u^ϕ and $T(n; \lambda)$ be defined as in Definition B.3. Then*

$$\frac{\|u^{\phi+\pi/2}\|^2_{L+1}}{\|u^\phi\|^2_{L+1}} \le 2\left(\frac{1}{L}\sum_{n=1}^{L}\|T(n;\lambda)\|^2\right)^2.\tag{B.33}$$

Proof. By Definition B.3, u^ϕ and $u^{\phi+\pi/2}$ satisfy the initial conditions (B.2) with $\alpha = \phi$ and it thus follows that

$$T^\phi(n; \lambda) := T(n; \lambda)R\left(-\phi\right) = \begin{pmatrix} u^\phi_{n+1} & u^{\phi+\pi/2}_{n+1} \\ u^\phi_n & u^{\phi+\pi/2}_n \end{pmatrix}\tag{B.34a}$$

and

$$1 = W_n[u^\phi, u^{\phi+\pi/2}] = -p_n \det T^\phi(n; \lambda),\tag{B.34b}$$

by (B.18), (B.20), (B.3b) and (B.3c). Let the first and second columns of (B.34a) be denoted by $\mathbf{u}^\phi_n$ and $\mathbf{v}^\phi_n$. In terms of these vectors, the Wronskian reads $W_n[u^\phi, u^{\phi+\pi/2}] = p_n\mathbf{u}^\phi_n \cdot \mathcal{J}\mathbf{v}^\phi_n$, with $\mathcal{J} = \begin{pmatrix} 0 & 1 \\ -1 & 0 \end{pmatrix}$ and by (B.34b), Cauchy–Schwarz inequality, $p_n \le 1$ and $\|\mathcal{J}\| = 1$

$$\|\mathbf{u}^\phi_n\|\|\mathbf{v}^\phi_n\| \ge p_n|\mathbf{u}^\phi_n \cdot \mathcal{J}\mathbf{v}^\phi_n| = 1.\tag{B.34c}$$

By (B.34a)

$$\frac{1}{L}\sum_{n=1}^{L}\|\mathbf{v}^\phi_n\|^2 \le \frac{1}{L}\sum_{n=1}^{L}\|T(n;\lambda)\|^2$$

holds (term-by-term), and

$$1 \leq \left(\frac{1}{L} \sum_{n=1}^{L} \|\mathbf{u}_n^\phi\| \|\mathbf{v}_n^\phi\| \right)^2 \leq \frac{1}{L} \sum_{n=1}^{L} \|\mathbf{u}_n^\phi\|^2 \frac{1}{L} \sum_{n=1}^{L} \|\mathbf{v}_n^\phi\|^2,$$

by (B.34c) and Cauchy–Schwarz inequality. From the last two equations, one concludes

$$\frac{\sum_{n=1}^{L} \|\mathbf{v}_n^\phi\|^2}{\sum_{n=1}^{L} \|\mathbf{u}_n^\phi\|^2} \leq \left(\frac{1}{L} \sum_{n=1}^{L} \|T(n;\lambda)\|^2 \right)^2, \tag{B.34d}$$

and this, together with

$$\sum_{n=1}^{L+1} |u_n^{\phi+\pi/2}|^2 \leq \sum_{n=1}^{L} \|\mathbf{v}_n^\phi\|^2$$

$$\sum_{n=1}^{L+1} |u_n^\phi|^2 \geq \frac{1}{2} \sum_{n=1}^{L} \|\mathbf{u}_n^\phi\|^2, \tag{B.34e}$$

yields (B.33). $\qquad\qquad\square$

Let $\mu = d\rho$ be the spectral measure of J^ϕ:

$$(e_1, f(J^\phi)e_1) = \int f(\lambda) d\rho(\lambda) \tag{B.35}$$

for every bounded measurable function with compact support in $(-2, 2)$. The vector e_1 is cyclic for J^ϕ in the sense that $\{(J^\phi)^k e_1 : k \in \mathbb{N}\}$ is a dense set in the Hilbert space $\mathcal{H}$ and any other spectral measure μ_Ψ is absolutely continuous with respect to μ. According to Proposition B.5,

$$\Sigma_{\mathrm{ac}} = \{\lambda \in [-2, 2]: \text{there is no subordinate solution}\} \tag{B.36}$$

is an essential support of μ_{ac} and has zero measure with respect to the singular part μ_{s}. Proposition B.8 gives one-half of the following.

Theorem B.1 ([163, Theorem 1.1]). *The essential support Σ_{ac} of the a.c. part μ_{ac} is given by*

$$\Sigma_{\mathrm{ac}} = \left\{ \lambda : \liminf_{L \to \infty} \frac{1}{L} \sum_{n=0}^{L-1} \|T(n; \lambda)\|^2 < \infty \right\}. \tag{B.37}$$

Proof. Let us denote by S the r.h.s. of (B.37). If J^ϕ has a subordinate solution at λ then the r.h.s. of (B.33) diverge. So, S is contained into Σ_{ac} and we quote [163, Proposition 3.3] for a proof that it is an essential support of $d\rho_{\text{ac}}$, i.e., for almost every $\lambda \in [-2, 2]$ with respect to μ_{ac}, we have $\lambda \in S$. $\qquad\square$

Now, we come back to Eqs. (B.28)–(B.32). Suppose that $f = f(\varphi, \theta)$ in (B.29) is replaced by its average $\bar{f} = f(\varphi)$ given by (B.31) and

$$\mathcal{E}_N(\varphi) = \frac{1}{N} \sum_{k=1}^{N} (f(\varphi, \theta_k) - \bar{f}(\varphi)) \longrightarrow 0, \quad \text{as } N \to \infty \text{ for a.e. } \varphi \in [0, \pi).$$
(B.38)

Since by definition (B.32), $r > 1$ if $v > 0$ and by (B.28)

$$\|T(n; \lambda)\|^2 \sim r^j$$

for $a_j < n < a_{j+1}$ a.e. $\lambda \in (-2, 2)$ (see Eqs. (B.42a) and (B.42c) below and (4.20b), for a derivation), we conclude by Theorem B.1 the following.

Theorem B.2. *Let $\mu = d\rho$ be the spectral measure of J^ϕ and suppose that the hypothesis (B.38) holds. Then the essential support of its a.c. part μ_{ac}, $\Sigma_{\text{ac}} = \emptyset$.*

The conclusions is not affected by the exclusion of a set $A \subset [-2, 2]$ of Lebesgue measure zero, by Definition B.1 of essential support Σ_{ac}. The presence of randomness ω on the sparse models in consideration allows the hypothesis to be rigorously proved (Theorem 4.3).

Metric properties of spectral measure. We state without proof an extension to Gilbert–Pearson theory, due Jitomirskaya–Last [133], which relates the Hausdorff decomposition of the spectral measure to a generalized subordinacy solution:

Theorem B.3 ([133, Theorem 1.2]). *Let $\mu = d\rho$ be given by (B.35), $\lambda \in [-2, 2]$ and $\alpha \in (0, 1)$. Then, the upper Hausdorff derivative*

$$D_\mu^\alpha(\lambda) := \limsup_{\varepsilon \to 0} \frac{\rho((\lambda - \varepsilon, \lambda + \varepsilon))}{(2\varepsilon)^\alpha} = \infty$$

iff

$$\liminf_{L \to \infty} \frac{\|u^\phi\|_L}{\|u^{\phi + \pi/2}\|_L^{\alpha/(2-\alpha)}} = 0.$$

One consequence of Theorem B.3 (see [133, Corollary 4.4]) may be restated in terms of the transfer matrix $T(n; \lambda)$.

Corollary B.1 ([63, Corollary 3.7]). *Suppose that for some $\alpha \in [0, 1)$ and every λ in a Borel set A,*

$$\limsup_{l \to \infty} \frac{1}{l^{2-\alpha}} \sum_{n=0}^{l} \|T(n; \lambda)\|^2 < \infty. \tag{B.39}$$

Then, the restriction $\rho(A \cap \cdot)$ of ρ to the set A is α-continuous.

The other consequence is (see [133, Corollary 4.5]) the following.

Corollary B.2 ([63, Corollary 3.8]). *Suppose that*

$$\liminf_{l \to \infty} \frac{\|u^\phi\|_l^2}{l^\alpha} = 0$$

holds for every λ in some Borel set A. Then, the restriction $\rho(A \cap \cdot)$ is α-singular.

The proof of both corollaries relates the behavior of the eigenvectors u_n^ϕ and $u_n^{\phi+\pi/2}$ with the norm of $T(n; \lambda)$ and the Prüfer radius R_k. Note that

$$R_j^2(\phi) = \left\| UT(n; \lambda) \begin{pmatrix} \cos \phi \\ \sin \phi \end{pmatrix} \right\|^2 = (u_{n+1}^\phi)^2 + (u_n^\phi)^2 - 2u_{n+1}^\phi u_n^\phi \cos \varphi \tag{B.40a}$$

holds with $a_j < n < a_{j+1}$, $\phi \in (0, \pi)$ and $\lambda = 2 \cos \varphi \in (-2, 2)$, by (B.23), (B.34a), (B.22) and (B.20). From (B.40a) we deduce

$$(1 - |\cos \varphi|)((u_{n+1}^\phi)^2 + (u_n^\phi)^2) \leq R_j^2(\phi) \leq (1 + |\cos \varphi|)((u_{n+1}^\phi)^2 + (u_n^\phi)^2). \tag{B.40b}$$

Problem B.5. *Let $\|A\|_U := \|UAU^{-1}\|$ be a matrix norm defined by (B.22) and $\|B\| = \sup_{\mathbf{v} \in \mathbb{C}^2} \|B\mathbf{v}\|/\|\mathbf{v}\|$. Show that*

$$C^{-1}\|A\| \leq \|A\|_U \leq C\|A\| \tag{B.41}$$

holds with $C = \sqrt{(1 + |\cos \varphi|)/(1 - |\cos \varphi|)}$.

Problem B.5 together with (B.28), imply

$$\|T(n;\lambda)\| \geq C^{-1}\|\mathbf{v}_0\|^{-1}\|UT(n;\lambda)U^{-1}\mathbf{v}_0\| = C'R_j(\phi) \qquad \text{(B.42a)}$$

and

$$\|T(n;\lambda)\| \leq C \sup_{\mathbf{v}:\|\mathbf{v}\|=1} \|UT(n;\lambda)U^{-1}\mathbf{v}\| = C''R_j(\bar{\phi}), \qquad \text{(B.42b)}$$

where $\bar{\mathbf{v}} = (\cos\bar{\theta}, \sin\bar{\theta})$ is the (unique) unit vector for which the supremum is attained and $\bar{\phi}$ solves $\theta_0(\phi) = \bar{\theta}$ for ϕ. To get rid of the sup, an artifact introduced in [152, Theorem 2.3] can be used (see (B.51d)) to replace equation (B.42b) by

$$\|T(n;\lambda)\| \leq \widetilde{C}\max(R_j(\phi^1), R_j(\phi^2)) \qquad \text{(B.42c)}$$

with $\widetilde{C} = C/|\sin(\theta_0^1 - \theta_0^2)/2|$ and $\theta_0^i = \theta_0(\phi^i)$, $i \in \{1,2\}$. See quoted reference for completion of the proof.

Proposition B.9. *If $(\beta_j)_{j\geq 1}$ satisfies a superexponential sparseness condition $\lim_{j\to\infty}\beta_{j-1}/\beta_j = 0$, then the spectrum $\sigma(J^\phi)$ of J^ϕ has Hausdorff dimension 1.*

Proof. It follows from (B.42c) and (4.20b) that, if $(\beta_j)_{j\geq 1}$ satisfies a super-exponential sparseness, then for any Borel set $A \subset [-2,2]$ and $\varepsilon > 0$ there exist $n_0 = n_0(\varepsilon, A)$ such that

$$\|T(n;\lambda)\|^2 \leq \tilde{C}_j r^j \leq n^\varepsilon$$

holds for $n \geq n_0$, uniformly in A. So,

$$\frac{1}{l^{2-\alpha}} \sum_{n=1}^{l} \|T(n;\lambda)\|^2 \leq Cl^{\alpha-1+\varepsilon}$$

holds for l sufficiently large and the lim sup is finite for any $\alpha > 1$. This, together with Corollary B.1 and definition of Hausdorff dimension (see Definition 3.1), concludes the proof. $\qquad\square$

We turn now to the point spectrum, which matters only for sequence $(\beta_j)_{j\geq 1}$ exponentially sparse

$$\beta_j = \beta^j \qquad \text{(B.43)}$$

for some integer $\beta > 1$, asymptotically as $j \to \infty$. By [163, Theorems 1.6 and 1.7] $(J^\phi - \lambda \mathbb{I})u = 0$ has no $l_2(\mathbb{N})$-solutions if

$$\sum_{n=1}^{\infty} \|T(n;\lambda)\|^{-2} = \infty \tag{B.44}$$

(see also [244, Theorem 2.1]) and an $l_2(\mathbb{N})$-solution if

$$\sum_{n=1}^{\infty} \|T(n;\lambda)\|^2 \left(\sum_{k=n}^{\infty} \|T(k;\lambda)\|^{-2} \right)^2 < \infty. \tag{B.45}$$

Condition (B.45) can be optimized for sparse models by improving Theorems 8.1 and 8.2 of [163] on the existence of subordinate and $l_2(\mathbb{N})$-solutions. By Eq. (B.19), for n and N s.t. $a_N < n < a_{N+1}$, we have

$$\|T(n,\lambda)\|_U \le \prod_{k=1}^{N} \|P_{+-}R(\beta_k \varphi)\| \le \|P_{+-}\|^N.$$

This together with (B.41) imply that $\|T(n;\lambda)\|$ cannot growth exponentially fast in n. As a consequence, the largest Lyapunov exponent, $\gamma = \lim_{n\to\infty} (1/n) \log \|T(n;\lambda)\|$, vanishes for $T(n;\lambda)$ as a product of one-step transfer matrices but it is strictly positive as a product of the $P_{+-}R(\beta_k \varphi)$: $\lim_{N\to\infty}(1/N) \log \|P_{+-}R(\beta_N \varphi) \cdots P_{+-}R(\beta_1 \varphi)\| > 0$ a.e. φ.

We follow closely Lemma 4.1 of [65] and Proposition 3.9 of [63], where decay at infinity of a subordinate solution were established (see also [279, Lemma 2.1]).

Proposition B.10. *Let J^ϕ be given by (B.17) with sparse sequence $(\beta_n)_{n\ge 1}$ and let t_n denote $\|T(a_n + 1; \lambda)\|$. Suppose that*

$$\sum_{n=1}^{\infty} t_n^{-2} < \infty \tag{B.46}$$

holds for some $\lambda \in [-2, 2]$. Then, there exists a unit vector $\mathbf{v}^* = (\cos\phi^*, \sin\phi^*)$ with $\phi^* \in [0, \pi)$ so that

$$\|T(n;\lambda)\mathbf{v}^*\|^2 \le B \left(\sum_{M=N}^{\infty} t_M^{-2} \right)^2 t_N^2 + t_N^{-2} \tag{B.47}$$

is satisfied for some constant $B < \infty$ and every $a_N < n < a_{N+1}$ with $N \in \mathbb{N}$.

Corollary B.3. *Under the assumptions of Proposition* B.10, *the vector*
$\mathbf{u}_n^{\phi^*} := T(n;\lambda)\mathbf{v}^* = \begin{pmatrix} u_{n+1}^{\phi^*} \\ u_n^{\phi^*} \end{pmatrix}$ *defines a strong subordinate solution* $u^{\phi^*} =$
$(u_n^{\phi^*})_{n\geq 0}$ *in the sense that*

$$\lim_{n\to\infty} \frac{\|\mathbf{u}_n^{\phi^*}\|}{\|\mathbf{u}_n^{\phi}\|} = 0$$

holds for any $\phi \in [0,\pi)$ *with* $\phi \neq \phi^*$.

Corollary B.4. *The strong subordinate solution* $u^{\phi^*} = (u_n^{\phi^*})_{n\geq 0}$ *is an*
$l^2(\mathbb{Z}_+)$-*solution of* $(J^\phi - \lambda I)u = 0$ *with* $\phi = \phi^*$, *provided*

$$\sum_{n=1}^{\infty} \beta_n t_n^{-2} < \infty \quad and \quad \sum_{n=1}^{\infty} \left(\sum_{m=n}^{\infty} t_m^{-2} \right)^2 \beta_n t_n^2 < \infty \qquad \text{(B.48)}$$

are verified in addition to the assumptions of Proposition B.10.

Corollary B.5. *Suppose, in addition to the assumptions of Proposition* B.10, *that*

$$C_n^{-1} r^n \leq t_n^2 \leq C_n r^n \qquad \text{(B.49)}$$

holds with $r > 1$ *and* $C_n^{1/n} \searrow 1$ *as* n *tends to* ∞. *Then*

$$\|\mathbf{u}_n^{\phi^*}\|^2 \leq \tilde{C}_n r^{-n} \qquad \text{(B.50)}$$

holds with $\tilde{C}_n^{1/n} \searrow 1$ *as* n *tends to* ∞.

Remark B.3. As the Prüfer angles $(\theta_n^\omega)_{n\geq 1}$ are uniformly distributed mod π (Theorem 4.3), Eq. (B.49) actually holds with r given by (B.32) a.e. λ. Theorem 4.5 excludes a set of λ's which is countable and independent of the initial Prüfer angle θ_0. These properties are crucial for proving pure point spectrum. Note that (B.42a) and (B.42c) hold for some θ_0^i, $i = 1, 2$, where θ_0, by (B.27), depends on λ and on the phase boundary ϕ.

Proof of Proposition B.10. Let $\lambda = 2\cos\varphi$ and write

$$T(a_n + 1; \lambda) = A_n(\lambda) \cdots A_1(\lambda)$$

where

$$U A_k U^{-1} = P_{+-} R(\beta_k \varphi).$$

Let $s_n = \|A_n(\lambda)\|$ denote the spectral norm of $A_n(\lambda)$. It follows from (B.5)

$$s_n \leq C\|A_n(\lambda)\|_U \leq C\|P_{+-}\| \leq D \tag{B.51a}$$

with $D = C\sqrt{1 + (p - 1/p)^2 \csc^2 \varphi}$, by (B.21), uniformly in n. As a consequence,

$$\sum_{n=1}^{\infty} \frac{s_{n+1}^2}{t_n^2} \leq D^2 \sum_{n=1}^{\infty} \frac{1}{t_n^2} < \infty, \tag{B.51b}$$

by hypothesis (B.46), verifies the assumption of [163, Theorem 8.1].

The transfer matrices A_k's are 2×2 unimodular real matrices. Since unimodular matrices form an algebra, the product of $T(a_n + 1; \lambda)$ with its adjoint $T^*(a_n + 1; \lambda)$ is a 2×2 unimodular symmetric real matrix with eigenvalues t_n^2 and t_n^{-2} and corresponding (orthonormal) eigenvectors $\mathbf{v}_n^+$ and $\mathbf{v}_n^-$: $\mathbf{v}_n^+ \cdot \mathbf{v}_n^- = 0$. Write $\mathbf{v}_\phi = \begin{pmatrix} \cos\phi \\ \sin\phi \end{pmatrix}$ and define ϕ_n by

$$\mathbf{v}_{\phi_n} = \mathbf{v}_n^-. \tag{B.51c}$$

Clearly, $\mathbf{v}_n^+ = \mathbf{v}_{\phi_n + \pi/2}$ and

$$\begin{aligned} \|T(a_n + 1; \lambda)\mathbf{v}_\phi\|^2 &= \mathbf{v}_\phi \cdot T^*(a_n + 1; \lambda)T(a_n + 1; \lambda)\mathbf{v}_\phi \\ &= t_n^2 |\mathbf{v}_\phi \cdot \mathbf{v}_n^+|^2 + t_n^{-2}|\mathbf{v}_\phi \cdot \mathbf{v}_n^-|^2 \\ &= t_n^2 \sin^2(\phi - \phi_n) + t_n^{-2}\cos^2(\phi - \phi_n). \end{aligned} \tag{B.51d}$$

by the spectral theorem.

The completion now follows exactly the steps of the proof of [163, Theorem 8.1]. The conclusion of Proposition B.10, Eq. (B.47), is the combination of [163, Eqs. (8.5) and (8.7)]. We review the main steps. Firstly, the sequence $(\phi_n)_{n \geq 1}$ converges to $\phi^* \in [0, \pi]$ under the condition (B.51b). The key estimate

$$|\phi_n - \phi_{n+1}| \leq \frac{\pi}{2} \frac{s_{n+1}^2}{t_n^2}, \tag{B.51e}$$

established using properties of a matrix norm, (B.51d) for $n+1$ and (B.51c), together with (B.51b) implies that $(\phi_n)_{n \geq 1}$ is a Cauchy sequence. Next, the telescope estimate

$$|\phi_n - \phi^*| \leq \sum_{m=n}^{\infty} |\phi_m - \phi_{m+1}| \leq \frac{\pi}{2} \sum_{m=n}^{\infty} \frac{s_{m+1}^2}{t_m^2}$$

replaced into (B.51d) yields

$$\|T(a_n + 1; \lambda)\mathbf{v}_{\phi^*}\|^2 \leq t_n^2(\phi^* - \phi_n)^2 + t_n^{-2}$$

$$\leq Bt_n^2 \left(\sum_{m=n}^{\infty} t_m^{-2}\right)^2 + t_n^{-2} \qquad (B.51f)$$

with $B = \frac{\pi}{2}D^4$, which concludes the proof. $\qquad \square$

Proof of Corollary B.3. Let $\mathbf{v}^* = \mathbf{v}_{\phi^*}$ and $\mathbf{v}_* = \mathbf{v}_{\phi^*+\pi/2}$. By (B.51d), $\|T(a_n + 1; \lambda)\mathbf{v}_*\|^2 \geq t_n^2/2$ is satisfied for sufficiently large n and for any $\phi \neq \phi^*$ we have $\mathbf{v}_\phi = a\mathbf{v}_* + b\mathbf{v}^*$, with $a^2 + b^2 = 1$, $a \neq 0$, and $\|T(a_n + 1; \lambda)\mathbf{v}_\phi\|^2 \geq a^2 t_n^2/2$ holds for some n sufficiently large. By (B.51d),

$$\frac{\|\mathbf{u}_n^{\phi^*}\|^2}{\|\mathbf{u}_n^{\phi}\|^2} \leq \frac{2}{a^2}\left(B\left(\sum_{m=n}^{\infty} t_m^{-2}\right)^2 + t_n^{-4}\right) \longrightarrow 0$$

by (B.46), concluding the proof. $\qquad \square$

Proof of Corollary B.4. We prove that the strongly subordinate solution is an $l_2(\mathbb{Z}_+)$-solution under assumptions (B.46) and (B.48). For any $k \in \mathbb{N}$ let n be such that $a_n + 1 \leq k < a_{n+1}$ holds. By (B.41),

$$\|T(k; \lambda)\mathbf{v}_{\phi^*}\|_U^2 \leq C\|\mathbf{v}_{\phi^*}\|^2$$

which, by (B.43), (B.40a) and (B.40b), together with (B.47), yields

$$\sum_{k=1}^{\infty} \|u_k^{\phi^*}\|^2 \leq \frac{1}{2(1 - |\cos\varphi|)} \sum_{k=1}^{\infty} \|UT(k; \lambda)\mathbf{v}_{\phi^*}\|^2$$

$$\leq B' \sum_{n=1}^{\infty} \beta^n t_n^2 \left(\sum_{m=n}^{\infty} t_m^{-2}\right)^2 + B'' \sum_{n=1}^{\infty} \beta^n t_n^{-2}$$

for some constants B' and B'', concluding the proof. $\qquad \square$

Proof of Corollary B.5. Equation (B.50) is a direct consequence of (B.47) and (B.49). $\qquad \square$

Appendix C

Symmetric Cantor Sets and Related Subjects

Let AB be a closed segment of length l and let

$$0 < \xi < 1/2 \tag{C.1}$$

be a number (the "ratio"). Following [144], consider a trisection of the segment AB in parts, respectively, equal to $l\xi$, $l(1 - 2\xi)$, and $l\xi$, and remove the *open* central interval (*"black"* interval) of length $l(1-2\xi)$; there remain two closed intervals (*"white"* intervals) of common length $l\xi$. Such a dissection of the given interval AB will be said to be a *dissection of type* $(2, \xi)$; the number 2 recalling that after dissection there remain two white intervals.

We now start from a fundamental interval $[a, b]$ (often the interval $[0, 2\pi]$). Let us operate a dissection of type $(2, \xi_1)$ on this interval; then one of type $(2, \xi_2)$ over each of the two remaining white intervals; further, a dissection of type $(2, \xi_3)$ over each of the 2^2 white intervals obtained, and so on, the infinite sequence $\{\xi_k\}_1^\infty$ being such that

$$0 < \xi_k < 1/2, \quad \text{for all } k = 1, 2, \ldots. \tag{C.2}$$

At the kth step, we shall have a set E_k consisting of 2^k white intervals of common length

$$(b - a)\xi_1 \cdots \xi_k. \tag{C.3}$$

The intersection

$$E \equiv \bigcap_{k=1}^{\infty} E_k \tag{C.4}$$

is a *perfect* set: E is closed and every point of E is a limit point of E_k. Further, it has no interior points, i.e., its closure is nowhere dense; it is a *Cantor set*. Its contiguous points are the edges of the black intervals obtained in the course of all the dissections. We have ($\mathcal{L}$ will denote Lebesgue measure in this appendix), by (C.3):

$$\mathcal{L}(E) = 0 \Leftrightarrow \lim_{k \to \infty} 2^k \xi_1 \cdots \xi_k = 0. \tag{C.5}$$

Special cases of the symmetric Cantor sets described above are prototypes of *self-similar sets*. In the case $\xi_k = \xi < 1/2$ above one has a *homogeneous perfect set* E *of type* $(2, \xi)$, but this notion may be generalized, see [144, p. 16]. This set E may be written as $E = \bigcup_{i=1}^{2} f_i(E)$, where $f_1(x) = x/3$ and $f_2(x) = x/3 + 2/3$. The generalization of this structure to $T = \bigcup_{i=1}^{N} f_i(T)$, where T is a nonempty compact set and $(f_1, \ldots, f_N)$ are similitudes, i.e., functions of type $f_i(x) = \lambda_i x + b_i$, with $0 < \lambda_i < 1$ is called an *iterated function system* [83] (IFS) for T, and T the attractor or the invariant set for the IFS. If there exists an open set V such that $f_i(V) \subset V$ and $f_i(V) \cap f_j(V) = \emptyset$ for $i \neq j$, we say that the IFS satisfies the open set condition. This is the case for the Cantor set of type $(2, \xi)$, with $V = (0, \xi)$. Even further generalizations to $\mathbb{R}^n$ exist: see [83]. Measures supported by these sets are called self-similar measures, and their construction has been described in detail by Guarneri and Schulz-Baldes in a beautiful review [103]: the latter authors also describe the construction of Schrödinger operators (one-dimensional Jacobi matrices) having the previously mentioned self-similar measures as spectral measures μ_Ψ with $\Psi = \delta_0$ as cyclic vector (see Theorem 1.4). This construction was used in models of quasicrystals in three dimensions, see Sec. 3.4.

Claim C.1. *The 2^k origins of the white intervals of which E_k consists are given by the basic formula*

$$a + l[\epsilon_1(1 - \xi_1) + \epsilon_2 \xi_1(1 - \xi_2) + \cdots + \epsilon_k \xi_1 \cdots \xi_{k-1}(1 - \xi_k)] \tag{C.6}$$

with $\epsilon_i \in \{0, 1\} \; \forall i = 1, \ldots, k$ and $l = b - a$.

Proof. By induction: for $k = 1$, $a + l[\epsilon_1(1 - \xi_1)]$, for $\epsilon_1 \in \{0, 1\}$ is s.t.

$$a + 0 = a \quad \text{and} \quad a + l(1 - \xi_1) = a + l\xi_1 + l(1 - 2\xi_1).$$

Suppose, now, $k \geq k_0 \geq 1$, and that the 2^{k_0} origins are given by

$$a + l[\epsilon_1(1 - \xi_1) + \cdots + \epsilon_{k_0} \xi_1 \cdots \xi_{k_0 - 1}(1 - \xi_{k_0})]. \tag{C.7a}$$

For $k = k_0 + 1$, one has, to each (C.7a), to add $l\xi_1 \cdots \xi_{k_0}(1 - \xi_{k_0+1})$, but this corresponds to build

$$a + l[\epsilon_1(1 - \xi_1) + \cdots + \epsilon_{k_0}\xi_1 \cdots \xi_{k_0-1}(1 - \xi_{k_0})$$
$$+ \epsilon_{k_0+1}\xi_1 \cdots \xi_{k_0}(1 - \xi_{k_0+1})] \tag{C.7b}$$

with $\epsilon_{k_0+1} \in \{0, 1\}$. This ends the induction.   □

Since $\lim_{k \to \infty} \xi_1 \cdots \xi_k = 0$ (because of (C.2)), the points of E given by (C.4) are given by the infinite series

$$x = a + l[\epsilon_1(1 - \xi_1 + \epsilon_2\xi_1(1 - \xi_2) + \cdots + \epsilon_k\xi_1 \cdots \xi_{k-1}(1 - \xi_k) + \cdots] \tag{C.8}$$

with $\epsilon_k \in \{0, 1\}$ for all integers $k \geq 1$. Define

$$r_k = \xi_1 \cdots \xi_{k-1}(1 - \xi_k) \tag{C.9}$$

Then, by (C.2),

$$r_k > r_{k+1} + r_{k+2} + \cdots . \tag{C.10}$$

We associate to each perfect symmetric set E the *Lebesgue function* which is constructed in the following way. Let $L_k(x)$ be the continuous function which equals zero for $x \leq a$, one for $x \geq b$, increasing linearly by $\frac{1}{2^k}$ on each white interval building E_k, and constant in each black interval contiguous to E_k. If $k \to \infty$, $L_k(x)$ tends uniformly (see below) to a function $L(x)$, continuous, constant in each interval contiguous to E, monotone increasing, with $L(a) = 0$, $L(b) = 1$: the *Lebesgue function constructed on E*. The same simple induction leading from (C.7a) to (C.7b) shows that for all $x \in E$, where E is given by (C.4),

$$L(x) = \frac{\epsilon_1}{2} + \frac{\epsilon_2}{2^2} + \cdots + \frac{\epsilon_k}{2^k} + \cdots . \tag{C.11}$$

In order to see that the convergence is uniform, let $L_p(x) = \sum_{i=1}^{p} \epsilon_i/2^i$; then $|L_p(x) - L_{p+1}(x)| \leq 2^{-p}$; thus the $\lim_{p \to \infty} L_p(x)$ is uniform.

The variation of $L(x)$ along an interval will be called the *L-measure* of this interval, denoted by dL; the Stieltjes integral of a continuous function $f(x)$ w.r.t. dL will be denoted by $\int_a^b f(x)dL(x)$; in particular, the

Fourier–Stieltjes transform of dL,

$$\Gamma(u) \equiv \int_{-\infty}^{\infty} \exp(iux)dL(x) = \int_{a}^{b} \exp(iux)dL(x). \qquad \text{(C.12)}$$

Observing that $L(x)$ grows by $1/2^k$ on every white interval building up E_k, we have

$$\Gamma(u) = \lim_{k\to\infty} \Gamma_k(u), \qquad \text{(C.13a)}$$

where

$$\Gamma_k(u) \equiv \frac{1}{2^k} \sum_{\epsilon_j \in \{0,1\} \forall 1 \leq j \leq k} \exp\left(iu[a + l(\epsilon_1 r_1 + \cdots + \epsilon_k r_k)]\right) \qquad \text{(C.13b)}$$

with r_k defined by (C.9). Thus

$$\Gamma_k(u) = \exp(iua) \prod_{j=1}^{k} \frac{1 + \exp(iulr_j)}{2}$$

$$= \exp\left[iu\left(a + l/2 \sum_{j=1}^{k} r_j\right)\right] \prod_{j=1}^{k} \cos(ulr_j/2) \qquad \text{(C.14a)}$$

from which, using $\sum_{j=1}^{\infty} r_j = 1$ and $b - a = l$,

$$\Gamma(u) = \exp[iu(a+b)/2] \prod_{j=1}^{\infty} \cos(ulr_j/2). \qquad \text{(C.14b)}$$

Problem C.1. *Prove the convergence of the infinite product on the r.h.s. of* (C.14b).

In particular, if $[a, b]$ is centered at the origin, with $a = -d$, $b = d$,

$$\Gamma(u) = \prod_{j=1}^{\infty} \cos(udr_j). \qquad \text{(C.15)}$$

Definition C.1. *If all ξ_j have a common value ξ, then $E = E_\xi$ is said to be a symmetric Cantor set of constant ratio ξ.*

Problem C.2. *Prove that for E_ξ a symmetric Cantor set of constant ratio ξ constructed over $[-d, d]$,*

$$\Gamma(u) = \prod_{j=1}^{\infty} \cos[ud(1 - \xi)\xi^{j-1}]. \qquad \text{(C.16)}$$

The Hausdorff dimension $\dim_H$ of the symmetric Cantor set of ratio ξ is

$$\dim_H(E) = |\log 1/2|/|\log \xi|.$$

(see [144]). There are several additional concepts of dimension which are relevant to describe sets of this type, such as the information dimension or the fractal dimension (see [78] for details). Let A be a set in $\mathbb{R}^n$, and $n(\epsilon)$ the minimal number of balls of radius ϵ which are necessary to cover A. The fractal dimension $D_F(A)$ is defined by

$$D_F(A) = \lim_{\epsilon \to 0} \frac{n(\epsilon)}{|\log \epsilon|}.$$

For the set E_ξ, considering the $E_k = E_{k,\xi}$ in (C.4), take $\epsilon = \xi^{-k}$; then $n(\epsilon) = 2^k$, and $D_F(E) = \lim_{k \to \infty} \frac{\log 2^k}{\lceil \log \xi^{-k} \rceil} = \frac{\log 2}{\lceil \log \xi \rceil}$. For the set E_ξ the Hausdorff and fractal dimensions are equal. For any countable set, it is easy to see that both dimensions yield zero, and for an interval in $\mathbb{R}$, they yield one. Hence, a nonzero Hausdorff or fractal dimension is a signal of uncountability (but there are uncountable sets of zero Hausdorff measure, see Chap. 4 for explicit examples). It is easy to see that E_ξ is uncountable; take, e.g., $\xi = 1/3$. The elements of $E_{1/3}$ are all the numbers which in base three representation $\sum_{i=1}^\infty \epsilon_i 3^{-i}$ have $\epsilon_i \in \{0, 2\}$ $\forall i$. We may, therefore, define a one-to-one map of $E_{1/3}$ to the set of numbers $\sum_{i=1}^\infty \epsilon_i 2^{-i}$ with $\epsilon_i \in \{0, 1\} \forall i$, i.e., to the whole interval $[0, 1]$, hence the set $E_{1/3}$ is uncountable.

We now show that the Lebesgue function L, given by (C.11), is UαH (see Definition 3.4):

Proposition C.1. *Let $\xi_k = \xi$ for $k = 1, 2, \ldots$ with ξ satisfying (C.1). Then L, given by (C.11), is UαH, i.e.,*

$$L(x) - L(x') \le C(x - x')^{\frac{|\log 1/2|}{\lceil \log \xi \rceil}} \tag{C.17a}$$

where, for definiteness,

$$0 < x - x' < 1. \tag{C.17b}$$

Proof. Suppose first that both x and x' satisfy (C.17b) and are of the form (C.8), then

$$x - x' = l[(\epsilon'_{p+1} - \epsilon_{p+1})\xi^p(1 - \xi) + (\epsilon'_{p+2} - \epsilon_{p+2})\xi^{p+1}(1 - \xi) + \cdots]. \tag{C.18}$$

If the first p ϵ_i are equal for x and x', for definiteness, then $\epsilon'_{p+1} - \epsilon_{p+1} = 1$, and $|\epsilon'_{p+k} - \epsilon_{p+k}| \le 1$. Thus, from (C.18),

$$
\begin{aligned}
x' - x &\ge l\xi^p(1-\xi) - l[\xi^{p+1}(1-\xi) + \xi^{p+2}(1-\xi) + \cdots] \\
&= l\xi^p(1-\xi) - l\xi^{p+1} = l\xi^p(1-2\xi) = A\xi^p,
\end{aligned}
\tag{C.19a}
$$

where, by (C.1),

$$
0 < A < \infty, \quad \text{with } A \text{ independent of } x \text{ and } x'. \tag{C.19b}
$$

By (C.11),

$$
L(x') - L(x) \le 2^{-p} + 2^{-(p+1)} + \cdots = \frac{2^{-p}}{1 - 1/2} = 22^{-p}. \tag{C.19c}
$$

By (C.19a), $(x-x')^{-1} \le A^{-1}\xi^{-p}$, whence $\log(x-x')^{-1} \le \log A^{-1} + p|\log \xi|$, and therefore

$$
p \ge \frac{-\log(x - x')}{|\log \xi|} + \frac{\log A}{|\log \xi|}. \tag{C.19d}
$$

By (C.19c) and (C.19d),

$$
L(x') - L(x) \le C2^{\frac{-\log(x'-x)}{|\log \xi|}} = C(x - x)^\alpha, \quad \alpha \equiv \frac{|\log 1/2|}{|\log \xi|} \tag{C.19e}
$$

which is (C.17a). If x or x' or both are not of the form (C.8), let x_0 be the right end point of the black interval containing x, and x'_0 the left-hand end point of the black interval containing x'. As $x - x' > x_0 - x'_0$, and $L(x') - L(x) = L(x_0) - L(x'_0)$ by the construction of L, (C.17a) subsists. $\qquad\square$

We have now the following proposition.

Proposition C.2. *Let $J_\alpha(L)$ be defined by (3.53b) (for $\mu = L$). Then, for any $0 < \epsilon < \alpha$,*

$$
0 < J_{\alpha-\epsilon}(L) < \infty. \tag{C.20}
$$

Proof. By Proposition (C.1), L is UαH, and then, by Lemma 3.1, (C.20) holds. $\qquad\square$

Proposition C.3. *Let $d = \pi$, and*

$$
\xi = 1/p \tag{C.21a}
$$

with p any odd integer such that

$$p \geq 3. \tag{C.21b}$$

Then (C.20) holds, with α given by (C.19e), but

$$\Gamma(u) \nrightarrow 0, \quad as \ u \to \infty. \tag{C.21c}$$

Proof. By (C.16) with $d = \pi$,

$$\Gamma(u) = \prod_{j=1}^{\infty} \cos[\pi u(1 - 1/p)p^{-(j-1)}] = \prod_{j=1}^{\infty} \cos\left[\frac{\pi u(p-1)}{p^j}\right]. \tag{C.21d}$$

Since p is an odd integer ≥ 3, it follows from (C.21d) that

$$\Gamma(n) = \Gamma(pn) \quad \forall n \in \mathbb{Z} \tag{C.21e}$$

and thus (C.21c) is proved. □

Proposition C.3 shows that there exist singular measures which are not Rajchman measures (see Sec. 3.3) and Propositions C.2 and C.3 together show that, also in one dimension, finiteness of the integral $J_{\alpha-\epsilon}(\mu)$ does not imply pointwise convergence (see also Sec. 3.3). The further problem of whether it could happen that $\Gamma(u) \to$ as $u \to \infty$ except for u in a set of zero Lebesgue measure, apparently left open by the proof of Proposition C.3, is negatively answered in Sec. 3.3 (see Proposition 3.1).

The amusing remark (C.21e) goes back at least to [144]. Lyons [175] also mentions it and remarks that, to our benefit, (C.21e) was not observed by F. Riesz, who, instead, constructed his famous Riesz products [213] for precisely the purpose of providing an example of a continuous measure which is not a Rajchman measure.

The proof of Proposition C.1 is patterned after Salem's paper [225].

We close this appendix with some important remarks. If, instead of (C.2),

$$1/2 < \xi_k < 1, \tag{C.22}$$

the support of dL is the interval $[-d, d]$. By a theorem of Jessen and Wintner [131], the corresponding measure may be either singular or a.c.; in the latter case, we have an a.c. measure with nowhere dense support (see Theorem 3.2). Such measures have the following property [16].

Proposition C.4. *Suppose μ is an absolutely continuous measure supported by a Cantor set C. Then $\hat{\mu} \notin L^1(\mathbb{R})$.*

Proof. (See [16; [54], Chap. 10, p. 214]). $d\mu(x) = f(x)dx$ for a function f supported by C. Since C is a Cantor set, f cannot be continuous; but, if $\hat{\mu} \in L^1$, f *would* be continuous. □

Example C.1. The famous Vieta's identity (A.1) may be rewritten as

$$\frac{\sin t}{t} = \left(\prod_{k=1}^{\infty} \cos(t/2^{2k-1})\right)\left(\prod_{k=1}^{\infty} \cos(t/2^{2k})\right). \qquad (C.23)$$

By [142], each product on the r.h.s. of (C.23) is the Fourier–Stieltjes transform of a s.c. measure. Indeed, the second term on the r.h.s. of (C.23) is the Fourier–Stieltjes transform $\hat{\mu}_E$ of a measure supported by a perfect symmetric set E of constant ratio $\xi = 1/4$ and is, therefore, s.c. [142, p. 15]. The first term on the r.h.s. of (C.23) equals $\hat{\mu}_E(2t) = \hat{\mu}_{E'}(t)$, where E' is the perfect symmetric set of constant ratio $\xi = 1/4$ but constructed on the interval $[-\pi/2, \pi/2]$, which is, of course, also s.c. by the same argument. Finally, the l.h.s. of (C.23) belongs to $L^2(\mathbb{R}, \mathcal{L})$ and it follows from Theorem 3.14 (for $p = 2$) that μ is a.c..

Definition C.2 ([16]). Let H be a self-adjoint operator on a separable Hilbert space $\mathcal{H}$. The vector $\phi \in \mathcal{H}$ is called a transient vector for H if

$$(\phi, \exp(-itH)\phi) = O(t^{-N}) \quad \forall N \in \mathbb{N}. \qquad (C.24)$$

The closure of the set of transient vectors is called $\mathcal{H}_{\text{tac}}$ (transient a.c. subspace). One has [16, 54] the following proposition.

Proposition C.5. $\mathcal{H}_{\text{tac}}$ *is a subspace of* $\mathcal{H}$ *such that* $\mathcal{H}_{\text{tac}} \subset \mathcal{H}_{\text{ac}}$, *and*

$$\mathcal{H}_{\text{tac}} = \overline{\{\phi : \hat{\mu}_\phi \in L^1(\mathbb{R})\}}, \qquad (C.25)$$

where the bar indicates closure. One defines

$$\mathcal{H}_{\text{rac}} = \mathcal{H}_{\text{tac}}^{\perp} \cap \mathcal{H}_{\text{ac}} \qquad (C.26)$$

which is called the recurrent a.c. subspace [16]: *both* $\mathcal{H}_{\text{tac}}$ *and* $\mathcal{H}_{\text{rac}}$ *are invariant subspaces under* H (*see* [16, 54]).

Section 3.4 discusses the connection of the above spectra with quantum dynamical stability and provides or mentions explicit nontrivial examples of transient and recurrent a.c. spectra.

Bibliography

1. M. Abramowitz and I. Stegun. *Handbook of mathematical functions.* Dover, 1965.
2. N. I. Achieser and I. M. Glasmann. *Theorie der Linearen Operatoren im Hilbert Raum.* Verlag Harri Deutsch, Thun, Frankfurt am Main, 1977.
3. M. Aizenman and S. Warzel. *Comm. Math. Phys.*, 264:371, 2006.
4. M. Aizenman and S. Warzel. *Phys. Rev. Lett.*, 106:136804, 2011.
5. N. I. Akhiezer. *The classical moment problem.* Oliver and Boyd, Edinburgh and London, 1965.
6. W. Amrein and V. Georgescu. *Helv. Phys. Acta*, 46:635, 1973.
7. W. O. Amrein, J. M. Jauch, and K. B. Sinha. *Scattering theory in quantum mechanics.* W. A. Benjamin, 1977.
8. P. W. Anderson. *Phys. Rev.*, 109:1492, 1958.
9. P. W. Anderson. *Proc. Natl. Acad. Sci.*, 69:1097, 1972.
10. D. V. Anosov. *Sov. Math. Dokl.*, 4:1153, 1963.
11. H. Araki. *Mathematical theory of quantum fields.* Oxford University Press, 1999.
12. H. Araki and E. Barouch. *J. Stat. Phys.*, 31:327, 1983.
13. H. Araki and R. Haag. *Comm. Math. Phys.*, 4:77–91, 1967.
14. V. I. Arnold. *Méthodes Mathématiques de la Mécanique Classique.* Editions Mir, Moscow, 1976.
15. V. I. Arnold and A. Avez. *Ergodic problems of classical mechanics.* W. A. Benjamin, 1968.
16. J. Avron and B. Simon. *J. Func. Anal.*, 43:1–31, 1981.
17. V. Bach, J. Fröhlich, and I. M. Sigal. *J. Math. Phys.*, 41:3985, 2000.
18. J. A. Baeta, H. Hey, and W. F. Wreszinski. *J. Stat. Phys.*, 76:1479, 1994.
19. V. Baladi and M. Viana. *Ann. Sci. École. Norm. Sup.*, 29:483–517, 1996.
20. J. C. A. Barata and D. H. U. Marchetti. *J. Stat. Phys.*, 88:231–267, 1997.
21. J. Bellissard. In S. Albeverio and Ph. Blanchard, editors, *Trends in the eighties.* World Scientific, Singapore, 1985.
22. J. Bellissard and H. Schulz-Baldes. *J. Stat. Phys.*, 99:587–594, 2000.
23. R. D. Benguria, P. Duclos, C. Fernandez, and C. Sing-Long. *J. Phys. A*, 43:474007, 2010.
24. M. V. Berry. *Eur. J. Phys.*, 2:91, 1981.

25. M. V. Berry and J. Goldberg. *Nonlinearity*, 1:1, 1988.
26. P. Blanchard and E. Brüning. *Mathematical methods in physics*. Birkhäuser, 2003.
27. P. Bleher, H. R. Jauslin, and J. L. Lebowitz. *J. Stat. Phys.*, 53:551, 1988.
28. C. Bluhm. *J. Four. Anal. Appl.*, 5:355–362, 1999.
29. R. Bowen and D. Ruelle. *Invent. Math.*, 29:181, 1975.
30. O. Bratelli, A. Kishimoto, and D. W. Robinson. *Comm. Math. Phys.*, 61:209, 1978.
31. O. Bratelli and D. W. Robinson. *Operator algebras and quantum statistical mechanics I*. Springer, 1987.
32. O. Bratelli and D. W. Robinson. *Operator algebras and quantum statistical mechanics II*. Springer, 2nd edition, 1997.
33. W. Brenig and R. Haag. *Fort. der Physik*, 7:183–242, 1959.
34. R. Brunetti and K. Fredenhagen. *Phys. Rev. A*, 66:044101, 2002.
35. R. Brunetti and K. Fredenhagen. *Rev. Math. Phys.*, 14:897–906, 2002.
36. R. Brunetti, K. Fredenhagen, and M. Hoge. *Found. Phys.*, 40:1368–1378, 2010.
37. L. Bunimovich. *Nonlinearity*, 21:T13–T17, 2008.
38. P. Busch. The time-energy uncertainty relation. In R. Sala Mayato, J. G. Muga, and I. L. Egusquiza, editors, *Time and quantum mechanics*. Springer, 2002.
39. G. Casati and L. Molinari. *Progr. Theoret. Phys. Suppl.*, 98:287, 1989.
40. C. Chandre and H. R. Jauslin. *J. Math. Phys.*, 39:5856, 1998.
41. S. R. Channon and J. L. Lebowitz. *Ann. N. Y. Acad. Sci.*, 357:108, 1980.
42. K. L. Chung. *A course in probability theory*. Academic Press, 1974.
43. K. L. Chung. *Elementary probability theory with stochastic processes*. Springer, 3rd edition, 1979.
44. C. Cohen-Tannoudji, J. Duport-Roc, and G. Arynberg. *Atom-photon interactions*. J. Wiley, 1992.
45. J. M. Combes. Connections between quantum dynamics and spectral properties of time evolution operators. In E. M. Harrell, W. F. Ames, and J. V. Herod, editors, *Differential equations with applications to mathematical physics*. Academic Press, Boston, 1993.
46. J. M. Combes and G. Mantica. In *Long time behavior of classical and quantum systems*, Ser. Concr. Appl. Math., pages 107–123, Bologna, 2001. World Scientific.
47. M. Combescure. *Ann. Inst. Henri Poincarè*, 47:63, 1987.
48. I. P. Cornfeld, S. V. Fomin, and Ya. G. Sinai. *Ergodic theory*. Springer-Verlag, 1981.
49. O. Costin. *Asymptotics and Borel summability*. Chapman and Hall/CRC, 2009.
50. O. Costin and M. Huang. *J. Stat. Phys.*, 144:846, 2011.
51. O. Costin, J. L. Lebowitz, and A. Rokhlenko. *J. Phys. A*, 33:6311, 2000.
52. O. Costin, J. L. Lebowitz, and C. Stucchio. *Rev. Math. Phys.*, 20:835, 2008.
53. O. Costin, J. L. Lebowitz, C. Stucchio, and S. Tanveer. *J. Math. Phys.*, 51:1, 2010.

54. H. L. Cycon, R. M. Froese, W. Kirsch, and B. Simon. *Schrödinger Operators.* Springer-Verlag, 1987.
55. D. Damanik. *Proc. Symp. Pure Math.*, 76 v2:505–538, 2007.
56. D. Damanik and S. Tcheremchantsev. *Comm. Math. Phys.*, 236:513, 2003.
57. H. Davenport, P. Erdös, and W. J. LeVeque. *Michigan Math. J.*, 10:311–314, 1963.
58. E. B. Davies. *Quantum theory of open systems.* Academic Press, 1976.
59. A. M. Ozorio de Almeida. *Hamiltonian systems, chaos and quantization.* Cambridge University Press, 1988.
60. S. de Bièvre and M. Degli Esposti. *Ann. Inst. Henri Poincarè*, 69:130, 1998.
61. S. de Bièvre and M. Merkli. *Class. Quan. Grav.*, 23:5227, 2006.
62. S. L. de Carvalho. Espectro e dimensão hausdorff de operadores bloco-jacobi com perturbações esparsas distribuidas aleatoriamente. Ph.D. thesis, IFUSP, 2010.
63. S. L. de Carvalho, D. H. U. Marchetti, and W. F. Wreszinski. *J. Math. Anal. Appl.*, 368:218–234, 2010.
64. S. L. de Carvalho, D. H. U. Marchetti, and W. F. Wreszinski. Pointwise decay of Fourier–Stieltjes transform of the spectral measure for Jacobi matrices with faster-than-exponential sparse perturbations. arXiv: 1010.5274, 2010.
65. S. L. de Carvalho, D. H. U. Marchetti, and W. F. Wreszinski. On the uniform distribution of Prüfer angles and its implication to a sharp spectral transition of Jacobi matrices with randomly sparse perturbations. arXiv: 1006.2849, 2011.
66. R. de la Madrid and M. Gadella. *Am. J. Phys.*, 70:626, 2002.
67. R. del Rio, S. Jitomirskaya, Y. Last, and B. Simon. *J. Anal. Math.*, 69:153–200, 1996.
68. M. Demuth and M. Krishna. *Determining spectra in quantum theory.* Birkhäuser Boston, 2005.
69. W. deRoeeck, T. Jacobs, C. Maes, and K. Netocny. *J. Phys. A*, 36:11547, 2003.
70. J. Dixmier. *Les C-ètoile algèbres et leurs rèpresentations.* Gauthiers-Villars, Paris, 1969.
71. J. D. Dollard. *J. Math. Phys.*, 5:729, 1964.
72. J. R. Dorfman. *An introduction to chaos in nonequilibrium statistical mechanics.* Cambridge University Press, 1999.
73. P. Duarte. *Ann. Inst. Henri Poincarè*, 11:359–459, 1994.
74. P. Duarte. *Erg. Th. Dyn. Syst.*, 29:1781–1813, 2008.
75. D. A. Dubin and G. L. Sewell. *J. Math. Phys.*, 11:2990, 1970.
76. D. Dürr, R. Grummt, and M. Kolb. On the time dependent analysis of gamow decay. arXiv: 1011.6084, 2011.
77. J. H. Eberly. *J. Phys. B*, 23:L619, 1990.
78. J. P. Eckmann and D. Ruelle. *Rev. Mod. Phys.*, 57:617, 1985.
79. G. G. Emch. *J. Math. Phys.*, 7:1198, 1966.
80. G. G. Emch, H. Narnhofer, G. L. Sewell, and W. Thirring. *J. Math. Phys.*, 35:5582, 1994.

81. V. Enss. In G. Velo and A. S. Wightman, editors, Geometric methods in spectral and scattering theory of Schrodinger operators, *Rigorous atomic and molecular physics*. Plenum Press, New York, 1981.

82. L. C. Evans and R. F. Gariepy. *Measure theory and fine properties of functions*. CRC Press, Boca Raton, Ann Arbor and London, 1992.

83. K. Falconer. *The geometry of fractal sets*. Cambridge University Press, 1985.

84. R. P. Feynman. *The character of the physical law*. MIT Press, Cambridge MA, 1967.

85. R. P. Feynman. *Lectures on physics III — Quantum Mechanics*. Addison-Wesley, 1965.

86. M. Floater and T. Lyche. *Math. Comp.*, 76:867–877, 2007.

87. K. Fredenhagen. *Comm. Math. Phys.*, 97:461, 1985.

88. A. Fring, V. Kostrykin, and R. Schrader. *J. Phys. A*, 30:8559, 1997.

89. A. Galves, A. C. R. Nogueira, and M. E. Vares. Introdução aos sistemas markovianos de partículas. Quinto simpósio nacional de probabilidade e estatística, IME/USP, 1982.

90. G. Gamow. *Z. Physik*, 51:204, 1928.

91. P. L. Garrido, S. Goldstein, J. Lukkarinen, and R. Tumulka. Paradoxical reflections in quantum mechanics. arXiv: 0808.0610, 2011.

92. D. Gilbert and D. P. Pearson. *J. Math. Anal. Appl.*, 128:30–58, 1987.

93. E. A. Gislason, N. H. Sabelli, and J. W. Wood. *Phys. Rev. A*, 31:2078, 1985.

94. J. Glimm and A. Jaffe. *Acta Math.*, 125:204, 1970.

95. I. Ya. Goldscheidt, S. Molchanov, and L. Pasthur. *Func. Anal. and Appl.*, 11:1–11, 1977.

96. S. Goldstein, J. L. Lebowitz, C. Mastrodonato, R. Tumulka, and N. Zanghi. Normal typicality and von Neumann's ergodic theorem. *Proc. Roy. Soc. A*, 466:3203–3224, 2010.

97. S. Goldstein, J. L. Lebowitz, C. Mastrodonato, R. Tumulka, and N. Zanghi. On the approach to thermal equilibrium of macroscopic quantum systems. *Phys. Rev. E*, 81:011109, 2010.

98. S. W. Graham and G. Kolesnik. *Van der Corput method of exponential sums*. Cambridge University Press, 1991.

99. V. Grecchi and A. Sacchetti. *J. Stat. Phys.*, 103:339, 2001.

100. A. Grigis and J. Sjöstrand. *Microlocal analysis for differential operators — an introduction*. Cambridge University Press, 1994.

101. I. Guarneri. *Europhys. Lett.*, 10:95, 1989.

102. I. Guarneri. *Europhys. Lett.*, 21:729, 1993.

103. I. Guarneri and H. Schulz-Baldes. *Rev. Math. Phys.*, 11:1249, 1999.

104. J. Guckenheimer and P. Holmes. *Nonlinear oscillations dynamical systems and bifurcations of vector fields*. Springer-Verlag, 1983.

105. S. J. Gustafson and I. M. Sigal. *Mathematical concepts in quantum mechanics*. Springer-Verlag, Berlin, Heidelberg, 2006.

106. R. Haag and D. Kastler. *J. Math. Phys.*, 5:548, 1964.

107. R. Haag, D. Kastler, and E. B. Trych-Pohlmeyer. *Comm. Math. Phys.*, 38:111, 1974.

108. G. Hagedorn and A. Joye. *Comm. Math. Phys.*, 207:439, 1999.

109. E. M. Harrell. *Comm. Math. Phys.*, 75:239, 1980.
110. E. M. Harrell. *Proc. Symp. Pure Math.*, 76:227–248, 2006.
111. F. Hausdorff. *Math. Ann.*, 79:157–179, 1919.
112. K. Hepp. *Helv. Phys. Acta*, 45:237, 1972.
113. E. Hille. *Ordinary differential equations in the complex domain.* Dover, 1997.
114. M. Hirsch and S. Smale. *Differential equations dynamical systems and linear algebra.* Academic Press, 1974.
115. H. Hochstadt. *The functions of mathematical physics.* Wiley, 1971.
116. M. Holschneider. *Comm. Math. Phys.*, 100:457, 1994.
117. L. Hörmander. *The analysis of linear partial differential operators I.* Springer, 2nd edition, 1983.
118. J. S. Howland. *J. Func. Anal.*, 74:52, 1980.
119. J. S. Howland. Quantum stability. In E. Balslev, editor, *Schrödinger operators.* Springer, Berlin, 1992.
120. X. Hu and S. J. Taylor. *Math. Proc. Cambridge Phil. Soc.*, 115:527, 1994.
121. M. Huang. *J. Stat. Phys.*, 137:569–592, 2009.
122. N. M. Hugenholtz. In R. F. Streater, editor, States and representations in Statistical Mechanics, *Mathematics of Contemporary Physics.* Academic Press, London, 17–76, 1972.
123. T. Ichinose. On the spectra of tensor products of linear operators in banach spaces. *J. Reine Angew. Math.*, 244:119–153, 1970.
124. C. Jäkel, H. Narnhofer, and W. F. Wreszinski. *J. Math. Phys.*, 51:052703, 2010.
125. V. Jaksic and Y. Last. *Duke Math. J.*, 133:185–204, 2006.
126. V. Jaksic and C. A. Pillet. *J. Stat. Phys.*, 108:787, 2002.
127. V. Jaksic. *Lect. Notes in Math.*, 1880:235–312, 2006.
128. H. R. Jauslin. *Stability and chaos in classical and quantum hamiltonian systems. II Granada seminar on computational physics*, World Scientific, 1993.
129. H. R. Jauslin and J. L. Lebowitz. *Chaos*, 1:114, 1991.
130. H. R. Jauslin, O. Sapin, S. Guérin, and W. F. Wreszinski. *J. Math. Phys.*, 45:4377, 2004.
131. B. Jessen and A. Wintner. *Trans. Amer. Math. Soc.*, 38:48–88, 1935.
132. S. Jitomirskaya. Ergodic Schrödinger operators. *Proc. Symp. Pure Math.*, 76 v2:613, 2007.
133. S. Jitomirskaya and Y. Last. *Comm. Math. Phys.*, 211:643, 2000.
134. S. Jitomirskaya and B. Simon. *Comm. Math. Phys.*, 165:201, 1994.
135. R. Jost. *The general theory of quantized fields.* American Mathematical Society, Providence, RI, 1965.
136. R. Jost. *Das Märchen vom Elfenbeinernen Turm.* Springer-Verlag, 1995.
137. W. F. Donoghue Jr. *Monotone matrix functions and analytic continuation.* Springer-Verlag, original edition, 1974.
138. M. Kac. *Statistical independence in probability, analysis and number theory*, volume 12. The Carus Mathematical Monographs, Mathematical Association of America. Distributed by John Wiley and Sons, Inc., 1959.

139. S. Kaczmarz and H. Steinhaus. *Theorie der Orthogonalreihen.* Warszawa-Lwow, 1935.
140. J. P. Kahane. *Ann. Inst. Fourier*, 14:519–526, 1964.
141. J. P. Kahane. *Some random series of functions.* Cambridge University Press, 2nd edition, 1985.
142. J. P. Kahane and R. Salem. *Colloquium Math.*, 6:193–202, 1958.
143. J. P. Kahane and R. Salem. *Bull. Amer. Math. Soc.*, 70:259–261, 1964.
144. J. P. Kahane and R. Salem. *Ensembles Parfaits et séries trigonométriques.* Hermann, Paris, 2nd edition, 1994.
145. A. Katok and B. Hasselblatt. *Introduction to the modern theory of dynamical systems.* Cambridge University Press, 1997.
146. Y. Katznelson. *An introduction to harmonic analysis.* Dover, 1976.
147. J. Keating. *Phys. World*, 46–50, April 1990.
148. R. Ketzmerick, G. Petschel, and T. Geisel. *Phys. Rev. Lett.*, 69:695, 1992.
149. S. Khan and D. B. Pearson. *Helv. Phys. Acta*, 65:505–527, 1992.
150. A. I. Khinchin. *Continued Fractions.* Dover, 1992.
151. C. King. *Lett. Math. Phys.*, 23:215, 1991.
152. A. Kiselev, Y. Last, and B. Simon. *Comm. Math. Phys.*, 194:1–45, 1998.
153. A. Klein. *Adv. Math.*, 133:163, 1998.
154. P. M. Koch and K. A. H. van Leeuwen. *Phys. Rep.*, 255:289, 1995.
155. D. Krutikov and C. Remling. *Comm. Math. Phys.*, 223:509–532, 2001.
156. L. Kuipers and H. Niederreiter. *Uniform distribution of sequences.* Dover, 1974.
157. H. Kunz and B. Souillard. *Comm. Math. Phys.*, 78:201, 1980–1981.
158. L. J. Landau. *Comm. Math. Phys.*, 17:156–176, 1970.
159. L. J. Landau. *J. London Math. Soc.*, 61:197, 2000.
160. G. Jona Lasinio, F. Martinelli, and E. Scoppola. *Comm. Math. Phys.*, 80:223, 1981.
161. A. Lasota and M. Mackey. *Probabilistic properties of deterministic systems.* Cambridge University Press, 1985.
162. Y. Last. *J. Func. Anal.*, 142:406, 1996.
163. Y. Last and B. Simon. *Invent. Math.*, 135:329–367, 1999.
164. Y. Last. Exotic spectra — a review of Barry Simon's central contributions. *Proc. Symp. Pure Math.*, 76 v2:697–712, 2007.
165. R. Lavine. Spectral density and sojourn times. In J. Nuttall, editor, *Atomic scattering theory.* University of Western Ontario, London, Ontario, 1978.
166. R. Lavine. *Rev. Math. Phys.*, 13:267, 2001.
167. J. L. Lebowitz. Time asymmetric macroscopic behavior — an overview. In *Boltzmann's legacy.* Eur. Math. Soc., 2008.
168. J. L. Lebowitz and O. Penrose. *Phys. Today*, 20, 1973.
169. E. H. Lieb and M. Loss. *Analysis.* American Mathematical Society, Providence, RI, 2000.
170. E. H. Lieb and R. Seiringer. *The stability of matter in quantum mechanics.* Cambridge University Press, 2010.
171. I. J. Lowe and R. E. Norberg. *Phys. Rev.*, 107:46, 1957.
172. J. H. Lowenstein. *J. Phys.*, 7:68–85, 2005.

173. M. Ludvigsen. *General relativity — a geometric approach.* Cambridge University Press, 1999.
174. R. Lyons. *Ann. Math.*, 122:155–170, 1985.
175. R. Lyons. *J. Fourier Anal. Appl.*, 45:363, 1995.
176. W. Magnus and R. Winkler. *Hill's equation.* Dover, N.Y., 2004.
177. H. D. Maison. *Comm. Math. Phys.*, 10:48, 1968.
178. R. Mañé. *Ergodic theory and differentiable dynamics.* Springer-Verlag, Berlin, 1987.
179. D. H. U. Marchetti and W. F. Wreszinski. *J. Stat. Phys.*, 146:885–899, 2012.
180. D. H. U. Marchetti, W. F. Wreszinski, L. F. Guidi, and R. M. Angelo. *Nonlinearity*, 20:765, 2007.
181. J. Marklof and F. Cellarosi. On the limit curlicue process for theta sums. In *INdAM meeting hyperbolic dynamical systems in the sciences*, Corinaldo, 1st June, 2010.
182. P. A. Martin and F. Rothen. *Many body problems and quantum field theory — an introduction.* Springer, 2nd edition, 1990.
183. P. Mattila. *Geometry of sets and measures in Euclidean spaces.* Cambridge University Press, 1995.
184. M. Merkli. *J. Math. Anal. Appl.*, 327:376, 2007.
185. E. Merzbacher. *Quantum mechanics.* J. Wiley, 3rd edition, 1998.
186. Y. Meyer. *Algebraic numbers and harmonic analysis.* North-Holland, Amsterdam, 1972.
187. G. Mockenhaupt. *Geom. Func. Anal.*, 10:1579–1587, 2000.
188. S. Molchanov. *Contemp. Math.*, 217:157–181, 1998.
189. S. Molchanov and Vainberg. *Appl. Anal.*, 71:167–185, 1998.
190. S. A. Molchanov. *Homogenization.* World Scientific, Singapore, 1999.
191. H. Narnhofer and W. Thirring. *J. Stat. Phys.*, 57:811, 1989.
192. H. Narnhofer and W. Thirring. *Int. J. Mod. Phys. A*, 17:2937–2970, 1991.
193. E. Nelson. *Operator differential equations.* Notes for Mathematics, Volume 520, Princeton University, 1964–1965.
194. H. M. Nussenzveig. *Introduction to quantum optics.* Gordon and Breach Scientific Publishers, 1973.
195. H. M. Nussenzveig. *Causality and dispersion relations.* Academic Press, 1972.
196. V. I. Oseledec. *Trans. Moscow Math. Soc.*, 19:197, 1968.
197. J. Palis and F. Takens. *Hyperbolicity and sensitive chaotic dynamics at homoclinic bifurcations.* Cambridge University Press, 1993.
198. W. Pauli. *General principles of quantum mechanics.* Springer-Verlag, Berlin, 1980.
199. D. B. Pearson. *Comm. Math. Phys.*, 40:125, 1975.
200. D. B. Pearson. *Comm. Math. Phys.*, 60:13–36, 1978.
201. D. B. Pearson. *J. Phys. A*, 26:4067–4080, 1993.
202. O. Penrose. *Rep. Progr. Phys.*, 42:1937, 1979.
203. T. Pereira and D. H. U. Marchetti. *Progr. Theor. Phys.*, 122:1137, 2009.
204. Y. Peres and P. Shmerkin. *Ergodic Theory Dynam. systems*, 29:201–221, 2009.

205. P. Pfeifer and J. Fröhlich. *Rev. Mod. Phys.*, 67:759, 1995.

206. C. Radin. *J. Math. Phys.*, 11:2945, 1970.

207. M. Reed and B. Simon. *Methods in modern mathematical physics — v.1, Functional Analysis*. Academic Press, 1st edition, 1972.

208. M. Reed and B. Simon. *Methods of modern mathematical physics — v.2, Fourier Analysis, Self-adjointness*. Academic Press, 1975.

209. M. Reed and B. Simon. *Methods of modern mathematical physics — v.3, Scattering theory*. Academic Press, 1978.

210. M. Reed and B. Simon. *Methods of modern mathematical physics — v.4, Analysis of operators*. Academic Press, 1978.

211. C. Remling. *Comm. Math. Phys.*, 183:313–323, 1997.

212. C. Remling. *Math. Rev.* MR2300896, 2008a:81067.

213. F. Riesz. *Math. Zeit.*, 2:312–315, 1918.

214. W. Rindler. *Relativity-special, general and cosmological*. Oxford University Press, 2006.

215. A. W. Roberts and D. E. Varberg. *Convex functions*. Academic Press, 1973.

216. D. W. Robinson. *Comm. Math. Phys.*, 31:171–189, 1973.

217. C. A. Rogers and S. J. Taylor. *Acta Math.*, 109:207, 1963.

218. W. Rudin. *Principles of mathematical analysis*. McGraw-Hill, 2nd edition, 1964.

219. W. Rudin. *Functional analysis*. McGraw-Hill, Inc., 1973.

220. W. Rudin. *Real and Complex Analysis*. McGraw-Hill, 2nd edition, 1974.

221. D. Ruelle. *Nuovo Cim.*, 59A:655, 1969.

222. D. Ruelle. *Comm. Math. Phys.*, 125:239, 1989.

223. S. Sachdev. *Phys. World*, 7:25, 1994.

224. J. J. Sakurai. *Advanced quantum mechanics*. Addison-Wesley, 1967.

225. R. Salem. *J. Math. and Phys.*, 21:69–81, 1942.

226. R. Salem. *Ark. Mat.*, 1:353–365, 1950.

227. S. R. Salinas and W. F. Wreszinski. *J. Stat. Phys.*, 41:299, 1985.

228. O. Sapin, H. R. Jauslin, and S. Weigert. *J. Stat. Phys.*, 127:699, 2007.

229. G. Scharf, K. Sonnenmoser, and W. F. Wreszinski. *Phys. Rev. A*, 44:3250, 1991.

230. D. Schechtman, I. Blech, D. Gratias, and J. V. Cahn. *Phys. Rev. Lett.*, 53:1951, 1984.

231. L. Schwartz, *Mathematics for the Physical Sciences*. Dover Publ. Inc., 2008.

232. E. Schrödinger. *The spirit of science, in What is life? and other scientific essays*. Doubleday, Anchor B, Garden City, 1965.

233. H. Schulz-Baldes and J. Bellissard. *Rev. Math. Phys.*, 10:1–46, 1998.

234. G. L. Sewell. *Ann. Phys.*, 85:336, 1974.

235. G. L. Sewell. *Phys. Rep.*, 57:307–342, 1980.

236. G. L. Sewell. *Ann. Phys.*, 141:201–224, 1982.

237. G. L. Sewell. *Quantum theory of collective phenomena*. Clarendon Press, Oxford, 1986.

238. D. L. Shepelianskii. *Theor. Math. Phys.*, 49:117–121, 1981.

239. B. Simon. An overview of rigorous scattering theory. In J. Nuttall, editor, *Atomic scattering theory*, pages 1–24. University of Western Ontario, 1978.

240. B. Simon. *Functional Integration and Quantum Physics*. Academic Press, New York, 1979.
241. B. Simon. *J. Func. Anal.*, 63:123, 1985.
242. B. Simon. *Comm. Math. Phys.*, 134:209, 1990.
243. B. Simon. *Comm. Math. Phys.*, 176:713–722, 1996.
244. B. Simon and G. Stolz. *Proc. Am. Math. Soc.*, 124:2073–2080, 1996.
245. B. Simon and T. Wolff. *Comm. Pure Appl. Math.*, 39:75, 1986.
246. Ya. G. Sinai. *Russ. Math. Surv.*, 27:21, 1972.
247. Ya. G. Sinai. *Topics in ergodic theory*. Princeton University Press, 1994.
248. K. B. Sinha. *Ann. Inst. Henri Poincaré*, 26:263–277, 1977.
249. E. Skibsted. *Comm. Math. Phys.*, 104:591, 1986.
250. E. M. Stein and T. Murphy. *Harmonic analysis: real-variable methods, orthogonality and oscillatory integrals*. Princeton University Press, 1993.
251. M. H. Stone. *Linear transformations in Hilbert space and their applications to analysis*, Coll. Publ., Volume 15, American Mathematical Society, New York, 1932.
252. R. Strichartz. *J. Func. Anal.*, 89:154, 1990.
253. S. M. Susskind, S. C. Cowley, and E. J. Valeo. *Phys. Rev. A*, 42:3090, 1990.
254. W. Szlenk. *An introduction to the theory of smooth dynamical systems*. Wiley, 1984.
255. H. Tasaki. The approach to thermal equilibrium and thermodynamic normality. arXiv: 1003.5424v4, 2010.
256. G. Teschl. *Jacobi operators and completely integrable nonlinear lattices*. Mathematical Surveys and Monographs, Volume 72, American Mathematical Society, original edition, 2000.
257. W. Thirring. *A course in mathematical physics I — classical dynamical systems*. Springer, Berlin, 1992.
258. W. Thirring. *A course in mathematical physics IV — quantum mechanics of large systems*. Springer, Berlin, 1994.
259. W. Thirring. What are the quantum mechanical Lyapunov exponents. pages 223–237, Berlin, 1996. Proceedings of the 34 Internationale Universitätswoche für Kern und Teilchenphysik Schladming, Springer.
260. W. Thirring and H. A. Posch. *Phys. Rev. E*, 48:4333, 1993.
261. A. C. D. van Enter and J. L. van Hemmen. *J. Stat. Phys.*, 32:141, 1983.
262. L. van Hemmen. *Fort. der Physik*, 26:397–439, 1978.
263. R. Venegeroles. *Phys. Rev. Lett.*, 102:064101, 2009.
264. F. Verhulst. *Nonlinear differential equations and dynamical systems*. Springer, 1990.
265. M. Viana. Stochastic dynamics of deterministic systems. IMPA lecture notes 1997.
266. P. Walters. *Ergodic theory — introductory lectures*, Lecture Notes in Mathematics, Volume 458, Springer, Berlin, Heidelberg, New York, 1965.
267. S. Weigert. *Z. Phys. B*, 80:3, 1990.
268. S. Weigert. *Phys. Rev. A*, 48:1780, 1993.
269. A. S. Wightman. *Nuovo Cim. B*, 110:751, 1995.

270. I. F. Wilde. Lecture notes on local quantum theory and operator algebras. available from Ivan F. Wilde homepage.ntl.world.com.

271. W. F. Wreszinski. *Fort. der Physik*, 35:379–413, 1987.

272. W. F. Wreszinski. *Helv. Phys. Acta*, 70:109, 1997.

273. W. F. Wreszinski. *Rev. Math. Phys.*, 17:1–14, 2005.

274. W. F. Wreszinski. *J. Stat. Phys.*, 138:567–578, 2010.

275. W. F. Wreszinski. *J. Stat. Phys.*, 146:118, 2012.

276. W. F. Wreszinski and S. Casmeridis. *J. Stat. Phys.*, 90:1061, 1998.

277. K. Yajima and Y. Kitada. *Ann. Inst. Henri Poinc. A*, 39:145, 1983.

278. J. Zinn-Justin. *J. de Physique*, 40:969, 1979.

279. A. Zlatoš. *J. Func. Anal.*, 207:216–252, 2004.

280. A. Zygmund. *Trigonometric series v. I and II*. Cambridge University Press, 1968.

Index